Vector Analysis
and
Cartesian Tensors

Vector Analysis and Cartesian Tensors

Third edition

D.E. Bourne

Department of Applied and Computational Mathematics
University of Sheffield, UK

and

P.C. Kendall

Department of Electronic and Electrical Engineering
University of Sheffield, UK

CHAPMAN & HALL

University and Professional Division

London · Glasgow · New York · Tokyo · Melbourne · Madras

Published by Chapman & Hall, 2-6 Boundary Row, London SE1 8HN

Chapman & Hall, 2-6 Boundary Row, London SE1 8HN, UK

Blackie Academic & Professional, Wester Cleddens Road,
Bishopbriggs, Glasgow G64 2NZ, UK

Chapman & Hall, 29 West 35th Street, New York NY10001, USA

Chapman & Hall Japan, Thomson Publishing Japan, Hirakawacho
Nemoto Building, 6F, 1-7-11 Hirakawa-cho, Chiyoda-ku, Tokyo 102,
Japan

Chapman & Hall Australia, Thomas Nelson Australia, 102 Dodds Street,
South Melbourne, Victoria 3205, Australia

Chapman & Hall india, R. Seshadri, 32 Second Main Road, CIT East,
Madras 600 035, India

First edition 1967
Reprinted 1982, 1983, 1984, 1985, 1986, 1988, 1990, 1991
Second edition 1977
Third edition 1992
Reprinted 1993

© 1967, 1977, 1992 D.E. Bourne and P.C. Kendall

Typeset in 10/12pt Times by Pure Tech Corporation, India
Printed in Great Britain by TJ Press (Padstow) Ltd, Cornwall

ISBN 0 412 42750 8

A catalogue record for this book is available from the British Library
Library of Congress Cataloging-in-Publication Data available

Contents

Preface

We are very grateful to Chapman & Hall for their offer to reset this book completely. This has given us the opportunity to include small but important teaching points which have accumulated over a long period, to improve the notation and the diagrams, and to introduce some new material. The kinematics section has been extended to deal with the existence and nature of angular velocity, including rotating frames of reference and the concept of the Coriolis force. A new section on the application of vector analysis in optimization theory has been added, giving a simple approach to the method of steepest descent, which students have found stimulating. New examples and exercises have been added, and some deleted.

D.E. BOURNE
P.C. KENDALL

Preface to second edition

The most popular textbook approach to vector analysis begins with the definition of a vector as an equivalence class of directed line segments – or, more loosely, as an entity having both magnitude and direction. This approach is no doubt appealing because of its apparent conceptual simplicity, but it is fraught with logical difficulties which need careful handling if they are to be properly resolved. Consequently, students often have difficulty in understanding fully the early parts of vector algebra and many rapidly lose confidence. Another disadvantage is that subsequent developments usually make frequent appeal to geometrical intuition and much care is needed if analytical requirements are not to be obscured or overlooked. For example, it is seldom made clear that the definitions of the gradient of a scalar field and the divergence and curl of a vector field imply that these fields are continuously differentiable, and hence that the mere existence of the appropriate first order partial derivatives is insufficient.

The account of vector analysis presented in this volume is based upon the definition of a vector in terms of rectangular cartesian components which satisfy appropriate rules of transformation under changes of axes. This approach has now been used successfully for ten years in courses given from the first year onwards to undergraduate mathematicians and scientists, and offers several advantages. The rules for addition and subtraction of vectors, for finding scalar and vector products and differentiation are readily grasped, and the ability to handle vectors so easily gives the student immediate confidence. The later entry into vector field theory takes place naturally with gradient, divergence and curl being defined in their cartesian forms. This avoids the alternative, more sophisticated, definitions involving limits of integrals. Another advantage of the direct treatment of vectors by components is that introducing the student at a later stage to tensor analysis is easier. At that stage tensors are seen as a widening of the vector concept and no mental readjustment is necessary.

The approach to vectors through rectangular cartesian components does not obscure the intuitive idea of a vector as an entity with magnitude and direction. The notion emerges as an almost immediate consequence of the definition and is more soundly based, inasmuch as both the magnitude and direction then have precise analytical interpretations. The familiar parallelogram law of addition also follows easily.

The essential background ideas associated with rotations of rectangular cartesian coordinate axes are introduced in Chapter 1 at a level suitable for undergraduates beginning their first year. The second and third chapters deal with the basic concepts of vector algebra and differentiation of vectors, respectively; applications to the differential geometry of curves are also given in preparation for later work.

Vector field theory begins in Chapter 4 with the definitions of gradient, divergence and curl. We also show in this chapter how orthogonal curvilinear coordinate systems can be handled within the framework of rectangular cartesian theory.

An account of line, surface and volume integrals is given in the fifth chapter in preparation for the integral theorems of Gauss, Stokes and Green which are discussed in Chapter 6. The basic approach to vectors that we have adopted enables rigorous proofs to be given which are nevertheless within the grasp of the average student. Chapter 7 deals with some applications of vector analysis in potential theory and presents proofs of the principal theorems.

Chapters 8 and 9, on cartesian tensors, have been added to this second edition in response to the suggestion that it would be useful to have between two covers most of the vector and tensor analysis that undergraduates require. The case for adding this material is strengthened by the fact that the approach to vectors in the early chapters makes the transition to tensors quite straightforward. Chapter 8 deals with the basic algebra and calculus of cartesian tensors, including an account of isotropic tensors of second, third and fourth order. Chapter 9 briefly discusses those properties of second order tensors which have risen to importance in continuum mechanics over the last twenty years. Some theorems on invariants and the representation of isotropic tensor functions are proved.

We warmly acknowledge the many useful comments from students and colleagues who have worked with the first edition. They have enabled us to make improvements to the original text. We particularly thank the following: Dr G.T. Kneebone and Professor L. Mirsky for their early interest in the first edition; Professor A. Jeffrey and Thomas Nelson and Sons Ltd without whom this new edition would not have appeared.

D.E. BOURNE
P.C. KENDALL

Rectangular cartesian coordinates and rotation of axes

1.1 RECTANGULAR CARTESIAN COORDINATES

From a fixed point O, which we shall call the **origin of coordinates**, draw three fixed lines Ox, Oy, Oz at right angles to each other as in Fig. 1.1. These will be called the x-axis, y-axis, z-axis respectively and they will be referred to collectively as rectangular cartesian axes $Oxyz$. The planes Oyz, Ozx and Oxy are called the coordinate planes and they may be referred to as the yz-plane, zx-plane and xy-plane, respectively.

It is customary to choose the axes in such a way that Ox, Oy and Oz form a right-handed set, in that order. This means that to an observer looking along Oz, the sense of the smaller arc from a point on Ox to a point on Oy is

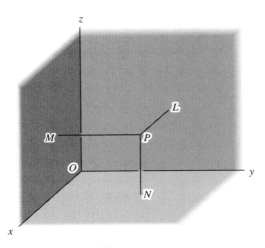

Fig. 1.1 Rectangular cartesian coordinates.

clockwise. Figure 1.2(a) illustrates this and Fig. 1.2(b) shows the relationship of Fig. 1.2(a) to a right hand. Notice that to an observer looking along Ox, the sense of the smaller arc from Oy to Oz is clockwise; and to an observer looking along Oy, the sense of the smaller arc from Oz to Ox is clockwise.

The three statements concerning observers looking along the respective axes exhibit **cyclic symmetry** in x, y, z, i.e. if, in any one of the three statements, we replace x by y, y by z and z by x, then one of the other two statements is produced. The operation of replacing x by y, y by z and z by x is called a **cyclic interchange** of x, y, z.

The position of a point P relative to a given set of rectangular cartesian axes may be specified in the following way. Draw the perpendiculars PL, PM, PN from P to the yz-, zx-, xy-planes respectively, as in Fig. 1.1. Let

$$x = \pm \text{ length of } PL$$

taking the positive sign if P lies on the same side of the yz-plane as Ox and the negative sign otherwise. Similarly, define

$$y = \pm \text{ length of } PM,$$
$$z = \pm \text{ length of } PN,$$

the positive or negative sign for y being taken according as P lies on the same or on the opposite side of the zx-plane as Oy, and the positive or negative sign for z being taken according as P lies on the same or on the opposite side of the xy-plane as Oz. The numbers x, y, z are called the x-coordinate, y-coordinate, z-coordinate of P. We may refer to P as the point (x, y, z).

It is an elementary observation that, when x, y, z are given, the position of

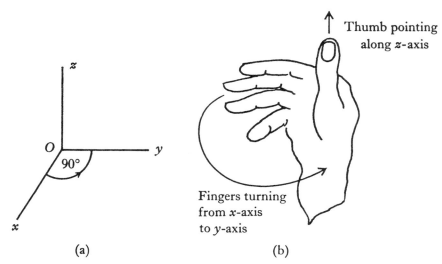

(a) (b)

Fig. 1.2 (a) The arrow turns in a clockwise sense as seen by an observer looking along Oz and (b) relation to a right hand.

P relative to the given axes is determined uniquely. Conversely, a given point P determines a unique triad of coordinates. In other words, there is a **one–one correspondence** between points P and triads of real numbers (x, y, z).

Distance from origin

To find the distance of P from the origin O, construct the rectangular paralle-lepiped which has PL, PM, PN as three edges (Fig. 1.3). Using Pythagoras's theorem, we have

$$OP^2 = ON^2 + PN^2$$
$$= PL^2 + PM^2 + PN^2.$$

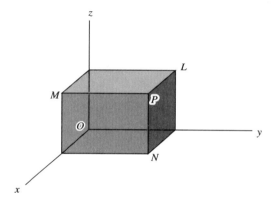

Fig. 1.3 Construction to find the distance OP.

Since the perpendicular distances of P from the coordinate planes are $|x|$, $|y|$, $|z|$, it follows that

$$OP = \sqrt{(x^2 + y^2 + z^2)}. \tag{1.1}$$

Distance between points

The distance between the points $P(x, y, z)$ and $P'(x', y', z')$ may be found in the following way. Through P construct three new coordinate axes PX, PY, PZ parallel to the original axes Ox, Oy, Oz as shown in Fig. 1.4. Let the coordinates of P' *relative to these new axes* be X, Y, Z. Then it is easily seen that

$$X = x' - x, \quad Y = y' - y, \quad Z = z' - z.$$

Applying result (1.1)

$$PP' = \sqrt{(X^2 + Y^2 + Z^2)}$$

and so, in terms of coordinates relative to the original axes,

$$PP' = \sqrt{\{(x' - x)^2 + (y' - y)^2 + (z' - z)^2\}}. \qquad (1.2)$$

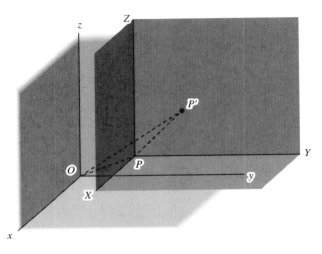

Fig. 1.4 Construction to find PP'.

EXERCISES

1.1 Show that the distance between the points (5, 4, 2) and (0, 3, 1) is $3\sqrt{3}$.

1.2 Show that the distance of the point $(a - b, a + b, c)$ from the origin is $\sqrt{(2a^2 + 2b^2 + c^2)}$.

1.3 Find the points in the xy-plane which are at unit distance from the origin and equidistant from the x-axis and y-axis.

1.4 Find the points which are at a distance of 5 units from the origin and whose distances from both the xy- and zx-planes are $2\sqrt{2}$ units.

1.5 Find the points which are at a distance of $\frac{1}{2}\sqrt{2}$ from every axis.

1.6 Find the distance between (i) the points $(1, -1, 0)$ and $(1, 2, 4)$, (ii) the points $(3, -1, 2)$ and $(-1, 5, -1)$.

1.7 The coordinates of a point O' relative to rectangular cartesian axes Ox, Oy, Oz are $(1, 1, -1)$. Through O' new axes $O'x'$, $O'y'$, $O'z'$ are taken such that they are respectively parallel to the original axes. Find the coordinates of O relative to the new axes. If a point P has coordinates $(-1, 2, 0)$ relative to the new axes, find its perpendicular distances from the xy-, xz-, yz-planes.

1.8 Find the perimeter of the triangle whose vertices lie at the points $(1, 0, 0)$, $(0, 1, 0)$ and $(0, 0, 1)$.

1.2 DIRECTION COSINES AND DIRECTION RATIOS

Direction cosines

Let OP be a line described in the sense from O (the origin) to a point P, and denote by α, β, γ the angles that OP makes with Ox, Oy, Oz (Fig. 1.5). We define the **direction cosines** of OP to be $\cos\alpha$, $\cos\beta$, $\cos\gamma$. For convenience we write

$$l = \cos\alpha, \quad m = \cos\beta, \quad n = \cos\gamma. \tag{1.3}$$

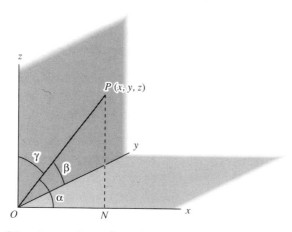

Fig. 1.5 The line OP makes angles α, β, γ with the axes.

The direction cosines of the x-axis, for example, are $1, 0, 0$.

Denote the foot of the perpendicular from P to the x-axis by N, let $OP = r$ and suppose that the coordinates of P are (x, y, z). From the triangle OPN we have $ON = |x| = r|\cos\alpha|$. Also, if α is an acute angle, $\cos\alpha$ and x are both positive, whilst if α is an obtuse angle, $\cos\alpha$ and x are both negative. It follows that $x = r\cos\alpha$, and similarly we may show that $y = r\cos\beta$, $z = r\cos\gamma$. The direction cosines of OP are therefore

$$l = x/r, \quad m = y/r, \quad n = z/r. \tag{1.4}$$

Since $r^2 = x^2 + y^2 + z^2$, we have

$$l^2 + m^2 + n^2 = 1. \tag{1.5}$$

This shows that the direction cosines of a line are not independent – they must satisfy (1.5).

The direction cosines of a line $\mathcal{L}$ not passing through the origin are defined to be the same as those of the parallel line drawn from the origin in the same sense as $\mathcal{L}$.

Direction ratios.

Any three numbers a, b, c such that

$$a : b : c = l : m : n \tag{1.6}$$

are referred to as **direction ratios** of OP. If (1.6) holds we have

$$l = a/d, \ m = b/d, \ n = c/d \tag{1.7}$$

where, by substituting into equations (1.5),

$$d = \pm \sqrt{(a^2 + b^2 + c^2)}. \tag{1.8}$$

The choice of sign in (1.8) indicates that there are two possible sets of direction cosines corresponding to a given set of direction ratios. These sets of direction cosines refer to oppositely directed parallel lines.

EXERCISES

1.9 Show that it is impossible for a line through the origin to be inclined at angles of $60°$, $120°$ and $30°$ to the x-axis, y-axis and z-axis, respectively, but that angles of $60°$, $120°$ and $135°$ are possible.

1.10 Find the direction cosines of the line joining the origin to the point $(6, 2, 5)$.

1.11 A line makes angles of $60°$ with both the x-axis and y-axis and is inclined at an obtuse angle to the z-axis. Show that its direction cosines are $\frac{1}{2}, \frac{1}{2}, -\frac{1}{2}\sqrt{2}$ and write down the angle it makes with the z-axis.

1.12 Find the direction cosines of the line which is equidistant from all three axes and is in the positive octant $x \geqslant 0$, $y \geqslant 0$, $z \geqslant 0$.

1.13 Find direction ratios for the line which makes an angle of $45°$ with the x-axis and an angle of $45°$ with the y-axis and which lies in the positive octant.

1.3 ANGLES BETWEEN LINES THROUGH THE ORIGIN

Consider two lines OA and OA' with direction cosines l, m, n and l', m', n'. To find the angle θ between them, denote by B, B' the points on OA, OA'

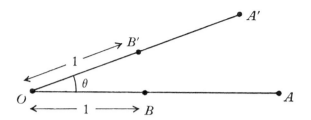

Fig. 1.6

(produced if necessary) such that $OB = OB' = 1$ (Fig. 1.6). Then the coordinates of B, B' are (l, m, n), (l', m', n'), using equation (1.4) with $r = 1$. Applying the cosine rule to the triangle OBB' gives

$$\cos \theta = \frac{OB^2 + OB'^2 - BB'^2}{2OB \cdot OB'} = 1 - \tfrac{1}{2} BB'^2.$$

But from (1.2)

$$BB'^2 = (l' - l)^2 + (m' - m)^2 + (n' - n)^2$$

$$= (l'^2 + m'^2 + n'^2) + (l^2 + m^2 + n^2) - 2(ll' + mm' + nn').$$

Using the results $l^2 + m^2 + n^2 = 1$, $l'^2 + m'^2 + n'^2 = 1$, we obtain

$$\cos \theta = ll' + mm' + nn'. \tag{1.9}$$

Note that, because $\cos (2\pi - \theta) = \cos \theta$, we still obtain equation (1.9) when the angle between OA and OA' is taken as $2\pi - \theta$.

Condition for perpendicular lines

Two lines through the origin are perpendicular if and only if

$$ll' + mm' + nn' = 0. \tag{1.10}$$

Proof

The two lines OA, OA' are at right angles if and only if $\theta = \tfrac{1}{2}\pi$ or $\theta = \tfrac{3}{2}\pi$, i.e. if and only if $\cos \theta = 0$. The result now follows from (1.9).

EXERCISES

1.14 Show that the angle between the lines whose direction cosines are $\tfrac{1}{2}\sqrt{2}$, $\tfrac{1}{2}\sqrt{2}$, 0 and $\tfrac{1}{3}\sqrt{3}$, $\tfrac{1}{3}\sqrt{3}$, $\tfrac{1}{3}\sqrt{3}$ is $\cos^{-1} \tfrac{1}{3}\sqrt{6}$.

1.15 Show that the lines whose direction cosines are $\tfrac{1}{2}\sqrt{2}$, 0, $\tfrac{1}{2}\sqrt{2}$ and $\tfrac{1}{2}\sqrt{2}$, 0, $-\tfrac{1}{2}\sqrt{2}$ are perpendicular.

1.16 Find the angle between any two of the diagonals of a cube. [*Hint.* Choose axes suitably with origin at the centre of the cube.]

1.4 THE ORTHOGONAL PROJECTION OF ONE LINE ON ANOTHER

Let two lines OP, OA meet at an angle θ. Then we define the orthogonal projection of OP on OA to be $OP \cos \theta$ (Fig. 1.7).

Note that if N is the foot of the perpendicular from P to OA (produced if necessary beyond O or beyond A), then $ON = OP \, |\cos \theta|$.

The work in Section 1.2 shows that the orthogonal projections of OP on rectangular cartesian axes with origin O are the x, y, z coordinates of P relative to these axes. We now extend this result to find the orthogonal projection of OP on a line OA which is not necessarily part of one of the coordinate axes.

Let the direction cosines of OA be l, m, n and let P be the point (x, y, z). Then the orthogonal projection of OP on OA is

$$lx + my + nz. \qquad (1.11)$$

Proof

By equations (1.4) the direction cosines of OP are x/r, y/r, z/r, where $r = OP$. Hence, by formula (1.9), the angle θ between OP and OA is given by

$$\cos \theta = (lx + my + nz)/r.$$

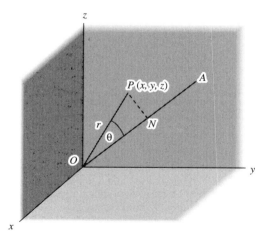

Fig. 1.7 ON is the orthogonal projection of OP on the line OA.

From the definition of the orthogonal projection of OP on OA, expression (1.11) follows at once.

EXERCISES

1.17 Points A, B have coordinates $(1, 4, -1), (-1, 3, 2)$ respectively. If O is the origin, find the point P on OA produced which is such that the orthogonal projection of OP on OB is of length $9\sqrt{14}/7$.

1.18 A line OP joins the origin O to the point $P(3, 1, 5)$. Show that the orthogonal projection of OP on the line in the positive octant making equal angles with all three axes is $3\sqrt{3}$.

1.19 The feet of the perpendiculars from the point $(4, -4, 0)$ to the lines through the origin whose direction cosines are $(\frac{1}{2}\sqrt{2}, \frac{1}{2}\sqrt{2}, 0), (\frac{1}{3}, \frac{2}{3}, \frac{2}{3})$ are denoted by N, N'. Find the lengths of ON, ON', where O is the origin, and explain why one of these lengths is zero.

1.5 ROTATION OF AXES

The transformation matrix and its properties

Consider two sets of right-handed rectangular cartesian axes $Oxyz$, $Ox'y'z'$. It is easily seen that, by a suitable continuous movement about O, the set of axes $Oxyz$ (with Ox, Oy, Oz always fixed relative to each other) may be brought into coincidence with the set $Ox'y'z'$. Such a movement will be called a **rotation** of the axes.

Note that if one set of axes is right-handed and the other left-handed, it is impossible to bring them into coincidence by a rotation. It will be convenient to refer to $Oxyz$ as the original axes and $Ox'y'z'$ as the new axes.

Let the direction cosines of Ox' relative to the axes $Oxyz$ be l_{11}, l_{12}, l_{13}. Further, denote the direction cosines of Oy' and Oz' by l_{21}, l_{22}, l_{23} and l_{31}, l_{32}, l_{33}. We may conveniently summarize this by the array

$$
\begin{array}{c|ccc}
O & x & y & z \\
\hline
x' & l_{11} & l_{12} & l_{13} \\
y' & l_{21} & l_{22} & l_{23} \\
z' & l_{31} & l_{32} & l_{33}
\end{array}
\qquad (1.12)
$$

In this array, the direction cosines of Ox' relative to the axes $Oxyz$ occur in the first row, the direction cosines of Oy' occur in the second row and those of Oz' in the third row. Furthermore, reading down the three columns in turn, it is easily seen that we obtain the direction cosines of the axes Ox, Oy, Oz

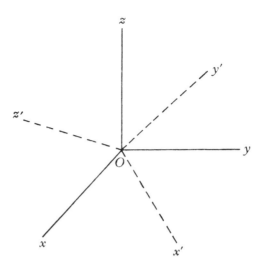

Fig. 1.8 A rotation of axes.

relative to the axes $Ox'y'z'$. The array of direction cosines in (1.12) is called the **transformation matrix**.

Since the axes Ox', Oy', Oz' are mutually perpendicular,

$$l_{11}l_{21} + l_{12}l_{22} + l_{13}l_{23} = 0,$$
$$l_{21}l_{31} + l_{22}l_{32} + l_{23}l_{33} = 0, \tag{1.13}$$
$$l_{31}l_{11} + l_{32}l_{12} + l_{33}l_{13} = 0.$$

Also, from Section 1.2, the sums of the squares of direction cosines are all unity and so

$$l_{11}^2 + l_{12}^2 + l_{13}^2 = 1,$$
$$l_{21}^2 + l_{22}^2 + l_{23}^2 = 1, \tag{1.14}$$
$$l_{31}^2 + l_{32}^2 + l_{33}^2 = 1.$$

The six equations in (1.13), (1.14) are called the **orthonormality conditions**: it should be observed how they are formed from the array (1.12).

Since the elements of the columns form the direction cosines of the axes Ox, Oy, Oz relative to the axes $Ox'y'z'$, it follows by a similar argument that

$$l_{11}l_{12} + l_{21}l_{22} + l_{31}l_{32} = 0,$$
$$l_{12}l_{13} + l_{22}l_{23} + l_{32}l_{33} = 0, \tag{1.15}$$
$$l_{13}l_{11} + l_{23}l_{21} + l_{33}l_{31} = 0,$$

and

$$l_{11}^2 + l_{21}^2 + l_{31}^2 = 1,$$
$$l_{12}^2 + l_{22}^2 + l_{32}^2 = 1, \qquad (1.16)$$
$$l_{13}^2 + l_{23}^2 + l_{33}^2 = 1.$$

Equations (1.15) and (1.16) are an important alternative form of the orthonormality conditions. They may be derived from equations (1.13) and (1.14) by a purely algebraic argument.

The transformation matrix satisfies one further condition which arises from the fact that the axes $Oxyz$, $Ox'y'z'$ are both right-handed. Consider the determinant

$$T = \begin{vmatrix} l_{11} & l_{12} & l_{13} \\ l_{21} & l_{22} & l_{23} \\ l_{31} & l_{32} & l_{33} \end{vmatrix}.$$

(For the reader unacquainted with determinants, an account of all the theory needed in this book is included in Appendix 1.)

Denoting the transpose of T by T' we have

$$T^2 = TT' = \begin{vmatrix} l_{11} & l_{12} & l_{13} \\ l_{21} & l_{22} & l_{23} \\ l_{31} & l_{32} & l_{33} \end{vmatrix} \times \begin{vmatrix} l_{11} & l_{21} & l_{31} \\ l_{12} & l_{22} & l_{32} \\ l_{13} & l_{23} & l_{33} \end{vmatrix}.$$

Hence, multiplying the two determinants and using the orthonormality conditions (1.13) and (1.14)

$$T^2 = \begin{vmatrix} 1 & 0 & 0 \\ 0 & 1 & 0 \\ 0 & 0 & 1 \end{vmatrix} = 1.$$

Thus $T = \pm 1$.

Now, when the axes $Oxyz$, $Ox'y'z'$ coincide, it is easily seen that the appropriate values of the direction cosines in the array (1.12) are $l_{ij} = 1$ when $i = j$, $l_{ij} = 0$ when $i \neq j$, and so for this particular case

$$T = \begin{vmatrix} 1 & 0 & 0 \\ 0 & 1 & 0 \\ 0 & 0 & 1 \end{vmatrix} = 1.$$

If the axes are rotated out of coincidence, the direction cosines l_{ij} will vary during the rotation in a continuous manner (i.e. with no 'sudden jumps' in value) and as the determinant T is the sum of products of the direction cosines its value will also vary continuously. But at all stages of the rotation $T = 1$ or -1 and so, for no discontinuity in value to occur, T must take the value 1 *throughout* the rotation, or else take the value -1. Since $T = 1$ when the two coordinate systems coincide, it follows that in all positions

$$T = \begin{vmatrix} l_{11} & l_{12} & l_{13} \\ l_{21} & l_{22} & l_{23} \\ l_{31} & l_{32} & l_{33} \end{vmatrix} = 1. \tag{1.17}$$

This is the additional condition to be satisfied by the transformation matrix.

We have shown that, if the components of the array (1.12) are the direction cosines of the new axes relative to the original axes, conditions (1.13), (1.14) and (1.17) are necessarily satisfied. These conditions are also sufficient for the array to represent a rotation of right-handed axes $Oxyz$. For, firstly, if equations (1.13) are satisfied the axes Ox', Oy', Oz' are mutually perpendicular; secondly, if equations (1.14) are satisfied the rows in the transformation matrix represent direction cosines of Ox', Oy', Oz'; and finally, if (1.17) is satisfied the system $Ox'y'z'$ is right-handed.

Transformation of coordinates

Let a point P have coordinates (x, y, z) and (x', y', z') relative to the axes $Oxyz$ and $Ox'y'z'$ respectively. The x'-, y'-, z'-coordinates of P are the orthogonal projections of OP on Ox', Oy', Oz'. Hence, using (1.11) to calculate these, we obtain

$$\begin{aligned} x' &= l_{11}\, x + l_{12}\, y + l_{13}\, z, \\ y' &= l_{21}\, x + l_{22}\, y + l_{23}\, z, \\ z' &= l_{31}\, x + l_{32}\, y + l_{33}\, z. \end{aligned} \tag{1.18}$$

Equations (1.18) show how the coordinates of P transform under a rotation of axes. It should be noted how these expressions are formed from the array (1.12).

We could, of course, regard the axes $Ox'y'z'$ as the original set and the axes $Oxyz$ as the new set, and determine the coordinates (x, y, z) in terms of (x', y', z'). Remembering that the elements of the columns in (1.12) are the direction cosines of the x-, y-, z-axes relative to the axes $Ox'y'z'$, it follows, by using (1.11) again, that

$$\begin{aligned} x &= l_{11}\, x' + l_{21}\, y' + l_{31}\, z', \\ y &= l_{12}\, x' + l_{22}\, y' + l_{32}\, z', \\ z &= l_{13}\, x' + l_{23}\, y' + l_{33}\, z'. \end{aligned} \tag{1.19}$$

The reader may verify as an exercise that equations (1.19) also follow algebraically from equations (1.18) by solving for x, y, z.

EXERCISES

1.20 Two sets of axes $Oxyz$, $Ox'y'z'$ are such that the first set may be placed in the position of the second set by a rotation of $180°$ about the x-axis. Write down in the form of array (1.12) the set of direction cosines which corresponds to this rotation. If a point has coordinates $(1,1,1)$ relative to the axes $Oxyz$, find its coordinates relative to the axes $Ox'y'z'$.

1.21 A set of axes $Ox'y'z'$ is initially coincident with a set $Oxyz$. The set $Ox'y'z'$ is then rotated through an angle θ about the z-axis, the direction of rotation being from the x-axis to the y-axis. Show that

$$x' = x \cos \theta + y \sin \theta,$$
$$y' = -x \sin \theta + y \sin \theta,$$
$$z' = z.$$

[*Hint.* Consider the direction cosines of the new axes and use equations (1.18).]

1.22 Show that the following equations represent a rotation of a set of axes about a fixed point:

$$x' = x \sin \theta \, \cos \varphi + y \sin \theta \, \sin \varphi + z \cos \theta,$$
$$y' = x \cos \theta \, \cos \varphi + y \cos \theta \, \sin \varphi - z \sin \theta,$$
$$z' = -x \sin \varphi + y \cos \varphi.$$

[*Hint.* Show that the coefficients of x, y, z satisfy (1.17) and the orthonormality conditions.]

1.23 Solve the equations of Exercise 1.22 for x, y, z in terms of x', y', z'. [*Hint.* Multiply the first equation by $\sin \theta \, \cos \varphi$, the second by $\cos \theta \, \cos \varphi$, the third by $-\sin \varphi$ and add to obtain

$$x = x' \sin \theta \, \cos \varphi + y' \cos \theta \, \cos \varphi - z' \sin \varphi.$$

Similarly for y and z.]

1.24 With reference to the transformation array (1.12) in the text, show that

$$l_{11} = l_{22} \, l_{33} - l_{23} \, l_{32},$$
$$l_{12} = l_{23} \, l_{31} - l_{21} \, l_{33},$$
$$l_{13} = l_{21} \, l_{32} - l_{22} \, l_{31}.$$

Write down two sets of three similar relations. [*Hint.* By using (1.17) show that if l_{11}, l_{12}, l_{13} have the values given above then $l_{11}^2 + l_{12}^2 + l_{13}^2 = 1$. The solution to this exercise is given with the answers at the end of the book.]

1.6 THE SUMMATION CONVENTION AND ITS USE

It is possible to simplify the statement of equations (1.18) and (1.19) by relabelling the coordinates (x, y, z) as (x_1, x_2, x_3) and the coordinates (x', y', z') as (x'_1, x'_2, x'_3). With this change of notation equations (1.18) become

$$x'_1 = l_{11}x_1 + l_{12}x_2 + l_{13}x_3 = \sum_{j=1}^{3} l_{1j} x_j \tag{1.20}$$

and

$$x'_2 = \sum_{j=1}^{3} l_{2j} x_j, \tag{1.21}$$

$$x'_3 = \sum_{j=1}^{3} l_{3j} x_j. \tag{1.22}$$

Even more briefly, we may write equations (1.20) to (1.22) in the form

$$x'_i = \sum_{j=1}^{3} l_{ij} x_j, \quad i = 1, 2, 3. \tag{1.23}$$

Similarly, equations (1.19) may be reduced to the form

$$x_i = \sum_{j=1}^{3} l_{ji} x'_j, \quad i = 1, 2, 3. \tag{1.24}$$

In equations (1.23), (1.24) the suffix j appears twice in the sums on the right-hand sides and we sum over all three possible values of j. This situation occurs so frequently that it is convenient to adopt a convention which often avoids the necessity of writing summation signs.

Summation convention

Whenever a suffix appears twice in the same expression, that expression is to be summed over all possible values of the suffix.

Using the summation convention, equations (1.23), (1.24) become simply

$$x'_i = l_{ij} x_j, \tag{1.25}$$

$$x_i = l_{ji} x'_j. \tag{1.26}$$

It is understood here that when a suffix is used alone in an equation (such as i on the left-hand and right-hand sides of (1.25) and (1.26)) the equation under consideration holds for each value of that suffix. The reader should notice that equations (1.25), (1.26) are much more elegant and more convenient than the original forms (1.18), (1.19).

Kronecker delta

The Kronecker delta is defined by

$$\delta_{ij} = \begin{cases} 0 & \text{when } i \neq j \\ 1 & \text{when } i = j. \end{cases} \tag{1.27}$$

By introducing this symbol and using the summation convention the orthonormality conditions (1.13), (1.14) are embodied in the single equation

$$l_{ik}\, l_{jk} = \delta_{ij}. \tag{1.28}$$

For example, taking $i = 1, j = 2$, this becomes

$$l_{1k}\, l_{2k} = 0;$$

that is

$$l_{11}l_{21} + l_{12}l_{22} + l_{13}l_{23} = 0,$$

which is the first of equations (1.13). Again, taking $i = j = 1$ in (1.28),

$$l_{1k}\, l_{1k} = 1;$$

that is

$$l_{11}^2 + l_{12}^2 + l_{13}^2 = 1,$$

which is the first of equations (1.14). Similarly, taking the other possible combinations of the suffixes i, j we may obtain the remaining four orthonormality conditions.

The alternative form of the orthonormality conditions as expressed by (1.15), (1.16) are embodied in the equation

$$l_{ki}\, l_{kj} = \delta_{ij}. \tag{1.29}$$

The Kronecker delta is a useful symbol in many contexts other than vector analysis.

Further remarks on the summation convention

1. A repeated suffix is known as a **dummy suffix** because it may be replaced by any other suitable symbol. For example,

$$l_{1j}\, l_{2j} = l_{1k}\, l_{2k} = l_{1\alpha}l_{2\alpha},$$

since in each expression summation over the repeated suffix is implied.
2. When the summation convention is in use, care must be taken to avoid using any suffix more than twice in the same expression. (The meaning of $l_{1j}\, l_{jj}$, for example, is not clear.)

3. As far as we are concerned in this book there are only three possible values for a suffix, namely 1, 2 and 3. The reader will appreciate, though, that elsewhere it might be convenient to increase or decrease the range of a suffix.

Some parts of vector analysis can be shortened considerably by using the convention. *We shall usually warn the reader when the summation convention is in use.*

EXERCISES

1.25 If

$$a_{11} = 1, \qquad a_{12} = -1, \qquad a_{13} = 0,$$
$$a_{21} = -2, \qquad a_{22} = 3, \qquad a_{23} = 1,$$
$$a_{31} = 2, \qquad a_{32} = 0, \qquad a_{33} = 4,$$

show that

$$a_{ii} = 8, \qquad a_{i1} a_{i2} = -7, \qquad a_{i2} a_{i3} = 3,$$
$$a_{1i} a_{2i} = -5, \qquad a_{2i} a_{3i} = 0, \qquad a_{i1} a_{2i} = -6.$$

1.26 If the numbers a_{ij} are as given in Exercise 1.25 and if

$$b_1 = 1, \quad b_2 = -1, \quad b_3 = 4,$$

show that

$$a_{1i} b_i = 2, \quad a_{j1} b_j = 11, \quad a_{ji} a_{i1} b_j = 49.$$

[*Hint.* For the last part, first evaluate $a_{j1} b_j$, $a_{j2} b_j$ and $a_{j3} b_j$.]

1.27 Show that

$$\delta_{ij} b_j = \delta_{ji} b_j = b_i.$$

1.28 If the numbers a_{ij} are as given in Exercise 1.25, evaluate

(i) $a_{1j} \delta_{1j}$, (ii) $a_{12} \delta_{ii}$, (iii) $a_{1i} a_{2k} \delta_{ik}$.

1.29 The suffix i may assume all integral values from 0 to ∞. If a_i, b_i are defined by

$$a_i = x^i, \quad b_i = \frac{1}{i!},$$

where x is a constant and, by definition, $0! = 1$, show that

$$a_i b_i = e^x.$$

1.30 If the quantities e_{ij}, e'_{ij} satisfy the relation

$$l_{mi} \, l_{nj} \, e_{ij} = e'_{mn},$$

and if

$$l_{ki} \, l_{kj} = \delta_{ij},$$

show that

$$e_{ij} = l_{mi} \, l_{nj} \, e'_{mn}.$$

[*Hint.* Multiply the first equation by $l_{mp} \, l_{nq}$.]

1.7 INVARIANCE WITH RESPECT TO A ROTATION OF THE AXES

Consider a function $f(\alpha_1, \alpha_2, \ldots)$ of several elements $\alpha_1, \alpha_2, \ldots$ such that given any set of rectangular cartesian axes $Oxyz$, the elements $\alpha_1, \alpha_2, \ldots$ are determined by a definite rule. Denote by $\alpha'_1, \alpha'_2, \ldots$ the elements corresponding to any other set of rectangular cartesian axes $Ox'y'z'$ with the same origin O. Then, if

$$f(\alpha'_1, \alpha'_2, \ldots) = f(\alpha_1, \alpha_2, \ldots)$$

the function f is said to be **invariant with respect to a rotation of the axes**.
  The examples which follow should clarify the idea of invariance.

Examples of invariants

1. The function $\sqrt{(x^2 + y^2 + z^2)}$ is invariant, since (1.19) gives

$$\begin{aligned}
\sqrt{(x^2 + y^2 + z^2)} = \sqrt{\{} &(l_{11}^2 + l_{12}^2 + l_{13}^2)x'^2 + (l_{21}^2 + l_{22}^2 + l_{23}^2)y'^2 \\
&+ (l_{31}^2 + l_{32}^2 + l_{33}^2)z'^2 + 2(l_{11} \, l_{21} + l_{12} \, l_{22} + l_{13} \, l_{23})x' \, y' \\
&+ 2(l_{21} \, l_{31} + l_{22} \, l_{32} + l_{23} \, l_{33})y' \, z' \\
&+ 2(l_{31} \, l_{11} + l_{32} \, l_{12} + l_{33} \, l_{13})z' \, x' \}.
\end{aligned}$$

Using the orthonormality conditions (1.13) and (1.14), this reduces to

$$\sqrt{(x^2 + y^2 + z^2)} = \sqrt{(x'^2 + y'^2 + z'^2)}.$$

This result has an immediate geometrical interpretation – it expresses the fact that the distance between the origin O and a point P does not depend upon the system of coordinates used in calculating the distance.
  The proof given above can be shortened considerably by using the summation convention (see Exercise 1.32 at the end of this section).
2. If OA and OB are two lines through the origin, the expression representing the cosine of the angle between them is clearly invariant with respect to a rotation of the axes. To verify this algebraically let (a_1, a_2, a_3), (b_1, b_2, b_3) be the coordinates of A, B relative to the axes $Oxyz$. If $OA = a$, $OB = b$, the direction cosines of OA and OB are

$$\frac{a_1}{a}, \frac{a_2}{a}, \frac{a_3}{a} \quad \text{and} \quad \frac{b_1}{b}, \frac{b_2}{b}, \frac{b_3}{b}.$$

If θ is the angle between OA, OB, formula (1.9) gives

$$\cos \theta = \frac{a_1 b_1 + a_2 b_2 + a_3 b_3}{ab} = \frac{a_i b_i}{ab},$$

using the summation convention. From Example 1 we see that a and b are invariant with respect to a rotation of the axes. Thus, to show that $\cos \theta$ is invariant, it only remains to show that $a_i b_i$ is invariant.

Using (1.25) we see that on transforming to new axes $Ox'y'z'$, the coordinates of A and B become (a_1', a_2', a_3') and (b_1', b_2', b_3'), where

$$a_i' = l_{ij} a_j, \quad b_i' = l_{ik} b_k.$$

Thus

$$a_i' b_i' = l_{ij} l_{ik} a_j b_k. \tag{1.30}$$

(Notice that before forming the expression for $a_i' b_i'$ different dummy suffixes must be used in the formulae for a_i' and b_i'; otherwise a suffix would appear more than twice in the right-hand side of (1.30).) By using the orthonormality conditions in the form (1.29) (with the suffixes changed to those required here), (1.30) becomes

$$\begin{aligned} a_i' b_i' &= \delta_{jk} a_j b_k \\ &= a_k b_k = a_i b_i. \end{aligned}$$

It follows that the quantity $a_i b_i \, (= a_1 b_1 + a_2 b_2 + a_3 b_3)$ is invariant under a rotation of the axes, as required, and that $\cos \theta$ is invariant also.

The concept of invariance with respect to rotation of the axes is important because the recognizable aspects of a physical system are usually invariant in this way.

For example, the distance between two points, the volume of a specified region and the resolute of a force along a given line are all independent of any special coordinate system, and the expressions which represent them are invariant with respect to a rotation of the axes.

EXERCISES

1.31 Find the x', y', z' coordinates of the points $x = 1$, $y = 1$, $z = 0$ and $x = 0$, $y = 1$, $z = 1$ for the rotation of axes given in Exercise 1.21, namely

$$x' = x \cos \theta + y \sin \theta,$$

$$y' = -x \sin \theta + y \cos \theta,$$
$$z' = z.$$

Verify that the angle between the lines joining the origin to these two points works out as $60°$ with either set of axes.

1.32 Show that the quantity $x_1^2 + x_2^2 + x_3^2 = x_i \, x_i$ is invariant under a rotation of axes.

1.8 MATRIX NOTATION

Another way of expressing some of the results obtained in this chapter is afforded by the use of matrices. It is not our intention to make much use of matrix notation in this book, but those readers familiar with matrices may welcome the following brief remarks.

The matrix of direction cosines in (1.12) may be denoted by

$$L = \begin{pmatrix} l_{11} & l_{12} & l_{13} \\ l_{21} & l_{22} & l_{23} \\ l_{31} & l_{32} & l_{33} \end{pmatrix}. \tag{1.31}$$

Its transpose is

$$L^T = \begin{pmatrix} l_{11} & l_{21} & l_{31} \\ l_{12} & l_{22} & l_{32} \\ l_{13} & l_{23} & l_{33} \end{pmatrix}. \tag{1.32}$$

With this notation the orthonormality conditions (1.13) and (1.14) may be expressed as

$$LL^T = I \tag{1.33}$$

where

$$I = \begin{pmatrix} 1 & 0 & 0 \\ 0 & 1 & 0 \\ 0 & 0 & 1 \end{pmatrix} \tag{1.34}$$

is the unit matrix. By pre-multiplying (1.33) by L^{-1}, which is the inverse of L, the result that

$$L^T = L^{-1} \tag{1.35}$$

is obtained.
Writing

$$\mathbf{x} = \begin{pmatrix} x_1 \\ x_2 \\ x_3 \end{pmatrix} \quad \text{and} \quad \mathbf{x}' = \begin{pmatrix} x_1' \\ x_2' \\ x_3' \end{pmatrix} \tag{1.36}$$

the transformation rules (1.25) and (1.26) may be expressed as

$$\mathbf{x}' = L\mathbf{x} \tag{1.37}$$

and

$$\mathbf{x} = L^T\mathbf{x}' \tag{1.38}$$

respectively. Expression (1.38) may be derived at once from (1.37) by pre-multiplying by L^{-1} and using (1.35).

Scalar and vector algebra | 2

2.1 SCALARS

Any mathematical entity or any property of a physical system which can be represented by a single real number is called a **scalar**. In the case of a physical property, the single real number will be a measure of the property in some chosen units (e.g. kilogrammes, metres, seconds).

Particular examples of scalars are: (i) the modulus of a complex number; (ii) mass; (iii) volume; (iv) temperature. Note that real numbers are themselves scalars.

Single letters printed in italics (such as a, b, c, etc.) will be used to denote real numbers representing scalars. For convenience statements like 'let a be a real number representing a scalar' will be abbreviated to 'let a be a scalar'.

Equality of scalars

Two scalars (measured in the same units if they are physical properties) are said to be equal if the real numbers representing them are equal.

It will be assumed throughout this book that in the case of physical entities the same units are used on both sides of any equality sign.

Scalar addition, subtraction, multiplication and division

The sum of two scalars is defined as the sum of the real numbers representing them. Similarly, scalar subtraction, multiplication and division are defined by the corresponding operations on the representative numbers. In the case of physical scalars, the operations of addition and subtraction are physically meaningful only for similar scalars such as two lengths or two masses.

Some care is necessary in the matter of units. For example, if a, b are two physical scalars it is meaningful to say their sum is $a + b$ only if the units of measurement are the same.

Again, consider the equation $T = \frac{1}{2} m v^2$ giving the kinetic energy T of a particle of mass m travelling with speed v. If T has the value 30 kg m^2 s^{-2} and v has the value 0.1 km s^{-1}, then to calculate $m = 2T/v^2$ consistent units for length and time must first be introduced. Thus, converting the given speed to m s^{-1} we find v has the value 100 m s^{-1}. Hence the value of m is $2 \times 30/10000 = 0.006$ kg.

Henceforth it is to be understood that consistent units of measurement are used in operations involving physical properties.

2.2 VECTORS: BASIC NOTIONS

From an elementary standpoint the reader will probably have already encountered properties of physical systems which require for their complete specification a scalar magnitude and a direction – the velocity of a moving point and the force on a body are particular examples. Such properties are called vectors. We define a vector formally below, but as the definition might otherwise seem strange we shall try to put it into perspective by continuing for the moment on an intuitive basis.

Consider the velocity of a point P moving relative to fixed rectangular cartesian coordinate axes $Oxyz$. Denote by v_1, v_2, v_3 the rate at which P is travelling away from the yz-, zx-, xy-planes in the directions x-, y-, z-increasing respectively. Then v_1, v_2, v_3 are called the components of the velocity; together they completely describe the instantaneous motion of P.

If the axes $Oxyz$ are moved to new fixed positions $Ox'y'z'$, without rotation, the velocity components of P relative to the new axes will be v_1, v_2, v_3 as before: for v_1 is the rate at which P moves away from the yz-plane and this will be the same as the rate at which it moves away from the $y'z'$-plane since the two planes are parallel; and similarly for the other two directions.

If the axes are rotated about O to new positions $Ox'y'z'$ the velocity components, v_1', v_2', v_3' say, of P relative to the new axes must be related in some way to the original components, for the two sets of components are each sufficient to define the motion of P. The relationship will depend upon the relative position of the two sets of axes (which is defined by the direction cosines of one set relative to the other), and by continuing this reasoning on an intuitive basis it could be obtained explicitly.

The formal definition which follows will be seen to fit into the pattern which the brief remarks above suggest. That vectors do indeed have an associated magnitude and direction will follow later in this section as a consequence of the definition and so the consistency of this with intuitive ideas will be confirmed.

Definition

A **vector** is any mathematical or physical entity which is such that:
1. when it is associated with a set of rectangular cartesian axes $Oxyz$ it can be represented completely by three scalars a_1, a_2, a_3 related, in turn, to the axes of x, y, and z;
2. the triad of scalars in 1 is invariant under a translation of the axes, i.e. if $Oxyz$, $O'x'y'z'$ are rectangular cartesian axes (with different origins O, O') such that Ox is parallel to $O'x'$, Oy is parallel to $O'y'$ and Oz is parallel to $O'z'$, and if the triads associated with the two coordinate systems are (a_1, a_2, a_3), (a_1', a_2', a_3') respectively, then $a_1 = a_1'$, $a_2 = a_2'$, $a_3 = a_3'$;
3. if the triads of scalars associated with two sets of axes $Oxyz$, $Ox'y'z'$ (with the same origin O) are (a_1, a_2, a_3), (a_1', a_2', a_3') respectively, and if the direction cosines of Ox', Oy', Oz' relative to the axes $Oxyz$ are given by the transformation matrix (1.12), then

$$a_1' = l_{11} a_1 + l_{12} a_2 + l_{13} a_3,$$
$$a_2' = l_{21} a_1 + l_{22} a_2 + l_{23} a_3, \qquad (2.1)$$
$$a_3' = l_{31} a_1 + l_{32} a_2 + l_{33} a_3.$$

Introducing the summation convention, equations (2.1) reduce to

$$a_i' = l_{ij} a_j \quad (i = 1, 2, 3). \qquad (2.2)$$

Vectors will be denoted by letters printed in heavy type. Thus

$$\mathbf{a} = (a_1, a_2, a_3). \qquad (2.3)$$

The scalars a_1, a_2, a_3 are called respectively the x-component, y-component, z-component of the vector $\mathbf{a}$.

To avoid repetition, components of all vectors are henceforth to be taken as referred to axes $Oxyz$ unless stated otherwise.

Equality of vectors

Two vectors $\mathbf{a} = (a_1, a_2, a_3)$ and $\mathbf{b} = (b_1, b_2, b_3)$, referred to the same coordinate system, are defined to be equal if and only if their respective components are equal, i.e.

$$\mathbf{a} = \mathbf{b} \iff a_1 = b_1, \ a_2 = b_2, \ a_3 = b_3. \qquad (2.4)$$

The zero vector

The vector whose components are all zero is called the zero vector or null vector and is written as

$$\mathbf{0} = (0, 0, 0). \qquad (2.5)$$

The position vector

Let A and B be points whose coordinates relative to axes $Oxyz$ are (a_1, a_2, a_3) and (b_1, b_2, b_3), respectively. The position of B relative to A is a vector written

$$\overrightarrow{AB} = (b_1 - a_1, \ b_2 - a_2, \ b_3 - a_3).$$

It is called the **position vector of B relative to A**.

Proof

To prove that $\overrightarrow{AB}$ is a vector it is necessary to establish that conditions 1 to 3 of the definition are satisfied.

1. The scalars $b_1 - a_1$, $b_2 - a_2$, $b_3 - a_3$ are the coordinates of B relative to axes $Ax'y'z'$ drawn through A and parallel to the original axes $Oxyz$ (Fig. 2.1). These scalars are clearly related to the axes Ox, Oy, Oz respectively and they define completely the position of B relative to A (within the coordinate system $Oxyz$).

2. Suppose the axes $Oxyz$ are moved parallel to themselves so that they pass through a new origin whose coordinates are $(-x_0, -y_0, -z_0)$ referred to the axes in their original position. The coordinates of A become $(x_0 + a_1, y_0 + a_2, z_0 + a_3)$ and those of B become $(x_0 + b_1, y_0 + b_2, z_0 + b_3)$. Thus, if $(\overrightarrow{AB})'$ denotes the position vector referred to the new axes,

$$(\overrightarrow{AB})' = [(x_0 + b_1) - (x_0 + a_1), \ (y_0 + b_2) - (y_0 + a_2), \ (z_0 + b_3) - (z_0 + a_3)]$$
$$= (b_1 - a_1, b_2 - a_2, b_3 - a_3).$$

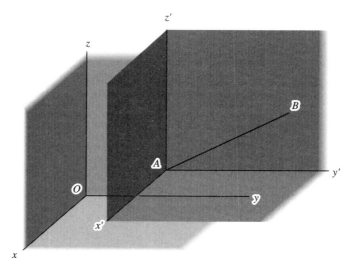

Fig. 2.1

It follows that the components of $\overrightarrow{AB}$ are invariant with respect to a translation of the axes.

3. Let Ox', Oy', Oz' be rectangular coordinate axes whose direction cosines relative to the axes $Oxyz$ are given by (1.12). Using (1.23), the coordinates (a_1', a_2', a_3') and (b_1', b_2', b_3') of A and B referred to the axes $Ox'y'z'$ are

$$a_i' = \sum_{j=1}^{3} l_{ij} a_j, \quad b_i' = \sum_{j=1}^{3} l_{ij} b_j \quad (i = 1, 2, 3)$$

giving

$$b_i' - a_i' = \sum_{j=1}^{3} l_{ij} (b_j - a_j) \quad (i = 1, 2, 3). \tag{2.6}$$

It follows that the components of $\overrightarrow{AB}$ obey the vector transformation law (2.2). As all three conditions of the definition are satisfied, $\overrightarrow{AB}$ is a vector.

The vector $\overrightarrow{OP}$ giving the position of a point $P(x, y, z)$ relative to an origin O is called the **position vector** of P. Later it will also be convenient to denote this vector by $\mathbf{r} = (x, y, z)$.

Notice that the position vector of a point O relative to itself is the zero vector $(0, 0, 0)$.

Examples of vectors occurring in physical systems

Many properties of physical systems are vectors. For example, velocities and accelerations of moving points, forces, angular velocities, angular accelerations and couples are all vectors. In the theory of electricity and magnetism, electric field strength, magnetic field strength and electric current density are also vectors.

In the next chapter (section 3.7), velocity, acceleration and angular velocity will be defined. We shall also refer occasionally to some of the other vectors mentioned above; their definitions may be found in appropriate reference books.

Geometrical representation of vectors

Let $\mathbf{a} = (a_1, a_2, a_3)$ be any non-zero vector and let A be the point whose x-, y-, z-coordinates are a_1, a_2, a_3. Then $\overrightarrow{OA} = (a_1, a_2, a_3)$, showing that $\mathbf{a}$ and $\overrightarrow{OA}$ have the same components. Thus, the **directed straight line segment** OA may be taken as a geometrical representation of $\mathbf{a}$. When a directed line segment such as OA represents a vector this may be shown in a diagram by drawing an arrow on the line pointing from O to A (Fig. 2.2). Any zero vector is appropriately represented geometrically by a single point.

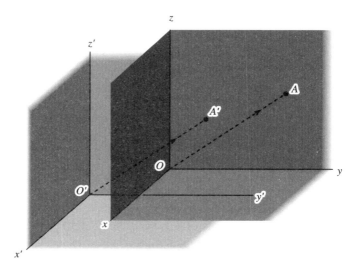

Fig. 2.2 $\overrightarrow{O'A'} = \overrightarrow{OA}$

Let $O'A'$ be a line segment equal in length to OA and drawn parallel to and in the same sense as OA. If $O'x'$, $O'y'$, $O'z'$ are axes through O' such that they are parallel to Ox, Oy, Oz respectively, the coordinates of A' referred to these new axes will be (a_1, a_2, a_3). Hence $\overrightarrow{O'A'}$ is a second geometrical representation of the vector **a**. It follows that the geometrical representation of **a** as a directed line segment is not unique. Any other directed line segment $O'A'$ which is parallel to OA and of equal length may also be used to represent **a**.

Direction of a vector

Because a non-zero vector can be represented as a directed line segment, a vector is said to have (or be associated with) a **direction**. Naturally this is taken to be the same as the direction of the directed line segments which represent the vector. Thus the direction of the vector **a** is the direction in which OA points. The direction of the null vector is not defined.

Two vectors are said to be **parallel** if they are in the same direction and **anti-parallel** if they are in opposite directions.

Magnitude of a vector

The **magnitude** of a vector $\mathbf{a} = (a_1, a_2, a_3)$ is defined as

$$a = \sqrt{(a_1^2 + a_2^2 + a_3^2)}. \tag{2.7}$$

It will also be convenient occasionally to denote the magnitude of a vector $\mathbf{a}$ by $|\mathbf{a}|$.

If $\mathbf{a}$ is represented geometrically as $\overrightarrow{OA} = (a_1, a_2, a_3)$ it is seen that a is just the length of the line segment OA. Since this length is invariant under a rotation of the axes (cf. section 1.7, Example 1) it follows that the magnitude of a vector is also invariant. The magnitude of a vector is also sometimes termed its **modulus** or its **norm**.

Unit vectors

A vector of unit magnitude is called a **unit vector**. These are frequently distinguished by a circumflex; thus $\hat{\mathbf{r}} = (\cos\theta, \sin\theta, 0)$ is a unit vector.

If $\mathbf{a}$ is any vector, the unit vector whose direction is that of $\mathbf{a}$ is denoted by $\hat{\mathbf{a}}$.

EXERCISES

2.1 Show that equations (2.2) are equivalent to

$$a_k = l_{ik} a_i'.$$

[*Hint.* Multiply (2.2) by l_{ik} and use the orthonormality conditions (1.29).]

2.2 Relative to axes $Oxyz$, points P, Q have coordinates $(1, 2, 3)$, $(0, 0, 1)$. Find the components, referred to the axes $Oxyz$, of: (i) the position vector of P relative to O; (ii) the position vector of O relative to P; (iii) the position vector of P relative to Q.

2.3 Find the magnitudes of the vectors $\mathbf{a} = (1, 3, 4)$ and $\mathbf{b} = (2, -1, 0)$.

2.4 Show that the vectors $\mathbf{a} = (0, -3, 3)$, $\mathbf{b} = (0, -5, 5)$ have the same direction. What is the ratio a/b?

2.5 The transformation matrix for a rotation from axes $Oxyz$ to axes $Ox'y'z'$ is

O	x	y	z
x'	0	1	0
y'	-1	0	0
z'	0	0	1.

Describe (in words or by a diagram) how the positions of the two sets of axes are related.

Relative to the set of axes $Oxyz$, a vector $\mathbf{a}$ has components $(2, 1, 2)$. Find the components of $\mathbf{a}$ relative to the axes $Ox'y'z'$.

2.6 Show that $\mathbf{a} = (\cos\theta, \sin\theta\cos\varphi, \sin\theta\sin\varphi)$ is a unit vector.

2.7 Find two unit vectors which are perpendicular to each of the vectors $\mathbf{a} = (0, 0, 1)$, $\mathbf{b} = (0, 1, 1)$. Are there any more unit vectors which are perpendicular to both $\mathbf{a}$ and $\mathbf{b}$?

2.3 MULTIPLICATION OF A VECTOR BY A SCALAR

If the components of a vector $\mathbf{a}$ in every rectangular cartesian coordinate system are each multiplied by a scalar λ, we define the ordered triads of scalars so formed to be components of $\lambda\mathbf{a}$(or $\mathbf{a}\lambda$). Thus, if $\mathbf{a} = (a_1, a_2, a_3)$ when referred to axes $Oxyz$,

$$\lambda\mathbf{a} = \mathbf{a}\lambda = (\lambda a_1, \lambda a_2, \lambda a_3). \tag{2.8}$$

It is easily seen that $\lambda\mathbf{a}$ is a vector. Condition 1 of the definition given in section 2.2 is obviously satisfied, as is condition 2, since the triad (a_1, a_2, a_3) is invariant under a translation of the axes. Further, referred to the axes $Ox'y'z'$ (defined in section 2.2),

$$\mathbf{a} = (a_1', a_2', a_3')$$

and hence the components of $\lambda\mathbf{a}$ referred to these axes are $(\lambda a_1', \lambda a_2', \lambda a_3')$. To satisfy 3 we require

$$\lambda a_i' = l_{ij}(\lambda a_j), \quad (i = 1, 2, 3)$$

and as these are the same as equations (2.2) multiplied by λ, the result follows. Note that when $\lambda = 0$, $\lambda\mathbf{a}$ is the null vector defined by equation (2.5).

From equation (2.8) and the definition of the magnitude of a vector

$$\begin{aligned}
|\lambda\mathbf{a}| &= \sqrt{(\lambda^2 a_1^2 + \lambda^2 a_2^2 + \lambda^2 a_3^2)} \\
&= |\lambda|\sqrt{(a_1^2 + a_2^2 + a_3^2)} \\
&= |\lambda|a.
\end{aligned}$$

Thus, when a vector is multiplied by a scalar λ, its magnitude is multiplied by $|\lambda|$.

Multiplication of a non-zero vector by a positive scalar leaves the direction of the vector unchanged, but multiplication by a negative scalar reverses its direction. To see this, let $\overrightarrow{OA} = (a_1, a_2, a_3)$ represent $\mathbf{a}$; $\lambda\mathbf{a}$ will then be represented by $\lambda\overrightarrow{OA} = (\lambda a_1, \lambda a_2, \lambda a_3) = \overrightarrow{OA'}$, say. Since A, A' have coordinates (a_1, a_2, a_3), $(\lambda a_1, \lambda a_2, \lambda a_3)$ respectively, the direction cosines of OA and OA' will be

$$\frac{a_1}{a}, \frac{a_2}{a}, \frac{a_3}{a} \quad \text{and} \quad \frac{\lambda a_1}{|\lambda|a}, \frac{\lambda a_2}{|\lambda|a}, \frac{\lambda a_3}{|\lambda|a}$$

where $a = \sqrt{(a_1^2 + a_2^2 + a_3^2)}$. If $\lambda > 0$, $\lambda/|\lambda| = 1$, whilst if $\lambda < 0$, $\lambda/|\lambda| = -1$. It follows that OA, OA' are in the same direction if $\lambda > 0$ and in opposite directions if $\lambda < 0$ (Fig. 2.3). The direction of $\lambda\mathbf{a}$ is therefore the same as or opposite to that of $\mathbf{a}$ according as λ is positive or negative.

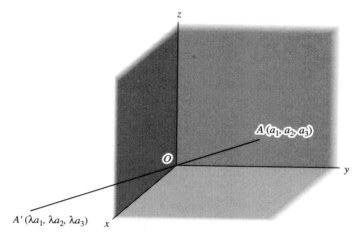

Fig. 2.3 The case $\lambda < 0$.

The vector $-\mathbf{a}$

We define

$$-\mathbf{a} = (-1)\mathbf{a}. \qquad (2.9)$$

If $\mathbf{a}$, $-\mathbf{a}$ are represented by $\overrightarrow{OA}$, $\overrightarrow{OA'}$ respectively, then OA, OA' will be oppositely directed line segments of equal length (Fig. 2.4).

Note that $\overrightarrow{OA} = -\overrightarrow{AO}$ since AO and OA', being parallel directed line segments of equal length, represent the same vector.

Fig. 2.4 Oppositely directed line segments OA and OA' of equal lengths.

EXERCISES

2.8 A point P on a line AB is such that $|AP| : |PB| = 3 : 2$. Show that
$$2\overrightarrow{AP} = \pm 3\overrightarrow{PB}.$$

2.9 Show that, for any vector $\mathbf{a}$,
$$\mathbf{a} = a\hat{\mathbf{a}}$$
where $\hat{\mathbf{a}}$ is a unit vector in the same direction as $\mathbf{a}$.

2.10 Show that the four points with position vectors

$$\mathbf{r}_1, \ \mathbf{r}_2, \ \frac{r_2}{r_1}\mathbf{r}_1, \ \frac{r_1}{r_2}\mathbf{r}_2,$$

where $r_1 \neq 0$, $r_2 \neq 0$, lie on a circle.

2.4 ADDITION AND SUBTRACTION OF VECTORS

Addition

The **sum** of two vectors $\mathbf{a} = (a_1, a_2, a_3)$ and $\mathbf{b} = (b_1, b_2, b_3)$ is defined as

$$\mathbf{a} + \mathbf{b} = (a_1 + b_1, a_2 + b_2, a_3 + b_3). \tag{2.10}$$

The set of all vectors is **closed** under the operation of addition, i.e. the sum of two vectors $\mathbf{a}$, $\mathbf{b}$ is a vector. For, considering the requirements 1–3 of the definition of a vector (section 2.2), it is seen at once that 1 and 2 are satisfied. Also, if (a_1', a_2', a_3'), (b_1', b_2', b_3') are the components of $\mathbf{a}$, $\mathbf{b}$ referred to the axes $Ox'y'z'$, equations (2.2) give

$$a_i' = l_{ij}\, a_j \ \text{ and } \ b_i' = l_{ij}\, b_j \ (i = 1, 2, 3).$$

Adding, we have

$$a_i' + b_i' = l_{ij}\,(a_j + b_j)$$

showing that the components of $\mathbf{a} + \mathbf{b}$ satisfy requirement 3.

As

$$a_i + b_i = b_i + a_i \quad (i = 1, 2, 3),$$

it follows from (2.10) that

$$\mathbf{a} + \mathbf{b} = \mathbf{b} + \mathbf{a}, \tag{2.11}$$

i.e. the operation of addition is **commutative**.

Also, if $\mathbf{c} = (c_1, c_2, c_3)$ is a third vector, then since

$$(a_i + b_i) + c_i = a_i + (b_i + c_i) \quad (i = 1, 2, 3)$$

we have

$$(\mathbf{a} + \mathbf{b}) + \mathbf{c} = \mathbf{a} + (\mathbf{b} + \mathbf{c}), \tag{2.12}$$

i.e. the operation of addition is **associative**.

It is important to appreciate that rules like the commutative and associative laws of addition cannot be taken for granted by analogy with similar laws for real numbers. A new mathematical system is being established involving entities which are not real numbers and there is no reason to suppose, until this has been proved, that such rules of manipulation are obeyed.

The triangle law of addition

From the definition of addition of vectors, we may deduce the triangle law of addition (also sometimes called the parallelogram law) as follows:

If two vectors are represented geometrically by $\overrightarrow{AB}$ and $\overrightarrow{BC}$, then their sum is represented by $\overrightarrow{AC}$ (Fig. 2.5). Thus

$$\overrightarrow{AB} + \overrightarrow{BC} = \overrightarrow{AC}.$$

$\overrightarrow{AC}$ is often called the **resultant** of $\overrightarrow{AB}$ and $\overrightarrow{BC}$.

Proof

Construct axes $Axyz$ through the point A and draw the line AD so as to complete the parallelogram $ABCD$ (Fig. 2.6). Let the coordinates of B, D be (b_1, b_2, b_3), (d_1, d_2, d_3) respectively. Then, as BC is parallel to AD, the coordinates of C will be $(b_1 + d_1, b_2 + d_2, b_3 + d_3)$. Thus, referred to the axes $Axyz$,

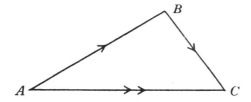

Fig. 2.5 $\overrightarrow{AB} + \overrightarrow{BC} = \overrightarrow{AC}$ (the triangle law of addition).

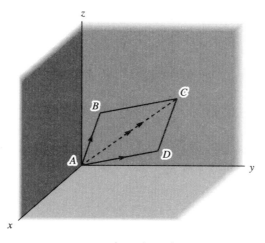

Fig. 2.6 If $ABCD$ is a parallelogram, $\overrightarrow{AB} + \overrightarrow{AD} = \overrightarrow{AC}$ (the parallelogram law).

$$\vec{AB} = (b_1, b_2, b_3),$$

$$\vec{BC} = \vec{AD} = (d_1, d_2, d_3),$$

$$\vec{AC} = (b_1 + d_1, b_2 + d_2, b_3 + d_3).$$

It follows at once from the law of addition of vectors that, as required,

$$\vec{AB} + \vec{BC} = \vec{AC}.$$

Subtraction

The **difference** of two vectors **a**, **b** is defined as

$$\mathbf{a} - \mathbf{b} = \mathbf{a} + (-\mathbf{b}). \tag{2.13}$$

Note that for any vector **a**

$$\mathbf{a} - \mathbf{a} = \mathbf{0}.$$

Figure 2.7 shows how the sum and difference of vectors **a**, **b** may be represented geometrically.

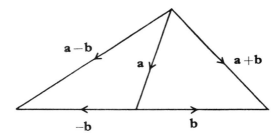

Fig. 2.7 The sum and difference of vectors **a** and **b**.

EXAMPLE 1

The position vectors of three points, A, B, P, relative to an origin O, are such that

$$\vec{OP} = \frac{\lambda \vec{OA} + \mu \vec{OB}}{\lambda + \mu}$$

where λ, μ are non-zero real numbers. Prove that P lies on AB (possibly produced) and that $AP : PB = |\mu| : |\lambda|$ (Fig. 2.8).

Solution

By the triangle law,

$$\vec{OA} = \vec{OP} + \vec{PA}$$

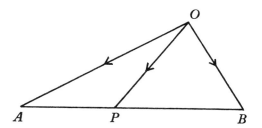

Fig. 2.8

and

$$\vec{OB} = \vec{OP} + \vec{PB}.$$

To satisfy the given relation between $\vec{OP}, \vec{OA}, \vec{OB}$ we require

$$\vec{OP} = \frac{\lambda(\vec{OP} + \vec{PA}) + \mu(\vec{OP} + \vec{PB})}{\lambda + \mu},$$

and hence

$$\mathbf{0} = \lambda \vec{PA} + \mu \vec{PB}.$$

Since $-\vec{PA} = \vec{AP}$ this may be expressed as

$$\vec{AP} = \frac{\mu}{\lambda} \vec{PB} \quad (\lambda, \mu \neq 0). \tag{2.14}$$

Equating the magnitudes of each side of (2.14) gives $AP = |\mu/\lambda| PB$. Thus

$$AP : PB = |\mu| : |\lambda|.$$

If λ, μ have the same sign, (2.14) shows that $\vec{AP}, \vec{PB}$ are in the same direction and so P lies on AB and between A and B. If λ, μ have opposite signs, $\vec{AP}, \vec{PB}$ are in opposite directions, hence P lies on AB produced beyond B (if $|\mu|/|\lambda| > 1$) or beyond A (if $|\mu|/|\lambda| < 1$).

EXERCISES

2.11 If **a** and **b** are vectors as given below, verify that their sums and differences are as shown.

	a	**b**	**a + b**	**a − b**
(i)	$(2, 2, 2)$	$(1, 0, 1)$	$(3, 2, 3)$	$(1, 2, 1)$
(ii)	$(3, 0, 0)$	$(5, 0, 0)$	$(8, 0, 0)$	$(-2, 0, 0)$
(iii)	$(1, -2, 6)$	$(-1, -3, 7)$	$(0, -5, 13)$	$(2, 1, -1)$.

2.12 Relative to axes $Oxyz$, points A, B are such that

$$\overrightarrow{OA} = (1, 1, 1), \quad \overrightarrow{AB} = (0, -1, 3).$$

What is the position vector of: (i) B relative to O; (ii) O relative to B?

2.13 On a flat horizontal plane an observer walks one mile N followed by one mile E, one mile S and one mile W. Explain vectorially why he finds himself back at the starting point.

2.14 If the angle between vectors $\mathbf{a}$ and $\mathbf{b}$ is 60°, and if $a = b = 3$, show that

$$|\mathbf{a} - \mathbf{b}| = 3.$$

2.15 From the property $AC \leqslant AB + BC$ of a triangle ABC, prove that

$$|\mathbf{a} + \mathbf{b}| \leqslant a + b.$$

For what particular cases is it true that

$$|\mathbf{a} + \mathbf{b}| = a + b ?$$

2.16 Prove that

$$|\mathbf{a} - \mathbf{b}| \leqslant a + b.$$

2.17 If $\hat{\mathbf{u}}, \hat{\mathbf{v}}$ are unit vectors with different directions, show that the vector $\hat{\mathbf{u}} + \hat{\mathbf{v}}$ bisects internally the angle between $\hat{\mathbf{u}}$ and $\hat{\mathbf{v}}$. Is $\frac{1}{2}(\hat{\mathbf{u}} + \hat{\mathbf{v}})$ a unit vector?

2.5 THE UNIT VECTORS i, j, k

Let $\mathbf{i}, \mathbf{j}, \mathbf{k}$ denote unit vectors in the directions of the x-axis, y-axis, z-axis respectively. Then

$$\mathbf{i} = (1, 0, 0), \quad \mathbf{j} = (0, 1, 0), \quad \mathbf{k} = (0, 0, 1).$$

Using the rules for multiplication of a vector by a scalar and for addition of vectors, a vector $\mathbf{a} = (a_1, a_2, a_3)$ may be written

$$\mathbf{a} = a_1 \mathbf{i} + a_2 \mathbf{j} + a_3 \mathbf{k}.$$

It is easily shown that, when the triad $\mathbf{i}, \mathbf{j}, \mathbf{k}$ is given, this representation of $\mathbf{a}$ is unique.

The three vectors $\mathbf{i}, \mathbf{j}, \mathbf{k}$ are unit vectors which are mutually perpendicular. Any set of three mutually perpendicular unit vectors is said to be an **orthonormal** set. Because any arbitrary vector can be represented as a linear combination of $\mathbf{i}, \mathbf{j}, \mathbf{k}$, these vectors are also said to form an **orthonormal basis** for the totality of all vectors. Orthonormal bases play an important role in vector analysis.

EXERCISES

2.18 A unit vector **a** in the positive quadrant of the xy-plane makes an angle of $45°$ with each of the axes Ox and Oy. Show that

$$\mathbf{a} = (\mathbf{i} + \mathbf{j})\,/\sqrt{2}.$$

2.19 The position vectors of points A and B relative to the origin O of axes $Oxyz$ are $\mathbf{i} - \mathbf{j} + 2\mathbf{k}$ and $5\mathbf{i} + \mathbf{j} + 6\mathbf{k}$ respectively. Show that $AB = 6$.

2.20 Find a, b, c if

$$(a + b - 2)\mathbf{i} + (c - 1)\,\mathbf{j} + (a + c)\mathbf{k} = \mathbf{0}.$$

2.6 SCALAR PRODUCTS

The **scalar product** (or **dot product**) of two vectors $\mathbf{a} = (a_1, a_2, a_3)$ and $\mathbf{b} = (b_1, b_2, b_3)$ is defined as

$$\mathbf{a}.\mathbf{b} = a_1 b_1 + a_2 b_2 + a_3 b_3. \tag{2.15}$$

This operation between two vectors is commutative because

$$\begin{aligned} \mathbf{b}.\mathbf{a} &= b_1 a_1 + b_2 a_2 + b_3 a_3 \\ &= \mathbf{a}.\mathbf{b}. \end{aligned} \tag{2.16}$$

The scalar product of **a** with itself is

$$\mathbf{a}.\mathbf{a} = a_1^2 + a_2^2 + a_3^2 = a^2 \,;$$

thus **a**.**a** is the square of the magnitude of **a**.

If $\mathbf{a} = (a_1, a_2, a_3)$, $\mathbf{b} = (b_1, b_2, b_3)$ and $\mathbf{c} = (c_1, c_2, c_3)$, then

$$\mathbf{a}.(\mathbf{b} + \mathbf{c}) = \mathbf{a}.\mathbf{b} + \mathbf{a}.\mathbf{c}; \tag{2.17}$$

this is the **distributive law**. It is easily proved, for we have

$$\begin{aligned} \mathbf{a}.(\mathbf{b} + \mathbf{c}) &= (a_1, a_2, a_3) . (b_1 + c_1,\; b_2 + c_2,\; b_3 + c_3) \\ &= (a_1 b_1 + a_1 c_1 + a_2 b_2 + a_2 c_2 + a_3 b_3 + a_3 c_3) \\ &= (a_1 b_1 + a_2 b_2 + a_3 b_3) + (a_1 c_1 + a_2 c_2 + a_3 c_3) \\ &= \mathbf{a}.\mathbf{b} + \mathbf{a}.\mathbf{c}\,. \end{aligned}$$

Scalar invariants

Any scalar which takes the same value in each coordinate system with which it may be associated is called a **scalar invariant**. Thus the components of a vector $\mathbf{a} = (a_1, a_2, a_3)$ are not scalar invariants because they may take different values in different coordinate systems. However the magnitude of **a**, i.e. $a = \sqrt{(a_1^2 + a_2^2 + a_3^2)}$, is a scalar invariant.

Since $\mathbf{a}.\mathbf{a} = a^2$, the scalar product of $\mathbf{a}$ with itself is a scalar invariant. This is a special case of the following more general result.

Scalar products are scalar invariants

To prove this, let (a_1', a_2', a_3'), (b_1', b_2', b_3') be the components of $\mathbf{a}$, $\mathbf{b}$ relative to the axes $Ox'y'z'$. From equation (2.2)

$$a_i' = l_{ij}\,a_j \quad \text{and} \quad b_i' = l_{ik}\,b_k \quad (i = 1, 2, 3)$$

and so

$$a_i'\,b_i' = l_{ij}\,l_{ik}\,a_j\,b_k.$$

But, from the orthonormality conditions (1.29)

$$l_{ij}\,l_{ik} = \delta_{jk}.$$

Hence

$$\begin{aligned} a_i'\,b_i' &= \delta_{jk}\,a_j\,b_k \\ &= a_k\,b_k, \end{aligned}$$

i.e. $a_1'\,b_1' + a_2'\,b_2' + a_3'\,b_3' = a_1 b_1 + a_2 b_2 + a_3 b_3$. This shows that the scalar product of $\mathbf{a}$ and $\mathbf{b}$ is invariant under a rotation of the axes. Since the scalar product is also obviously invariant under a translation of the axes (for the components of $\mathbf{a}$, $\mathbf{b}$ are themselves invariant under such a translation) it now follows that it is a scalar invariant.

Geometrical representation

Let two non-zero vectors $\mathbf{a} = (a_1, a_2, a_3)$, $\mathbf{b} = (b_1, b_2, b_3)$ be represented by $\overrightarrow{OA}$, $\overrightarrow{OB}$ and let the direction cosines relative to the axes $Oxyz$ of the directed line segments OA and OB be $\alpha_1, \alpha_2, \alpha_3$ and $\beta_1, \beta_2, \beta_3$ respectively. Then, since the coordinates of A, B are (a_1, a_2, a_3), (b_1, b_2, b_3), we have by equations (1.4)

$$\alpha_i = \frac{a_i}{a}, \quad \beta_i = \frac{b_i}{b} \quad (i = 1, 2, 3).$$

Thus, using the summation convention,

$$ab\alpha_i\beta_i = a_i b_i.$$

But if θ is the angle between OA, OB (Fig. 2.9), equation (1.9) shows that

$$\cos\theta = \alpha_i\,\beta_i.$$

Hence

$$\mathbf{a}\,.\,\mathbf{b} = ab\cos\theta. \tag{2.18}$$

This is an important relationship, often used as the definition of scalar product.

Note that since $\cos(2\pi - \theta) = \cos\theta$, no ambiguity arises if $2\pi - \theta$ is taken as the angle between OA, OB.

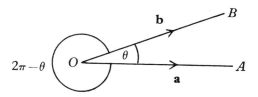

Fig. 2.9

Two non-zero vectors **a** and **b** are at right angles to each other if and only if $\mathbf{a} \cdot \mathbf{b} = 0$

Proof

If **a** and **b** are at right angles, $\theta = \frac{1}{2}\pi$ (or $\frac{3}{2}\pi$) so, by equation (2.18), $\mathbf{a} \cdot \mathbf{b} = 0$. Also, if $\mathbf{a} \cdot \mathbf{b} = 0$ and $a \neq 0$, $b \neq 0$ then $\cos\theta = 0$; hence $\theta = \frac{1}{2}\pi$ (or $\frac{3}{2}\pi$) showing that **a** and **b** are at right angles.

Scalar products of pairs of i, j, k

The unit vectors **i**, **j**, **k**, introduced in the previous section, are such that the scalar product of any one with itself is unity and the scalar product of any one with any other is zero, for these vectors are each of unit magnitude and are mutually perpendicular. Thus

and
$$\left. \begin{aligned} \mathbf{i} \cdot \mathbf{i} = \mathbf{j} \cdot \mathbf{j} = \mathbf{k} \cdot \mathbf{k} = 1, \\ \mathbf{i} \cdot \mathbf{j} = \mathbf{j} \cdot \mathbf{k} = \mathbf{k} \cdot \mathbf{i} = 0. \end{aligned} \right\} \quad (2.19)$$

Using these relations and the distributive law, the scalar product of two vectors

$$\mathbf{a} = a_1\mathbf{i} + a_2\mathbf{j} + a_3\mathbf{k} \quad \text{and} \quad \mathbf{b} = b_1\mathbf{i} + b_2\mathbf{j} + b_3\mathbf{k}$$

may be evaluated as follows:

$$\begin{aligned} \mathbf{a} \cdot \mathbf{b} &= (a_1\mathbf{i} + a_2\mathbf{j} + a_3\mathbf{k}) \cdot (b_1\mathbf{i} + b_2\mathbf{j} + b_3\mathbf{k}) \\ &= a_1 b_1 \mathbf{i} \cdot \mathbf{i} + a_1 b_2 \mathbf{i} \cdot \mathbf{j} + a_1 b_3 \mathbf{i} \cdot \mathbf{k} \\ &\quad + a_2 b_1 \mathbf{j} \cdot \mathbf{i} + a_2 b_2 \mathbf{j} \cdot \mathbf{j} + a_2 b_3 \mathbf{j} \cdot \mathbf{k} \\ &\quad + a_3 b_1 \mathbf{k} \cdot \mathbf{i} + a_3 b_2 \mathbf{k} \cdot \mathbf{j} + a_3 b_3 \mathbf{k} \cdot \mathbf{k} \\ &= a_1 b_1 + a_2 b_2 + a_3 b_3. \end{aligned}$$

Of course, the definition (2.15) gives this at once.

Resolution of vectors

The **resolute** of a vector $\mathbf{a}$ in the direction of (or along) a unit vector $\hat{\mathbf{n}}$ is defined as

$$a_n = \mathbf{a} . \hat{\mathbf{n}} . \tag{2.20}$$

If θ is the angle between $\mathbf{a}$ and $\hat{\mathbf{n}}$, resolving $\mathbf{a}$ in the direction of $\hat{\mathbf{n}}$ gives

$$a_n = \mathbf{a} . \hat{\mathbf{n}} = a |\hat{\mathbf{n}}| \cos \theta = a \cos \theta.$$

Figure 2.10 shows the geometrical interpretation of this. The resolute of $\mathbf{a}$ in the direction of $\hat{\mathbf{n}}$ is the sensed projection ON of $\mathbf{a}$ on $\hat{\mathbf{n}}$ (produced if necessary).

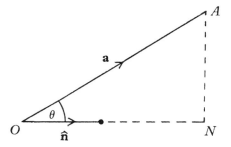

Fig. 2.10 The resolute of $\mathbf{a}$ in the direction $\hat{\mathbf{n}}$ is the sensed projection ON of $\mathbf{a}$ on $\hat{\mathbf{n}}$.

The resolute of $\mathbf{a}$ along any non-zero vector $\mathbf{b}$ is defined as $\mathbf{a} . \hat{\mathbf{b}}$ where

$$\hat{\mathbf{b}} = \mathbf{b}/b = (b_1, b_2, b_3)/(b_1^2 + b_2^2 + b_3^2)^{\frac{1}{2}}$$

is the unit vector in the same direction as $\mathbf{b}$.
    If

$$\mathbf{a} = a_1 \mathbf{i} + a_2 \mathbf{j} + a_3 \mathbf{k}$$

then the resolutes of $\mathbf{a}$ in the directions $\mathbf{i}$, $\mathbf{j}$, $\mathbf{k}$ are respectively

$$\mathbf{a} . \mathbf{i} = a_1, \quad \mathbf{a} . \mathbf{j} = a_2, \quad \mathbf{a} . \mathbf{k} = a_3.$$

Thus the resolutes of $\mathbf{a}$ along the x-, y-, z-axes are identical to the corresponding components of $\mathbf{a}$. It should be noted however that this result is not true if $\mathbf{a}$ is referred to a non-orthogonal coordinate system (see Exercise 2.45, section 2.8).
    Using the above results, we can write down at once the components, relative to given orthogonal axes, of a given vector of given magnitude and direction. For example, suppose that $\mathbf{a}$ is of magnitude 2 and is in the direction making an angle of 60° with Ox, 120° with Oy and 135° with Oz (cf Exercise 1.9). Then the components of $\mathbf{a}$ are

$$a_1 = 2 \cos 60° \ = 1,$$

$$a_2 = 2 \cos 120° = - 1,$$

$$a_3 = 2 \cos 135° = - \sqrt{2}.$$

The vector is then said to have been **resolved into components**.

EXERCISES

2.21 If **a** and **b** are vectors as given in the table below, verify that the scalar products are as shown. Which of the three pairs of vectors is perpendicular?

	a	**b**	**a . b**
(i)	$(1, -1, 0)$	$(3, 4, 5)$	-1
(ii)	$(4, 1, -3)$	$(-1, 3, -7)$	20
(iii)	$(3, 1, 4)$	$(2, -2, -1)$	$0.$

2.22 Using the formula $\mathbf{a} . \mathbf{b} = ab \cos \theta$, find the angle between the vectors $\mathbf{a} = (0, -1, 1)$, $\mathbf{b} = (3, 4, 5)$.

2.23 A set of rectangular cartesian axes is so arranged that the x-axis points east, the y-axis points north and the z-axis points vertically upwards. Evaluate the scalar products of the vectors **a** and **b** in the following cases.
  (i) **a** is of magnitude 3 and points SE, **b** is of magnitude 2 and points E;
  (ii) **a** is of unit magnitude and points NE, **b** is of magnitude 2 and points vertically upwards;
  (iii) **a** is of unit magnitude and points NE, **b** is of magnitude 2 and points W.

2.24 If

$$\mathbf{a} = \mathbf{i} - \mathbf{j}, \quad \mathbf{b} = -\mathbf{j} + 2\mathbf{k},$$

show that

$$(\mathbf{a} + \mathbf{b}) . (\mathbf{a} - 2\mathbf{b}) = -9.$$

2.25 Show that the vectors $\mathbf{i} + \mathbf{j} + \mathbf{k}$, $\lambda^2 \mathbf{i} - 2\lambda \mathbf{j} + \mathbf{k}$ are perpendicular if and only if $\lambda = 1$.

2.26 Find the component of **i** in the direction of the vector $\mathbf{i} + \mathbf{j} + 2\mathbf{k}$.

2.27 Resolve the vector $3\mathbf{i} + 4\mathbf{j}$ in the directions of the vectors $4\mathbf{i} - 3\mathbf{j}, 4\mathbf{i} + 3\mathbf{j}$, and **k**.

2.28 If **a** and **b** are such that $a = b$, use Fig. 2.11 to interpret geometrically the relation

$$(\mathbf{a} + \mathbf{b}) . (\mathbf{a} - \mathbf{b}) = a^2 - b^2 = 0.$$

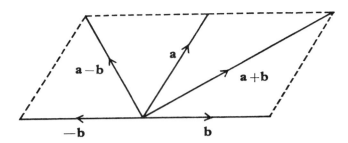

Fig. 2.11

2.29 Prove vectorially that the perpendiculars onto the sides of a triangle from the opposite vertices are concurrent. [*Hint.* Draw the perpendiculars from the vertices A, B of a triangle ABC and let them intersect at O. Let the position vectors of A, B, C relative to O be $\mathbf{a}$, $\mathbf{b}$, $\mathbf{c}$. Show that

$$\mathbf{a} \cdot (\mathbf{b} - \mathbf{c}) = \mathbf{b} \cdot (\mathbf{c} - \mathbf{a}) = 0.$$

Deduce that $\mathbf{c} \cdot (\mathbf{a} - \mathbf{b}) = 0$, and give the interpretation.]

2.7 VECTOR PRODUCTS

The **vector product** of a vector $\mathbf{a} = (a_1, a_2, a_3)$ with a vector $\mathbf{b} = (b_1, b_2, b_3)$ is defined as

$$\mathbf{a} \times \mathbf{b} = (a_2 b_3 - a_3 b_2, \ a_3 b_1 - a_1 b_3, \ a_1 b_2 - a_2 b_1). \tag{2.21}$$

Alternatively the vector product of

$$\mathbf{a} = a_1\mathbf{i} + a_2\mathbf{j} + a_3\mathbf{k} \ \text{ and } \ \mathbf{b} = b_1\mathbf{i} + b_2\mathbf{j} + b_3\mathbf{k}$$

is defined to be

$$\mathbf{a} \times \mathbf{b} = \begin{vmatrix} \mathbf{i} & \mathbf{j} & \mathbf{k} \\ a_1 & a_2 & a_3 \\ b_1 & b_2 & b_3 \end{vmatrix}. \tag{2.22}$$

The notation used in (2.21) anticipates that the vector product of two vectors is a vector and this result will now be established.

It should be noted first that $\mathbf{a} \times \mathbf{b}$ clearly satisfies conditions 1 and 2 of the definition of a vector, given in section 2.2, and so it only remains to be shown that 3 is satisfied.

For convenience put

$$\begin{aligned} \mathbf{a} \times \mathbf{b} &= (a_2 b_3 - a_3 b_2, \ a_3 b_1 - a_1 b_3, \ a_1 b_2 - a_2 b_1) \\ &= (c_1, c_2, c_3). \end{aligned}$$

Referred to the axes $Ox'y'z'$ (defined in section 2.2),

$$\mathbf{a} = (a_1', a_2', a_3'), \quad \mathbf{b} = (b_1', b_2', b_3')$$

and so, relative to these axes,

$$\mathbf{a} \times \mathbf{b} = (a_2' b_3' - a_3' b_2', \ a_3' b_1' - a_1' b_3', \ a_1' b_2' - a_2' b_1')$$
$$= (c_1', c_2', c_3').$$

Condition 3 will be satisfied if it can be proved that

$$\begin{aligned}
c_1' &= l_{11}c_1 + l_{12}c_2 + l_{13}c_3, \\
c_2' &= l_{21}c_1 + l_{22}c_2 + l_{23}c_3, \\
c_3' &= l_{31}c_1 + l_{32}c_2 + l_{33}c_3.
\end{aligned} \qquad (2.23)$$

Using the transformation law (2.1) for the vectors $\mathbf{a}$, $\mathbf{b}$ the quantity $c_1' = a_2' b_3' - a_3' b_2'$ becomes

$$\begin{aligned}
c_1' &= (l_{21}a_1 + l_{22}a_2 + l_{23}a_3)(l_{31}b_1 + l_{32}b_2 + l_{33}b_3) \\
&\quad - (l_{31}a_1 + l_{32}a_2 + l_{33}a_3)(l_{21}b_1 + l_{22}b_2 + l_{23}b_3) \\
&= (l_{22}l_{33} - l_{23}l_{32})(a_2 b_3 - a_3 b_2) \\
&\quad + (l_{23}l_{31} - l_{21}l_{33})(a_3 b_1 - a_1 b_3) \\
&\quad + (l_{21}l_{32} - l_{22}l_{31})(a_1 b_2 - a_2 b_1).
\end{aligned}$$

Substituting the results of Exercise 1.24, section 1.5, and using the definitions of c_1, c_2, c_3, this reduces to

$$c_1' = l_{11}c_1 + l_{12}c_2 + l_{13}c_3$$

which is the first of relations (2.23). The other two relations may be verified in a similiar way, and the proof that $\mathbf{a} \times \mathbf{b}$ is a vector is then complete.

Interchanging $\mathbf{a}$ and $\mathbf{b}$ in (2.21) gives

$$\mathbf{b} \times \mathbf{a} = (b_2 a_3 - b_3 a_2, \ b_3 a_1 - b_1 a_3, \ b_1 a_2 - b_2 a_1);$$

thus

$$\mathbf{b} \times \mathbf{a} = -\mathbf{a} \times \mathbf{b} \qquad (2.24)$$

showing that *the operation of vector multiplication is non-commutative. It is essential therefore to preserve the order of the vectors in a vector product.*

If $\mathbf{a}$, $\mathbf{b}$, $\mathbf{c}$ are any three vectors it is easily verified that

$$\mathbf{a} \times (\mathbf{b} + \mathbf{c}) = \mathbf{a} \times \mathbf{b} + \mathbf{a} \times \mathbf{c}; \qquad (2.25)$$

thus the vector product obeys the **distributive law**.

Vector products of pairs of i, j, k

Since

$$\mathbf{i} = (1, 0, 0), \ \mathbf{j} = (0, 1, 0), \ \mathbf{k} = (0, 0, 1)$$

we have

$$\mathbf{i} \times \mathbf{i} = \begin{vmatrix} \mathbf{i} & \mathbf{j} & \mathbf{k} \\ 1 & 0 & 0 \\ 1 & 0 & 0 \end{vmatrix} = \mathbf{0},$$

$$\mathbf{i} \times \mathbf{j} = \begin{vmatrix} \mathbf{i} & \mathbf{j} & \mathbf{k} \\ 1 & 0 & 0 \\ 0 & 1 & 0 \end{vmatrix} = \mathbf{k}$$

and four similar relations obtained by cyclic permutation of $\mathbf{i}$, $\mathbf{j}$, $\mathbf{k}$. In all

$$\left. \begin{array}{l} \mathbf{i} \times \mathbf{i} = \mathbf{j} \times \mathbf{j} = \mathbf{k} \times \mathbf{k} = \mathbf{0}, \\ \mathbf{i} \times \mathbf{j} = \mathbf{k}, \ \mathbf{j} \times \mathbf{k} = \mathbf{i}, \ \mathbf{k} \times \mathbf{i} = \mathbf{j}. \end{array} \right\} \tag{2.26}$$

These identities should be compared with the corresponding identities (2.19) involving scalar products of pairs of the vectors $\mathbf{i}$, $\mathbf{j}$, $\mathbf{k}$. Note that because of (2.24), interchanging the two vectors on the left-hand side of any one of the second set of identities (2.26) changes the sign in that identity; for example $\mathbf{j} \times \mathbf{i} = -\mathbf{k}$.

Equations (2.26) may be used in conjunction with the distributive law (2.25) to evaluate vector products. For example if

$$\mathbf{u} = \mathbf{i} + 3\mathbf{j} + \mathbf{k}, \ \mathbf{v} = 2\mathbf{i} - \mathbf{j} + 2\mathbf{k}$$

then

$$\begin{aligned} \mathbf{u} \times \mathbf{v} &= (\mathbf{i} + 3\mathbf{j} + \mathbf{k}) \times (2\mathbf{i} - \mathbf{j} + 2\mathbf{k}) \\ &= (2\mathbf{i} \times \mathbf{i} - \mathbf{i} \times \mathbf{j} + 2\mathbf{i} \times \mathbf{k}) + (6\mathbf{j} \times \mathbf{i} - 3\mathbf{j} \times \mathbf{j} + 6\mathbf{j} \times \mathbf{k}) \\ &\quad + (2\mathbf{k} \times \mathbf{i} - \mathbf{k} \times \mathbf{j} + 2\mathbf{k} \times \mathbf{k}) \\ &= (-\mathbf{k} - 2\mathbf{j}) + (-6\mathbf{k} + 6\mathbf{i}) + (2\mathbf{j} + \mathbf{i}) \\ &= 7\mathbf{i} - 7\mathbf{k}. \end{aligned}$$

It is however much more convenient to use (2.22). Thus

$$\mathbf{u} \times \mathbf{v} = \begin{vmatrix} \mathbf{i} & \mathbf{j} & \mathbf{k} \\ 1 & 3 & 1 \\ 2 & -1 & 2 \end{vmatrix} = 7\mathbf{i} - 7\mathbf{k}.$$

Geometrical interpretation of the vector product

Let $\mathbf{a}$, $\mathbf{b}$ be given vectors and choose rectangular axes $Oxyz$ such that $\mathbf{a}$, $\mathbf{b}$ are parallel to the xy-plane and Ox is in the same direction as $\mathbf{a}$. Let α be the angle between $\mathbf{a}$ and $\mathbf{b}$, measured in the sense turning from Ox into the positive quadrant of the xy-plane (Fig. 2.12). Then

$$\mathbf{a} = a_1\mathbf{i}, \ \mathbf{b} = b_1\mathbf{i} + b_2\mathbf{j}$$

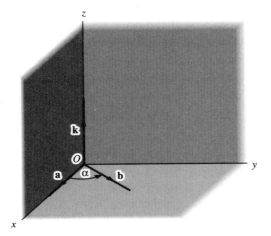

Fig. 2.12 The vector product $\mathbf{a} \times \mathbf{b}$ is of magnitude $ab \sin \alpha$ and in the direction of $\mathbf{k}$.

giving

$$\mathbf{a} \times \mathbf{b} = a_1 b_2 \, \mathbf{k}.$$

Also, for this special configuration,

$$a_1 = a, \quad b_2 = b \sin \alpha.$$

Hence

$$\mathbf{a} \times \mathbf{b} = ab \sin \alpha \, \mathbf{k}. \tag{2.27}$$

If $0 < \alpha < \pi$, $\sin \alpha > 0$ and so $\mathbf{a} \times \mathbf{b}$ is in the direction Oz. If $\pi < \alpha < 2\pi$, $\sin \alpha < 0$ so that $\mathbf{a} \times \mathbf{b}$ is in the opposite direction to Oz. However, if the angle

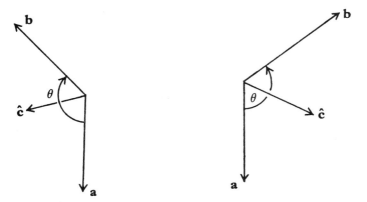

Fig. 2.13 The angle θ is always chosen in the range $0 \leqslant \theta \leqslant \pi$. In each diagram $\hat{\mathbf{c}}$ is perpendicular to both $\mathbf{a}$ and $\mathbf{b}$.

between **a** and **b** is defined suitably, the geometrical interpretation may be expressed more conveniently, as follows.

Let θ be the angle between **a** and **b**, measured in the sense turning from **a** to **b** and chosen so that $0 \leqslant \theta \leqslant \pi$. Then the vector product $\mathbf{a} \times \mathbf{b}$ is the vector $ab \sin \theta \; \hat{\mathbf{c}}$, where $\hat{\mathbf{c}}$ is a unit vector at right angles to both **a** and **b**, and such that, to an observer looking in the direction of $\hat{\mathbf{c}}$, the sense in which θ is measured is clockwise. Fig. 2.13 makes the situation clear. Since $0 \leqslant \theta \leqslant \pi$, the magnitude of $\mathbf{a} \times \mathbf{b}$ is $ab \sin \theta$ and its direction is that of $\hat{\mathbf{c}}$.

Two non-zero vectors **a**, **b** are parallel or anti-parallel if and only if $\mathbf{a} \times \mathbf{b} = 0$

Proof

If **a** and **b** are parallel, $\theta = 0$, and if **a** and **b** are anti-parallel, $\theta = \pi$. In either case $\mathbf{a} \times \mathbf{b} = \mathbf{0}$ since $\sin \theta = 0$. Also, if $\mathbf{a} \times \mathbf{b} = \mathbf{0}$ and if $a \neq 0, b \neq 0$, then $\sin \theta = 0$; hence $\theta = 0$ or π, showing that **a** and **b** are either parallel or anti-parallel.

EXAMPLE 2

Show that the area of a parallelogram with adjacent sides **a** and **b** is $|\mathbf{a} \times \mathbf{b}|$.

Solution

Denote by θ the smaller angle between **a** and **b** (Fig. 2.14). Drop a perpendicular from the end of **b** onto **a**. This perpendicular will be of length $b \sin \theta$. Thus the area of the parallelogram is base × height = $ab \sin \theta$ which is $|\mathbf{a} \times \mathbf{b}|$.

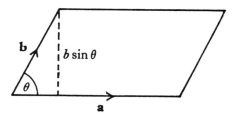

Fig. 2.14

EXAMPLE 3

Find the most general form for the vector **r** satisfying the equation

$$\mathbf{r} \times (1, 1, 1) = (2, -4, 2).$$

Solution

Let $\mathbf{r} = (a, b, c)$. Substituting this into the equation given, we have

$$(a, b, c) \times (1, 1, 1) = (2, -4, 2).$$

Thus

$$(b - c, c - a, a - b) = (2, -4, 2).$$

Using the definition of equality of vectors, this gives

$$b - c = 2,$$
$$c - a = -4,$$
$$a - b = 2.$$

These equations are not independent but they are consistent, for adding the first two gives

$$b - a = -2$$

which is the third equation.

Let

$$a = \lambda.$$

Then it follows at once that

$$b = \lambda - 2, \quad c = \lambda - 4.$$

Hence the general solution of the equation given may be represented in the form

$$\mathbf{r} = (\lambda, \ \lambda - 2, \ \lambda - 4)$$

where λ is arbitrary.

Remarks

If

$$\mathbf{r} \times \mathbf{a} = \mathbf{b}$$

the geometrical interpretation of vector products shows that $\mathbf{r}$ and $\mathbf{a}$ must both be perpendicular to $\mathbf{b}$. Hence, if $\mathbf{a}$ and $\mathbf{b}$ are given, the equation can have no solution for $\mathbf{r}$ unless $\mathbf{a}$ is perpendicular to $\mathbf{b}$.

It can be shown (see Exercise 2.54 at the end of this chapter) that the general solution for $\mathbf{r}$ is

$$\mathbf{r} = \lambda \mathbf{a} + (\mathbf{a} \times \mathbf{b})/a^2.$$

The reader may verify these remarks by reference to the example solved above.

EXAMPLE 4

Show vectorically that the bisectors of the angles of a triangle are concurrent.

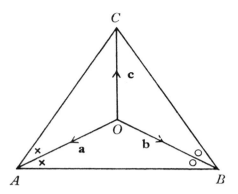

Fig. 2.15

Solution

Let the bisectors of the angles A and B of a triangle ABC intersect at O. Let the position vectors of A, B, C relative to O be $\mathbf{a}$, $\mathbf{b}$, $\mathbf{c}$, respectively. Then (see Fig. 2.15)

$$\overrightarrow{AC} = \mathbf{c} - \mathbf{a}, \quad \overrightarrow{CB} = \mathbf{b} - \mathbf{c}, \quad \overrightarrow{BA} = \mathbf{a} - \mathbf{b}.$$

Now if $\hat{\mathbf{u}}$ and $\hat{\mathbf{v}}$ are unit vectors, the internal bisector of the angle between them is parallel to the vector $\hat{\mathbf{u}} + \hat{\mathbf{v}}$ (see Exercise 2.17, section 2.4). Unit vectors in the directions of $\overrightarrow{CA}$ and $\overrightarrow{BA}$ are

$$\frac{\mathbf{a} - \mathbf{c}}{|\mathbf{a} - \mathbf{c}|} \quad \text{and} \quad \frac{\mathbf{a} - \mathbf{b}}{|\mathbf{a} - \mathbf{b}|}.$$

Thus $\mathbf{a}$ is parallel to

$$\frac{\mathbf{a} - \mathbf{c}}{|\mathbf{a} - \mathbf{c}|} + \frac{\mathbf{a} - \mathbf{b}}{|\mathbf{a} - \mathbf{b}|}.$$

Similarly $\mathbf{b}$ is parallel to

$$\frac{\mathbf{b} - \mathbf{a}}{|\mathbf{b} - \mathbf{a}|} + \frac{\mathbf{b} - \mathbf{c}}{|\mathbf{b} - \mathbf{c}|}.$$

These conditions may be expressed as

$$\mathbf{a} \times \left[\frac{\mathbf{a} - \mathbf{c}}{|\mathbf{a} - \mathbf{c}|} + \frac{\mathbf{a} - \mathbf{b}}{|\mathbf{a} - \mathbf{b}|} \right] = \mathbf{0}$$

and

$$\mathbf{b} \times \left[\frac{\mathbf{b} - \mathbf{a}}{|\mathbf{b} - \mathbf{a}|} + \frac{\mathbf{b} - \mathbf{c}}{|\mathbf{b} - \mathbf{c}|} \right] = \mathbf{0}.$$

Since $\mathbf{a} \times \mathbf{a} = \mathbf{0}$ and $\mathbf{b} \times \mathbf{b} = \mathbf{0}$ these simplify to

$$\mathbf{a} \times \left[\frac{\mathbf{c}}{|\mathbf{a} - \mathbf{c}|} + \frac{\mathbf{b}}{|\mathbf{a} - \mathbf{b}|} \right] = \mathbf{0} \tag{2.28}$$

and

$$\mathbf{b} \times \left[\frac{\mathbf{a}}{|\mathbf{b} - \mathbf{a}|} + \frac{\mathbf{c}}{|\mathbf{b} - \mathbf{c}|} \right] = \mathbf{0}. \tag{2.29}$$

The condition that CO should be the bisector of the angle at C can now be written down (by cyclic symmetry) as

$$\mathbf{c} \times \left[\frac{\mathbf{b}}{|\mathbf{c} - \mathbf{b}|} + \frac{\mathbf{a}}{|\mathbf{c} - \mathbf{a}|} \right] = \mathbf{0}. \tag{2.30}$$

The result (2.30) follows by adding (2.28) and (2.29), and observing that:

(i) $|\mathbf{a} - \mathbf{c}| = |\mathbf{c} - \mathbf{a}|$, with similar results for $|\mathbf{a} - \mathbf{b}|$ and $|\mathbf{b} - \mathbf{c}|$; and
(ii) $\mathbf{a} \times \mathbf{b} + \mathbf{b} \times \mathbf{a} = \mathbf{0}$, $\mathbf{a} \times \mathbf{c} = -\mathbf{c} \times \mathbf{a}$, $\mathbf{b} \times \mathbf{c} = -\mathbf{c} \times \mathbf{b}$.

EXERCISES

2.30 Show that, if vectors $\mathbf{a}$ and $\mathbf{b}$ are as given in the first two columns of the table below, $\mathbf{a} \times \mathbf{b}$ is as given in the third column.

	$\mathbf{a}$	$\mathbf{b}$	$\mathbf{a} \times \mathbf{b}$
(i)	$(3, 7, 2)$	$(1, 3, 1)$	$(1, -1, 2)$
(ii)	$(1, -3, 0)$	$(-2, 5, 0)$	$(0, 0, -1)$
(iii)	$(8, 8, -1)$	$(5, 5, 2)$	$(21, -21, 0)$.

2.31 Axes $Oxyz$ are positioned so that Ox points east, Oy points north and Oz points vertically upwards. Find the vector product $\mathbf{a} \times \mathbf{b}$ of vectors $\mathbf{a}$ and $\mathbf{b}$ in the following cases:

(i) $\mathbf{a}$ is of unit magnitude and points E, $\mathbf{b}$ is of magnitude 2 and points 30° N of E;
(ii) $\mathbf{a}$ is of unit magnitude and points E, $\mathbf{b}$ is of magnitude 2 and points SW;
(iii) $\mathbf{a}$ is of unit magnitude and points vertically upwards, $\mathbf{b}$ is of unit magnitude and points NE.

2.32 By taking components, prove the distributive law for vector products, namely

$$\mathbf{a} \times (\mathbf{b} + \mathbf{c}) = \mathbf{a} \times \mathbf{b} + \mathbf{a} \times \mathbf{c}.$$

2.33 Prove that for any scalar λ,

$$\mathbf{a} \times (\lambda \mathbf{b}) = (\lambda \mathbf{a}) \times \mathbf{b} = \lambda (\mathbf{a} \times \mathbf{b}).$$

2.34 If

$$\mathbf{a} \times \mathbf{b} = \mathbf{a} - \mathbf{b},$$

prove that

$$\mathbf{a} = \mathbf{b}.$$

2.35 Find the most general form for the vector $\mathbf{u}$ satisfying the equation

$$\mathbf{u} \times (2, 1, -1) = (1, 0, 0) \times (2, 1, -1).$$

2.36 Find a, b if

$$(a\mathbf{i} + b\mathbf{j} + \mathbf{k}) \times (2\mathbf{i} + 2\mathbf{j} + 3\mathbf{k}) = \mathbf{i} - \mathbf{j}.$$

2.37 By constructing an example, show that in general the associative law for vector products does not hold; that is, there exist $\mathbf{a}$, $\mathbf{b}$, $\mathbf{c}$ such that

$$\mathbf{a} \times (\mathbf{b} \times \mathbf{c}) \neq (\mathbf{a} \times \mathbf{b}) \times \mathbf{c}.$$

2.38 Prove vectorially that the medians of a triangle are concurrent. [*Hint.* Let E, F, G be the mid-points of the sides BC, CA, AB of a triangle ABC. Let AE, BF meet at O and denote the position vectors of A, B, C relative to O by $\mathbf{a}$, $\mathbf{b}$, $\mathbf{c}$. Find the position vectors of E and F, and deduce that

$$\mathbf{a} \times (\mathbf{b} + \mathbf{c}) = \mathbf{0} \text{ and } \mathbf{b} \times (\mathbf{c} + \mathbf{a}) = \mathbf{0}.$$

Hence show that $\mathbf{c} \times (\mathbf{a} + \mathbf{b}) = \mathbf{0}$ and deduce the result required.]

2.8 THE TRIPLE SCALAR PRODUCT

The scalar $\mathbf{a} \cdot (\mathbf{b} \times \mathbf{c})$ is called a triple scalar product. If

$$\mathbf{a} = (a_1, a_2, a_3), \quad \mathbf{b} = (b_1, b_2, b_3), \quad \mathbf{c} = (c_1, c_2, c_3)$$

we have

$$\mathbf{a} \cdot (\mathbf{b} \times \mathbf{c}) = \mathbf{a} \cdot \begin{vmatrix} \mathbf{i} & \mathbf{j} & \mathbf{k} \\ b_1 & b_2 & b_3 \\ c_1 & c_2 & c_3 \end{vmatrix}.$$

Thus

$$\mathbf{a} \cdot (\mathbf{b} \times \mathbf{c}) = \begin{vmatrix} a_1 & a_2 & a_3 \\ b_1 & b_2 & b_3 \\ c_1 & c_2 & c_3 \end{vmatrix}. \tag{2.31}$$

It is easily verified that

$$\mathbf{a} \cdot (\mathbf{b} \times \mathbf{c}) = (\mathbf{a} \times \mathbf{b}) \cdot \mathbf{c}, \tag{2.32}$$

that is, *the 'dot' and 'cross' may be interchanged in a triple scalar product.*

Geometrical interpretation

Consider the parallelepiped with adjacent edges representing vectors **a**, **b**, **c**, as shown in Fig. 2.16, with **b** and **c** horizontal. The volume V of the parallelepiped is 'area of base × height'. That is

$$V = |(bc \sin \theta)(a \cos \varphi)|,$$

where θ is the angle between **b** and **c** and φ is the angle between **a** and the upward vertical. But

$$\mathbf{b} \times \mathbf{c} = bc \sin \theta \, \mathbf{k},$$

where **k** is a unit vector vertically upwards. Also

$$\mathbf{a} \cdot \mathbf{k} = a \cos \varphi.$$

It follows that

$$V = |\mathbf{a} \cdot (\mathbf{b} \times \mathbf{c})|. \tag{2.33}$$

Condition for coplanar vectors

Three non-zero vectors **a**, **b**, **c** are coplanar if and only if

$$\mathbf{a} \cdot (\mathbf{b} \times \mathbf{c}) = 0.$$

Proof

As the three vectors have non-vanishing magnitudes, the volume V of a parallelepiped with adjacent edges **a**, **b**, **c** is zero if and only if the vectors are coplanar. Thus (2.33) shows that **a**, **b**, **c** are coplanar if and only if $\mathbf{a} \cdot (\mathbf{b} \times \mathbf{c}) = 0$.

Note

The triple scalar product $\mathbf{a} \cdot (\mathbf{b} \times \mathbf{c})$ vanishes if any two of the three vectors are parallel or anti-parallel, but the converse of this statement is not true.

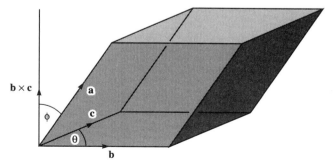

Fig. 2.16 The volume of a parallelepiped is $\mathbf{a} \cdot (\mathbf{b} \times \mathbf{c})$.

EXAMPLE 5

If the non-zero vectors **a**, **b**, **c** are not coplanar show that any other vector **A** may be expressed uniquely in the form

$$\mathbf{A} = \lambda\mathbf{a} + \mu\mathbf{b} + \nu\mathbf{c},$$

where λ, μ, ν are scalars.

Solution

Let $\mathbf{A} = (A_1, A_2, A_3)$, $\mathbf{a} = (a_1, a_2, a_3)$, $\mathbf{b} = (b_1, b_2, b_3)$, $\mathbf{c} = (c_1, c_2, c_3)$.
Then

$$\mathbf{A} = \lambda\mathbf{a} + \mu\mathbf{b} + \nu\mathbf{c},$$

if and only if

$$A_1 = \lambda a_1 + \mu b_1 + \nu c_1,$$
$$A_2 = \lambda a_2 + \mu b_2 + \nu c_2,$$
$$A_3 = \lambda a_3 + \mu b_3 + \nu c_3.$$

These simultaneous equations for λ, μ, ν have a unique solution if and only if

$$\begin{vmatrix} a_1 & a_2 & a_3 \\ b_1 & b_2 & b_3 \\ c_1 & c_2 & c_3 \end{vmatrix} \neq 0,$$

i.e.

$$\mathbf{a} \cdot (\mathbf{b} \times \mathbf{c}) \neq 0.$$

This condition is satisfied because **a**, **b**, **c** are not coplanar and are non-zero, and so the required result follows.

EXERCISES

2.39 By writing out components, or otherwise, prove that

$$\mathbf{a} \cdot (\mathbf{b} \times \mathbf{c}) = (\mathbf{a} \times \mathbf{b}) \cdot \mathbf{c}.$$

2.40 If (x, y, z) is any point on the plane through the points (x_1, y_1, z_1), (x_2, y_2, z_2) and the origin, show that

$$\begin{vmatrix} x & y & z \\ x_1 & y_1 & z_1 \\ x_2 & y_2 & z_2 \end{vmatrix} = 0.$$

2.41 Show that for all scalars λ,

$$(\mathbf{a} + \lambda\mathbf{b}) \cdot (\mathbf{b} \times \mathbf{c}) = \mathbf{a} \cdot (\mathbf{b} \times \mathbf{c}).$$

2.42 Show that

$$(\mathbf{a} + \mathbf{b} + \mathbf{c}) \cdot (\mathbf{b} \times \mathbf{c}) = \mathbf{a} \cdot (\mathbf{b} \times \mathbf{c}).$$

2.43 If

$$\mathbf{A}_1 = \lambda_1 \mathbf{a} + \mu_1 \mathbf{b} + \nu_1 \mathbf{c},$$
$$\mathbf{A}_2 = \lambda_2 \mathbf{a} + \mu_2 \mathbf{b} + \nu_2 \mathbf{c},$$
$$\mathbf{A}_3 = \lambda_3 \mathbf{a} + \mu_3 \mathbf{b} + \nu_3 \mathbf{c},$$

show that

$$\mathbf{A}_1 \cdot (\mathbf{A}_2 \times \mathbf{A}_3) = \begin{vmatrix} \lambda_1 & \mu_1 & \nu_1 \\ \lambda_2 & \mu_2 & \nu_2 \\ \lambda_3 & \mu_3 & \nu_3 \end{vmatrix} \mathbf{a} \cdot (\mathbf{b} \times \mathbf{c}).$$

2.44 Show that, given any four non-zero vectors $\mathbf{a}$, $\mathbf{b}$, $\mathbf{c}$, $\mathbf{d}$, there exist scalars p, q, r, s, not all zero, such that

$$p\mathbf{a} + q\mathbf{b} + r\mathbf{c} + s\mathbf{d} = \mathbf{0}.$$

[*Hint.* Consider the cases (i) when three of the vectors are not coplanar, (ii) when all four vectors are coplanar.]

2.45 Let OX, OY, OZ be a system of oblique rectilinear axes (i.e. axes such that OX, OY, OZ are straight lines which are not mutually perpendicular and not coplanar) and let $\mathbf{I}$, $\mathbf{J}$, $\mathbf{K}$ denote unit vectors in the three coordinate directions. If a vector $\mathbf{A}$ is expressed in the form

$$\mathbf{A} = A_1 \mathbf{I} + A_2 \mathbf{J} + A_3 \mathbf{K},$$

then A_1, A_2, A_3 are called the **components** of $\mathbf{A}$. Show that these components are not identical with the resolutes of $\mathbf{A}$ along OX, OY, OZ.

2.9 THE TRIPLE VECTOR PRODUCT

Vectors such as $(\mathbf{a} \times \mathbf{b}) \times \mathbf{c}$ and $\mathbf{a} \times (\mathbf{b} \times \mathbf{c})$ are called triple vector products. The following identities (proved below) are often needed:

$$(\mathbf{a} \times \mathbf{b}) \times \mathbf{c} = (\mathbf{a} \cdot \mathbf{c}) \mathbf{b} - (\mathbf{b} \cdot \mathbf{c}) \mathbf{a}; \tag{2.34}$$

$$\mathbf{a} \times (\mathbf{b} \times \mathbf{c}) = (\mathbf{a} \cdot \mathbf{c}) \mathbf{b} - (\mathbf{a} \cdot \mathbf{b}) \mathbf{c}. \tag{2.35}$$

Proof

Choose axes $Oxyz$ with the x-axis in the same direction as $\mathbf{a}$ and such that $\mathbf{b}$ is parallel to the xy-plane (as in Fig. 2.12, section 2.7). Then

$$\mathbf{a} = (a_1, 0, 0), \quad \mathbf{b} = (b_1, b_2, 0), \quad \mathbf{c} = (c_1, c_2, c_3).$$

Thus

$$\mathbf{a} \times \mathbf{b} = (0, \ 0, \ a_1 b_2)$$

giving

$$(\mathbf{a} \times \mathbf{b}) \times \mathbf{c} = (- a_1 b_2 c_2, \ a_1 b_2 c_1, \ 0). \tag{2.36}$$

Also

$$(\mathbf{a} \cdot \mathbf{c}) \mathbf{b} - (\mathbf{b} \cdot \mathbf{c}) \mathbf{a} = a_1 c_1 \mathbf{b} - (b_1 c_1 + b_2 c_2) \mathbf{a}$$
$$= (- a_1 b_2 c_2, \ a_1 b_2 c_1, \ 0). \tag{2.37}$$

Comparing (2.36) and (2.37), the identity (2.34) is established.

The second identity (2.35) is easily proved likewise, or by making use of (2.34).

Remark

To remember (2.34) and (2.35) observe that each of the vectors inside the brackets on the left appears once outside the brackets on the right and the 'middle' vector **b** appears first. Each term contains **a**, **b**, **c** once only.

EXERCISES

2.46 Adjacent sides of a triangle represent vectors **a** and **b**. Show that the area of the triangle is $\frac{1}{2} |\mathbf{a} \times \mathbf{b}|$.

2.47 Establish formula (2.35) by making use of (2.34). [*Hint*. Use the result that for any two vectors **A** and **B**, $\mathbf{A} \times \mathbf{B} = - \mathbf{B} \times \mathbf{A}$.]

2.48 Prove that, if **a**, **b**, **c** are non-zero and

$$(\mathbf{a} \times \mathbf{b}) \times \mathbf{c} = \mathbf{a} \times (\mathbf{b} \times \mathbf{c}) \ ,$$

then either (i) **b** is perpendicular to both **a** and **c**, or (ii) **a** and **c** are parallel or anti-parallel. [*Hint*. Expand using (2.34) and (2.35).]

2.10 PRODUCTS OF FOUR VECTORS

It is sometimes necessary to manipulate products of four vectors. This often involves the formulae (2.34) and (2.35) together with the knowledge that the dot and cross in the triple scalar product are always interchangeable. Thus for example

$$(\mathbf{a} \times \mathbf{b}) \cdot (\mathbf{c} \times \mathbf{d}) = \mathbf{a} \cdot [\mathbf{b} \times (\mathbf{c} \times \mathbf{d})]. \tag{2.38}$$

Expanding the triple vector product gives

$$\mathbf{b} \times (\mathbf{c} \times \mathbf{d}) = (\mathbf{b} \cdot \mathbf{d})\,\mathbf{c} - (\mathbf{b} \cdot \mathbf{c})\,\mathbf{d}.$$

Substituting into equation (2.38) we obtain

$$(\mathbf{a} \times \mathbf{b}) \cdot (\mathbf{c} \times \mathbf{d}) = (\mathbf{b} \cdot \mathbf{d})(\mathbf{a} \cdot \mathbf{c}) - (\mathbf{b} \cdot \mathbf{c})(\mathbf{a} \cdot \mathbf{d})$$

so that

$$(\mathbf{a} \times \mathbf{b}) \cdot (\mathbf{c} \times \mathbf{d}) = \begin{vmatrix} \mathbf{a} \cdot \mathbf{c} & \mathbf{a} \cdot \mathbf{d} \\ \mathbf{b} \cdot \mathbf{c} & \mathbf{b} \cdot \mathbf{d} \end{vmatrix}. \tag{2.39}$$

Other exercises on products of four vectors are given below.

EXERCISES

2.49 Show that

$$|\mathbf{a} \times \mathbf{b}|^2 = a^2 b^2 - (\mathbf{a} \cdot \mathbf{b})^2.$$

2.50 Given two vectors **a** and **r** through the origin, show on a diagram the vector $(\mathbf{a} \times \mathbf{r}) \times \mathbf{a}$. Deduce that the length of the perpendicular onto **a** from the point with position vector **r** is

$$|\mathbf{a} \times \mathbf{r}|^2 / |(\mathbf{a} \times \mathbf{r}) \times \mathbf{a}|.$$

2.51 If $\hat{\mathbf{a}}$ and $\hat{\mathbf{b}}$ are unit vectors show that

$$|\hat{\mathbf{a}} \times \hat{\mathbf{b}}|^2 = 1 - (\hat{\mathbf{a}} \cdot \hat{\mathbf{b}})^2.$$

Show that this is another form of the trigonometrical identity $\sin^2 \theta = 1 - \cos^2 \theta$.

2.52 Show that

$$\mathbf{a} \times [\mathbf{b} \times (\mathbf{c} \times \mathbf{a})] = (\mathbf{a} \cdot \mathbf{b})\,\mathbf{a} \times \mathbf{c}.$$

2.11 BOUND VECTORS

In mechanics the **point of application** of a force, or its **line of action**, may be important. The force, together with its point of application or its line of action, is then sometimes said to be a **bound vector**. We shall not enter into a detailed discussion which belongs more properly to a course on mechanics.

EXERCISES

2.53 Let $\mathcal{L}$ be the line of action of a force **F** and let O be a given point. Then the **moment** of **F** about O is defined as

$$\mathbf{G} = \mathbf{r} \times \mathbf{F}$$

where $\mathbf{r} = \overrightarrow{OP}$ is the position vector of any point P on the line $\mathcal{L}$. Show that $\mathbf{G}$ is independent of the particular point P chosen on $\mathcal{L}$.

2.54 As in the previous exercise, the moment $\mathbf{G}$ of a force $\mathbf{F}$ about a point O is given by the equation

$$\mathbf{G} = \mathbf{r} \times \mathbf{F}$$

where $\mathbf{r}$ is the position vector relative to O of any point on $\mathcal{L}$, the line of action of $\mathbf{F}$. By direct substitution (or otherwise) show that the position vectors of points on $\mathcal{L}$ are given by

$$\mathbf{r} = \lambda \mathbf{F} + (\mathbf{F} \times \mathbf{G})/F^2$$

where λ is a parameter. What is the perpendicular distance of O from $\mathcal{L}$?

Vector functions of a real variable. Differential geometry of curves

3.1 VECTOR FUNCTIONS AND THEIR GEOMETRICAL REPRESENTATION

The reader should already be familiar with the idea of a real function $f(x)$, say, of a real variable x. In this chapter the properties of vector functions $\mathbf{F}(t)$ of a real variable t will be discussed.

Suppose that the components of the vector

$$\mathbf{F}(t) = (f_1(t), f_2(t), f_3(t)) \qquad (3.1)$$

are single-valued functions of a real variable t. Then $\mathbf{F}(t)$ is called a **vector function** of t. In most applications t is a continuous variable and $f_1(t)$, $f_2(t)$, $f_3(t)$ are continuous* over some interval of t. If this is so $\mathbf{F}(t)$ is said to be a **continuous vector function** of t. Examples of such functions are

$$\mathbf{F}(t) = (2, \; t^{\frac{1}{2}}, \; \sin t) \qquad 0 \leqslant t < \infty$$

and

$$\mathbf{F}(t) = \begin{cases} (t^3, t, 3) & \text{for } -\infty < t \leqslant 2 \\ (2t^2, \, 2, \, 6t^{-1}) & \text{for } 2 < t < \infty. \end{cases}$$

The vector function

$$\mathbf{F}(t) = (1, t, t^{-1}) \qquad -1 \leqslant t \leqslant 1$$

is not continuous, because as t increases through zero the z-component t^{-1} changes value from $-\infty$ to ∞.

---

* Roughly speaking, a function is continuous if its value does not change suddenly at any point. However, for a precise definition of continuity in an interval see, for example, Hardy, G. H. (1952) *Pure Mathematics*, Cambridge University Press, p. 186.

Geometrical representation of vector functions

Let a continuous vector function $\mathbf{F}(t)$ be represented by the position vector $\overrightarrow{OP}$, where O is the origin and P is the point $(f_1(t), f_2(t), f_3(t))$. Then, as t varies over its permissible range of values, P describes a continuous curve (in three dimensions, see Fig. 3.1). It is clear that in general both the magnitude and direction of $\mathbf{F}(t)$ will vary with t. (A vector is constant only if both its magnitude and direction are fixed.) The equation

$$\overrightarrow{OP} = \mathbf{r} = \mathbf{F}(t), \tag{3.2}$$

where $\mathbf{r} = (x, y, z)$, is called the **parametric equation** of the curve described by P.

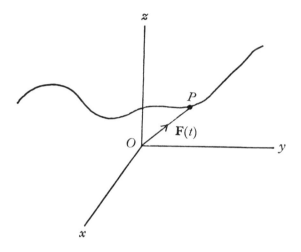

Fig. 3.1 A curve in three dimensions described by a point P whose position is given by an equation of the type (3.2).

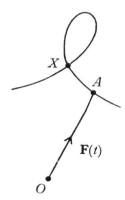

Fig. 3.2 A curve which intersects itself at X.

It should be noted that, although $\mathbf{F}(t)$ is taken to be a single-valued function of t, two (or possibly more) values of t may correspond to the same vector $\mathbf{F}$. In other words, there may fail to be a one-to-one correspondence between the vectors $\mathbf{F}(t)$ and the variable t. A simple instance of such a situation occurs when $\mathbf{F} = \overrightarrow{OA}$ is the position vector of a point A in motion. In this case t may be taken to denote time. If the point describes a curve which intersects itself at a point X, as in Fig. 3.2, then there will be two times t_1, t_2, say, at which A coincides with X. Thus

$$\mathbf{F}(t_1) = \mathbf{F}(t_2) = \overrightarrow{OX}.$$

A similar situation arises when a point retraces part (or all) of its path.

EXAMPLE 1

A point P has position vector

$$\overrightarrow{OP} = a(\cos\theta,\ 0,\ \sin\theta)$$

relative to rectangular cartesian axes $Oxyz$. Find the locus of P as θ varies and a remains constant.

Solution

Taking components,

$$x = a\cos\theta,\quad y = 0,\quad z = a\sin\theta.$$

Observing that $\cos^2\theta + \sin^2\theta = 1$, it is seen that the required locus is

$$x^2 + z^2 = a^2,\ y = 0.$$

Thus, as θ varies, P traces out the circle $x^2 + z^2 = a^2$ in the zx-plane.

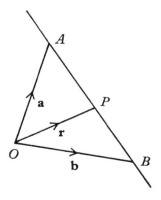

Fig. 3.3

EXAMPLE 2

Let **a**, **b** be the position vectors relative to the origin O of points A, B. Show that the equation of the straight line through A, B can be expressed in the form

$$\mathbf{r} = \mathbf{a} + (\mathbf{b} - \mathbf{a})t, \qquad (3.3)$$

where t is a parameter (Fig. 3.3).

Solution

The position vector of the point B relative to A is

$$\overrightarrow{AB} = \mathbf{b} - \mathbf{a}.$$

The point P with position vector **r** lies on the line through A and B (Fig. 3.3) if and only if

$$\overrightarrow{AP} = (\mathbf{b} - \mathbf{a})t,$$

where t is some real number. Noting that

$$\overrightarrow{OP} = \overrightarrow{OA} + \overrightarrow{AP}$$

we have

$$\mathbf{r} = \mathbf{a} + (\mathbf{b} - \mathbf{a})t.$$

This is the parametric equation of the straight line through A and B, because the position vectors of all points P on the line can be represented in this form.

Remark

If A is the point (x_0, y_0, z_0), B the point (x_1, y_1, z_1) and P the point (x, y, z) then the components of equation (3.3) are

$$x = x_0 + (x_1 - x_0)t, \ \ y = y_0 + (y_1 - y_0)t, \ \ z = z_0 + (z_1 - z_0)t.$$

Eliminating t gives

$$\frac{x - x_0}{x_1 - x_0} = \frac{y - y_0}{y_1 - y_0} = \frac{z - z_0}{z_1 - z_0}.$$

This is the common form of the equation in rectangular cartesian coordinates of the straight line through $A(x_0, y_0, z_0)$ and $B(x_1, y_1, z_1)$.

EXERCISES

3.1 The continuous parameter t can take all real values. Sketch the curves whose parametric equations are respectively

(i) $\mathbf{r} = (2 \cos \pi t, \sin \pi t, 0)$,

(ii) $\mathbf{r} = (\sin \pi t, 0, 0)$,

(iii) $\mathbf{r} = (t, |t|, 0)$,

(iv) $\mathbf{r} = (t^2, t^3 - t, 0)$,

(v) $\mathbf{r} = \begin{cases} (t, -t, 0) & \text{for } -\infty < t \le 0 \\ (t, -t^2, 0) & \text{for } 0 \le t < \infty. \end{cases}$

3.2 The points P and Q have position vectors

$$\mathbf{r}_P = (s^2 + c, s, 1),$$

$$\mathbf{r}_Q = (2t, t, t),$$

where s and t are parameters and c is a constant. Find the value of c for which the loci of P and Q intersect and show that the point of intersection is then $(2,1,1)$. What are the geometric forms of the two loci?

3.3 Show that the straight lines whose parametric equations are

$$\mathbf{r} = (1, 2, 5) + \lambda(0, 1, 0)$$

and

$$\mathbf{r} = (0, -2, 4) + \mu(1, 2, 1),$$

where λ, μ are parameters, have one point in common. Determine the coordinates of this point.

3.4 If λ, μ are parameters, show that the curves with parametric equations

$$\mathbf{r} = (1 + \lambda, 1 + 2\lambda, 1 + \lambda)$$

and

$$\mathbf{r} = (2\mu, \mu, 2 - 4\mu)$$

are straight lines and that they intersect at right angles.

3.5 The vectors $\mathbf{a}, \hat{\mathbf{u}}, \hat{\mathbf{v}}$ are constant and s, t are parameters which take all real values. Show that the locus of a point P with position vector

$$\mathbf{r} = \mathbf{a} + s\hat{\mathbf{u}} + t\hat{\mathbf{v}}$$

(relative to the origin) is a plane through the point with position vector $\mathbf{a}$ and parallel to the plane of the vectors $\hat{\mathbf{u}}, \hat{\mathbf{v}}$.

3.6 If θ, φ are parameters taking all real values show that the locus of a point P with position vector

$$\mathbf{r} = (\cos \theta, \sin \theta \cos \varphi, \sin \theta \sin \varphi)$$

is a sphere with centre at the origin and of unit radius.

3.2 DIFFERENTIATION OF VECTORS

Suppose that, in some interval of t, the functions $f_1(t)$, $f_2(t)$, $f_3(t)$ are differentiable once with respect to t. Then the **first derivative** of $\mathbf{F}(t)$ is defined in the same interval to be

$$\frac{d\mathbf{F}}{dt} = \left(\frac{df_1}{dt}, \frac{df_2}{dt}, \frac{df_3}{dt} \right). \tag{3.4}$$

The derivative of a vector is also a vector

To establish this we must verify that conditions 1, 2 and 3 of the definition given in section 2.2 are satisfied.

It is at once obvious that the first condition is satisfied. Also, because $f_1(t)$, $f_2(t)$, $f_3(t)$ are unchanged by a translation of the axes, it follows that

$$df_1/dt, \ df_2/dt, \ df_3/dt$$

are unchanged by such a translation and hence the second condition is satisfied.

To verify that 3 is satisfied let the components of $\mathbf{F}$ relative to fixed axes $Ox_1' x_2' x_3'$ be (f_1', f_2', f_3'). As in Chapter 1 (section 1.6), let l_{ij} denote the cosine of the angle between Ox_i' and Ox_j. Since the two sets of axes are taken to be fixed, each l_{ij} is independent of t, hence differentiating the equations of transformation (2.2),

$$\frac{df_i'}{dt} = l_{ij} \frac{df_j}{dt}. \tag{3.5}$$

But df_1'/dt, df_2'/dt, df_3'/dt are the components of $d\mathbf{F}/dt$ referred to the new axes $Ox_1' x_2' x_3'$, and comparing (3.5) with (2.2) it is seen that condition 3 holds. It follows that $d\mathbf{F}/dt$ is a vector.

The definition of higher order derivatives of $\mathbf{F}$ presents no difficulty. For example, if f_1, f_2, f_3 are twice differentiable functions of t in some range, then the second derivative of $\mathbf{F}$ with respect to t is defined in that range as

$$\frac{d^2\mathbf{F}}{dt^2} \equiv \frac{d}{dt}\left(\frac{d\mathbf{F}}{dt} \right) = \left(\frac{d^2f_1}{dt^2}, \frac{d^2f_2}{dt^2}, \frac{d^2f_3}{dt^2} \right). \tag{3.6}$$

By applying the theorem already proved to the vector $d\mathbf{F}/dt$ it follows at once that $d^2\mathbf{F}/dt^2$ is also a vector.

EXAMPLE 3

Find the values of λ for which the vector

$$\mathbf{A} = (\cos \lambda x, \sin \lambda x, 0)$$

satisfies the differential equation

$$\frac{d^2\mathbf{A}}{dx^2} = -9\mathbf{A}.$$

Solution

Using the formula for differentiation,

$$\frac{d\mathbf{A}}{dx} = (-\lambda \sin \lambda x, \; \lambda \cos \lambda x, \; 0)$$

and

$$\frac{d^2\mathbf{A}}{dx^2} = (-\lambda^2 \cos \lambda x, \; -\lambda^2 \sin \lambda x, \; 0).$$

Thus the given differential equation is satisfied if $\lambda^2 = 9$, that is if $\lambda = \pm 3$.

EXERCISES

3.7 Write down the derivatives $d\mathbf{r}/dt$ and $d^2\mathbf{r}/dt^2$ for the following vectors:

- (i) $\mathbf{r} = (2 \cos \pi t, \; \sin \pi t, \; 0)$,
- (ii) $\mathbf{r} = (t, t, e^t)$,
- (iii) $\mathbf{r} = (|t|, t, 0) \quad (t \neq 0)$.

3.8 Given that

$$\frac{d\mathbf{r}}{dt} = \{-e^{-t}(\cos t + \sin t), \; e^{-t}(\cos t - \sin t), \; 0\}$$

and that, when $t = 0$, $\mathbf{r} = (1, 0, 0)$, determine $\mathbf{r}$. Sketch the locus of the point with position vector $\mathbf{r}$ for values of $t \geq 0$.

3.9 Given that the general solution of the differential equation

$$\frac{d^2x}{dt^2} + \omega^2 x = 0 \quad (\omega \text{ constant})$$

is

$$x = A \cos \omega t + B \sin \omega t$$

where A, B are arbitrary constants, show that the general solution of the differential equation

$$\frac{d^2\mathbf{r}}{dt^2} + \omega^2 \mathbf{r} = 0$$

is

$$\mathbf{r} = \mathbf{A} \cos \omega t + \mathbf{B} \sin \omega t$$

where $\mathbf{A}$, $\mathbf{B}$ are arbitrary constant vectors.

If the motion of a point is such that its position vector $\mathbf{r}$ satisfies the differential equation above, show that the motion is confined to one plane.

3.3 DIFFERENTIATION RULES

The rules for differentiation of sums and products of vector functions are similar to the corresponding rules for differentiation of ordinary functions. If λ, $\mathbf{a}$, $\mathbf{b}$ are differentiable functions of t, the following identities hold:

$$\frac{d}{dt}(\mathbf{a} + \mathbf{b}) \equiv \frac{d\mathbf{a}}{dt} + \frac{d\mathbf{b}}{dt}, \tag{3.7}$$

$$\frac{d}{dt}(\lambda\mathbf{a}) \equiv \frac{d\lambda}{dt}\mathbf{a} + \lambda\frac{d\mathbf{a}}{dt}, \tag{3.8}$$

$$\frac{d}{dt}(\mathbf{a} \cdot \mathbf{b}) \equiv \frac{d\mathbf{a}}{dt} \cdot \mathbf{b} + \mathbf{a} \cdot \frac{d\mathbf{b}}{dt}, \tag{3.9}$$

$$\frac{d}{dt}(\mathbf{a} \times \mathbf{b}) \equiv \frac{d\mathbf{a}}{dt} \times \mathbf{b} + \mathbf{a} \times \frac{d\mathbf{b}}{dt}. \tag{3.10}$$

These identities are easily established by writing the vectors in component form.

For example,

$$\frac{d}{dt}(\mathbf{a} \cdot \mathbf{b}) \equiv \frac{d}{dt}(a_1 b_1 + a_2 b_2 + a_3 b_3)$$

$$\equiv \frac{da_1}{dt}b_1 + \frac{da_2}{dt}b_2 + \frac{da_3}{dt}b_3 + a_1\frac{db_1}{dt} + a_2\frac{db_2}{dt} + a_3\frac{db_3}{dt}$$

$$\equiv \frac{d\mathbf{a}}{dt} \cdot \mathbf{b} + \mathbf{a} \cdot \frac{d\mathbf{b}}{dt},$$

which proves (3.9).

Note that in (3.10) the order of $\mathbf{a}$ and $\mathbf{b}$ must be strictly observed because of the non-commutative property of the vector products.

EXAMPLE 4

Show that the first derivative of a unit vector $\hat{\mathbf{a}}(t)$ is always perpendicular to $\hat{\mathbf{a}}(t)$ provided that the derivative is not zero.

Solution

We have

$$\hat{\mathbf{a}} \cdot \hat{\mathbf{a}} = 1,$$

and hence

$$\frac{d\hat{\mathbf{a}}}{dt} \cdot \hat{\mathbf{a}} + \hat{\mathbf{a}} \cdot \frac{d\hat{\mathbf{a}}}{dt} = 0,$$

which gives

$$\hat{\mathbf{a}} \cdot d\hat{\mathbf{a}}/dt = 0.$$

Since neither $\hat{\mathbf{a}}$ nor $d\hat{\mathbf{a}}/dt$ is zero it follows that they must be perpendicular.

EXERCISES

3.10 Establish the identities (3.7), (3.8) and (3.10) in the text. Also verify that (3.9) holds for the case in which

$$\mathbf{a} = (1, t, t^2), \quad \mathbf{b} = (t^2, t, 1).$$

3.11 By writing $\mathbf{r} = r\hat{\mathbf{r}}$ show that, for any differentiable vector function $\mathbf{r} = \mathbf{r}(t)$,

$$\frac{dr}{dt} = \hat{\mathbf{r}} \cdot \frac{d\mathbf{r}}{dt}.$$

3.12 Prove that

$$\frac{d}{dt}\{(\mathbf{a} \times \mathbf{b}) \cdot \mathbf{c}\} \equiv \left(\frac{d\mathbf{a}}{dt} \times \mathbf{b}\right) \cdot \mathbf{c} + \left(\mathbf{a} \times \frac{d\mathbf{b}}{dt}\right) \cdot \mathbf{c} + (\mathbf{a} \times \mathbf{b}) \cdot \frac{d\mathbf{c}}{dt}.$$

3.13 If

$$\mathbf{a} \times \frac{d\mathbf{b}}{dt} = \mathbf{b} \times \frac{d\mathbf{a}}{dt}$$

for all values of t, what can be deduced about $\mathbf{a}$ and $\mathbf{b}$?

3.4 THE TANGENT TO A CURVE. SMOOTH, PIECEWISE SMOOTH AND SIMPLE CURVES

In this section, and the two following sections, we discuss some important concepts relating to curves.

Suppose that a continuous curve $\mathscr{C}$ is the locus of the point P whose position vector (relative to the origin O of fixed axes $Oxyz$) is

$$\overrightarrow{OP} = \mathbf{r} = \mathbf{r}(t) *$$
$$= (x(t), y(t), z(t)). \tag{3.11}$$

* We could have written $\mathbf{r} = \mathbf{f}(t)$, but the common practice of using $\mathbf{r}$ to symbolize the function as well as the dependent variable is convenient and should cause no confusion.

Let P' be a particular point on $\mathscr{C}$ at which $d\mathbf{r}/dt$ exists and is not zero. Then, at this point, *$d\mathbf{r}/dt$ lies along the tangent to the curve and is directed in the sense in which the curve is described by P as t increases.* To show this, let $t = t'$ at the point P'. Then

$$\overrightarrow{OP'} = \mathbf{r}(t').$$

At P'

$$\frac{d\mathbf{r}}{dt} = \lim_{t \to t'} \left(\frac{x(t) - x(t')}{t - t'} , \frac{y(t) - y(t')}{t - t'} , \frac{z(t) - z(t')}{t - t'} \right)$$

$$= \lim_{t \to t'} \frac{\mathbf{r}(t) - \mathbf{r}(t')}{t - t'}$$

$$= \lim_{t \to t'} \frac{\overrightarrow{P'P}}{t - t'} .$$

It is clear that as $t \to t'$, P approaches P' and so $\overrightarrow{P'P}/(t - t')$ ultimately lies along the tangent at P' (Fig. 3.4). Thus $d\mathbf{r}/dt$ is directed along the tangent to $\mathscr{C}$.

To determine the sense in which $d\mathbf{r}/dt$ points, choose axes with origin O at P' and with the x-axis parallel to $d\mathbf{r}/dt$. Then at O

$$\frac{d\mathbf{r}}{dt} = \left(\frac{dx}{dt} , 0, 0 \right).$$

As t increases, the point P moves through O in the positive or negative x-direction according as dx/dt is greater than or less than zero. It follows that $d\mathbf{r}/dt$ points in the direction in which P moves along $\mathscr{C}$ as t increases.

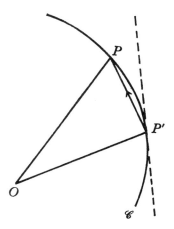

Fig. 3.4 As $t \to t'$, $\overrightarrow{P'P}/(t - t')$ ultimately lies along the tangent to $\mathscr{C}$ at P'.

The unit tangent

Suppose that at the point P', with parameter t', on the curve with parametric equation

$$\mathbf{r} = \mathbf{r}(t) \qquad t_0 \leqslant t \leqslant t_1, \tag{3.12}$$

$d\mathbf{r}/dt$ exists and is not zero.* Then the vector

$$\hat{\mathbf{T}} = \frac{d\mathbf{r}/dt}{|d\mathbf{r}/dt|} \tag{3.13}$$

is defined as the **unit tangent** at P'. If $|d\mathbf{r}/dt| \to 0$ or ∞ as $t \to t'$, then we define

$$\hat{\mathbf{T}} = \lim_{t \to t'} \frac{d\mathbf{r}/dt}{|d\mathbf{r}/dt|}, \tag{3.14}$$

provided the limit exists.

It is clear that $\hat{\mathbf{T}}$ is a unit vector, and the reasoning at the beginning of this section shows that it is directed along the tangent to the curve.

Smooth curve

The curve with parametric equation (3.12) is said to be **smooth** if, at all points in the interval $t_0 \leqslant t \leqslant t_1$, $\hat{\mathbf{T}}$ exists and is continuous. In rather less precise terms, smoothness may be taken to mean that the curve does not undergo a sudden change in direction at any point.

Piecewise smooth curve

Let

$$t_0 < t_1 < t_2 < \ldots < t_{n-1} < t_n .$$

The curve with parametric equation

$$\mathbf{r} = \mathbf{r}(t) \qquad t_0 \leqslant t \leqslant t_n$$

is said to be **piecewise smooth** if (i) $\mathbf{r}(t)$ is continuous in the interval $t_0 \leqslant t \leqslant t_n$ and (ii) the unit tangent $\hat{\mathbf{T}}$ is continuous in the interval $t_0 \leqslant t \leqslant t_n$ except at the points $t_1, t_2, \ldots, t_{n-1}$. Thus a piecewise smooth curve consists of a finite number of smooth curves, linked end to end (Fig. 3.5b).

Simple open curve

A piecewise smooth curve

$$\mathbf{r} = \mathbf{r}(t) \qquad t_0 \leqslant t \leqslant t_n$$

* For the definition of a derivative at an end point of an interval see, for example, Hardy, G. H. (1952) *Pure Mathematics*, Cambridge University Press, p. 286.

is said to be **simple and open** if each point on it corresponds to just one value of t. Thus a simple open curve does not cross or meet itself at any point.

Simple closed curve

A piecewise smooth curve with parametric equation

$$\mathbf{r} = \mathbf{r}(t) \qquad t_0 \leqslant t \leqslant t_n$$

is said to be **simple and closed** if its end points (corresponding to $t = t_0$ and $t = t_n$) coincide and each other point corresponds to just one value of t.

An elementary example of a simple closed curve is the unit circle in the xy-plane, with parametric equation

$$\mathbf{r} = (\cos t, \sin t, 0) \qquad 0 \leqslant t \leqslant 2\pi.$$

The reader should verify that this satisfies all the conditions of the above definition. Observe also that if the range of t were $0 \leqslant t \leqslant 4\pi$, the circle would be described twice as t covered the range, and so it would no longer be a simple curve.

Some examples of the various possible types of curve are shown in Fig. 3.5.

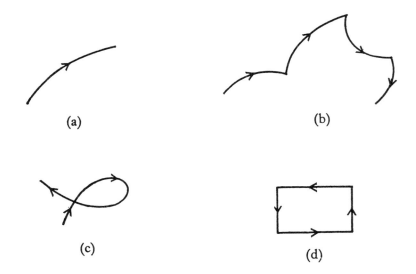

(a)

(b)

(c)

(d)

Fig. 3.5 Classification of curves. (a) Smooth curve. Also a simple curve if covered once only. (b) Piecewise smooth. Simple if covered once only. (c) Smooth curve. Not simple. (d) Simple closed curve (if covered once only).

Change of parameter

The curve with parametric equation

$$\mathbf{r} = \mathbf{r}(t) \qquad t_0 \leqslant t \leqslant t_1$$

has a **sense** (or **orientation**) defined as the direction in which it is described as t increases from t_0 to t_1. The sense is often indicated in a diagram by an arrow on the curve (Fig. 3.5). It is desirable that under a change of parameter the sense of description should be preserved. To ensure this the parametric transformation

$$t = t(u)$$

is only considered allowable if it is such that dt/du is non-negative at all points in the interval $u_0 \leqslant u \leqslant u_1$, where u_0, u_1 are the values of u corresponding to t_0, t_1, respectively. With this restriction u is nowhere decreasing as t is increasing and hence the sense of description of the curve is preserved.

EXAMPLE 5

Show that the unit tangent to the curve

$$\mathbf{r} = \begin{cases} (t^2, 2t, 0) & -1 \leqslant t \leqslant 1 \\ (1, 4 - 2t, 0) & 1 \leqslant t \leqslant 2 \end{cases}$$

is discontinuous at the point $t = 1$. Verify that the curve is piecewise smooth and indicate its sense in a diagram.

Solution

We have

$$\frac{d\mathbf{r}}{dt} = \begin{cases} (2t, 2, 0) & \text{for} \quad -1 \leqslant t \leqslant 1 \\ (0, -2, 0) & \text{for} \quad 1 \leqslant t \leqslant 2. \end{cases}$$

Thus

$$\hat{\mathbf{T}} = \begin{cases} \left(\dfrac{t}{(1 + t^2)^{\frac{1}{2}}}, \dfrac{1}{(1 + t^2)^{\frac{1}{2}}}, 0 \right) & \text{for} \quad -1 \leqslant t \leqslant 1 \\ (0, -1, 0) & \text{for} \quad 1 \leqslant t \leqslant 2. \end{cases}$$

It follows that, at the end points $t = 1$ of the intervals $-1 \leqslant t \leqslant 1$ and $1 \leqslant t \leqslant 2$,

$$\hat{\mathbf{T}} = \frac{1}{\sqrt{2}} (1, 1, 0)$$

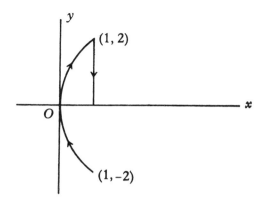

Fig. 3.6 A curve consisting of part of the parabola $y^2 = 4x$ and the straight line $x = 1$.

and

$$\hat{\mathbf{T}} = (0, -1, 0),$$

respectively. Thus $\hat{\mathbf{T}}$ is discontinuous at $t = 1$.

The curve is smooth in each of the separate intervals $-1 \leqslant t \leqslant 1$, $1 \leqslant t \leqslant 2$, because throughout each interval $\hat{\mathbf{T}}$ is continuous. Further, $\mathbf{r}(t)$ is clearly continuous at the point $t = 1$. It follows that the curve is piecewise smooth.

The curve lies entirely in the xy-plane. In the interval $-1 \leqslant t \leqslant 1$ it consists of part of the parabola $y^2 = 4x$; and in the interval $1 \leqslant t \leqslant 2$ it is part of the straight line $x = 1$ (Fig. 3.6). The sense of the curve is indicated by the arrows in the diagram.

EXERCISES

3.14 Show in a diagram the direction of $d\mathbf{r}/dt$ at the points $t = 0, t = 1, t = -1$ for the curves whose parametric equations are:

    (i) $\mathbf{r} = (2 \cos \frac{1}{2} \pi t, \ \sin \frac{1}{2} \pi t, \ 0) \ (-2 \leqslant t \leqslant 2)$,

    (ii) $\mathbf{r} = (t^2, t^3 - t, 0) \qquad\qquad (-\infty < t < \infty)$.

Classify the curves according to the definitions in this section.

3.15 Show that the unit tangent to the curve

$$\mathbf{r} = (3, t, t^2)$$

is

$$\hat{\mathbf{T}} = (0, 1, 2t)/(1 + 4t^2)^{\frac{1}{2}}.$$

3.16 Sketch the curve with parametric equation

$$\mathbf{r} = (t, |\sin t|, 0) \qquad 0 \leqslant t \leqslant 3\pi,$$

and show that it is piecewise smooth.

3.17 Sketch the curve with parametric equation

$$\mathbf{r} = \begin{cases} (\sin t, \cos t, 0) & \text{for} \quad 0 \leqslant t \leqslant \tfrac{1}{2}\pi \\ (1, \tfrac{1}{2}\pi - t, 0) & \text{for} \quad \tfrac{1}{2}\pi \leqslant t \leqslant \tfrac{3}{2}\pi, \end{cases}$$

and show that it is smooth. [*Hint.* Show that $\mathbf{r}$ and the unit tangent $\hat{\mathbf{T}}$ are continuous at the point $t = \tfrac{1}{2}\pi$.]

3.5 ARC LENGTH

Let

$$\mathbf{r} = \mathbf{r}(t) = (x(t), y(t), z(t)) \qquad t_0 \leqslant t \leqslant t_1 \tag{3.15}$$

be the parametric equation of a piecewise smooth curve $\mathscr{C}$. Define

$$\frac{ds}{dt} = \left| \frac{d\mathbf{r}}{dt} \right| = (\dot{x}^2 + \dot{y}^2 + \dot{z}^2)^{\frac{1}{2}}, \tag{3.16}$$

where $\dot{x}$, $\dot{y}$ and $\dot{z}$ denote dx/dt, dy/dt and dz/dt, respectively. Then

$$s(t) = \int_{t_0}^{t} (\dot{x}^2 + \dot{y}^2 + \dot{z}^2)^{\frac{1}{2}} \, dt \tag{3.17}$$

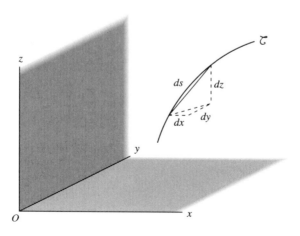

Fig. 3.7 Geometrical interpretation of the element of arc length $ds = \sqrt{(dx^2 + dy^2 + dz^2)}$.

is defined as the **arc length** of $\mathscr{C}$ from the point with parameter t_0 to the (variable) point with parameter t. The total length of the curve is defined as $s(t_1) = l$. Observe that $s(t_0) = 0$.

At points where ds/dt is finite, non-zero and continuous, the element of arc ds corresponding to a positive increment dt in t is defined as

$$ds = (\dot{x}^2 + \dot{y}^2 + \dot{z}^2)^{\frac{1}{2}} \, dt = (dx^2 + dy^2 + dz^2)^{\frac{1}{2}} .$$

This has an obvious intuitive geometrical interpretation as indicated in Fig. 3.7, and the motivation for the definitions (3.16) and (3.17) also becomes clear.

By the definition (3.16), $ds/dt \geqslant 0$, hence the substitution $t = t(s)$ is an allowable change of parameter. In terms of s,

$$\mathbf{r} = \mathbf{r}(s) = (x(s), y(s), z(s)) \qquad 0 \leqslant s \leqslant l, \tag{3.18}$$

and this is called the **intrinsic equation** of the curve.

The unit tangent $\hat{\mathbf{T}}$ may be obtained from the intrinsic equation quite simply. For, upon substituting (3.16), the expressions (3.13) and (3.14) both reduce to

$$\hat{\mathbf{T}} = \frac{d\mathbf{r}}{ds} = \left(\frac{dx}{ds}, \frac{dy}{ds}, \frac{dz}{ds} \right). \tag{3.19}$$

EXERCISE

3.18 Find the intrinsic equation of the curve with parametric equation

$$\mathbf{r} = (a \cos t, a \sin t, bt) \qquad 0 \leqslant t \leqslant 2\pi.$$

3.6 CURVATURE AND TORSION

Let $\hat{\mathbf{T}}$ be the unit tangent to the curve with intrinsic equation $\mathbf{r} = \mathbf{r}(s)$ and write

$$\frac{d\hat{\mathbf{T}}}{ds} = \kappa \hat{\mathbf{N}} \qquad \kappa \geqslant 0, \tag{3.20}$$

where κ is a positive function of s and $\hat{\mathbf{N}}$ is a unit vector. The worked example at the end of section 3.3 shows that $\hat{\mathbf{N}}$ is perpendicular to $\hat{\mathbf{T}}$: it is called the **principal unit normal vector**. The proportionality factor κ is defined to be the **curvature** and is a measure of the rate at which the direction of the tangent changes with s. For example, if the curve is a straight line, $\hat{\mathbf{T}}$ is constant in direction as well as in magnitude so $\kappa = 0$. The quantity $\rho = \kappa^{-1}$ is defined to be the **radius of curvature**.

Another vector of importance in the differential theory of curves is the **unit binormal vector**, defined as

$$\hat{\mathbf{B}} = \hat{\mathbf{T}} \times \hat{\mathbf{N}}. \tag{3.21}$$

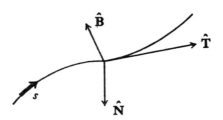

Fig. 3.8 The unit tangent $\hat{\mathbf{T}}$, unit principal normal $\hat{\mathbf{N}}$ and unit binormal $\hat{\mathbf{B}}$ at a particular point on a curve.

The three unit vectors $\hat{\mathbf{T}}$, $\hat{\mathbf{N}}$, $\hat{\mathbf{B}}$ form an orthonormal right-handed triad (Fig. 3.8).

The derivative of $\hat{\mathbf{B}}$ with respect to s is parallel or antiparallel to $\hat{\mathbf{N}}$. To show this we first differentiate (3.21) and use (3.20), thus obtaining the equation

$$\frac{d\hat{\mathbf{B}}}{ds} = \kappa\hat{\mathbf{N}} \times \hat{\mathbf{N}} + \hat{\mathbf{T}} \times \frac{d\hat{\mathbf{N}}}{ds} = \hat{\mathbf{T}} \times \frac{d\hat{\mathbf{N}}}{ds}.$$

Now $d\hat{\mathbf{B}}/ds$ is normal to $\hat{\mathbf{B}}$ and hence lies in the plane of $\hat{\mathbf{N}}$ and $\hat{\mathbf{T}}$. But $\hat{\mathbf{T}} \times d\hat{\mathbf{N}}/ds$ is normal to $\hat{\mathbf{T}}$ so

$$\frac{d\hat{\mathbf{B}}}{ds} = -\tau\hat{\mathbf{N}}, \tag{3.22}$$

where τ is a function of s. This proportionality factor is called the **torsion** of the curve and is a measure of the rate at which the direction of the binormal changes with s.

For curves which lie in a plane Π, $\hat{\mathbf{T}}$ clearly lies in Π, therefore $\hat{\mathbf{N}}$ (which is proportional to the rate of change of $\hat{\mathbf{T}}$) also lies in Π. Thus, for a plane curve, $\hat{\mathbf{B}}$ is a constant vector at right angles to Π and the torsion is zero.

EXAMPLE 6

Consider the **circular helix**, defined parametrically as

$$\mathbf{r} = (a \cos t, a \sin t, bt) \tag{3.23}$$

where a, b are constants. Equating the components,

$$x = a \cos t, \quad y = a \sin t, \quad z = bt,$$

and so for all t

$$x^2 + y^2 = a^2.$$

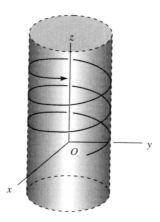

Fig. 3.9 Circular helix.

The curve therefore lies on the surface of a circular cylinder of radius a and axis Oz. It spirals around the z-axis, as shown in Fig. 3.9. For this curve

$$\frac{ds}{dt} = \left\{\left(\frac{dx}{dt}\right)^2 + \left(\frac{dy}{dt}\right)^2 + \left(\frac{dz}{dt}\right)^2\right\}^{\frac{1}{2}} = (a^2 + b^2)^{\frac{1}{2}}.$$

Hence

$$\hat{\mathbf{T}} = \frac{d\mathbf{r}}{ds} = \frac{d\mathbf{r}}{dt}\frac{dt}{ds} = \frac{1}{(a^2 + b^2)^{\frac{1}{2}}}(-a\sin t, a\cos t, b). \tag{3.24}$$

Also

$$\kappa\hat{\mathbf{N}} = \frac{d\hat{\mathbf{T}}}{ds} = \frac{-a}{a^2 + b^2}(\cos t, \sin t, 0). \tag{3.25}$$

It follows that the principal unit normal $\hat{\mathbf{N}}$ is always parallel to the xy-plane. Also, equating the magnitudes of the two sides of (3.25), the curvature is

$$\kappa = |a|/(a^2 + b^2). \tag{3.26}$$

From equations (3.24), (3.25)

$$\hat{\mathbf{B}} = \hat{\mathbf{T}} \times \hat{\mathbf{N}} = \frac{a}{\kappa(a^2 + b^2)^{3/2}}(b\sin t, -b\cos t, a).$$

Thus

$$\tau\hat{\mathbf{N}} = -\frac{d\hat{\mathbf{B}}}{ds} = \frac{-ab}{\kappa(a^2 + b^2)^2}(\cos t, \sin t, 0)$$

and using (3.25) this reduces to

$$\tau\hat{\mathbf{N}} = \frac{b}{a^2 + b^2}\,\hat{\mathbf{N}}.$$

The torsion of the helix is therefore

$$\tau = b/(a^2 + b^2). \tag{3.27}$$

Notice that when $b = 0$ the curve reduces to a circle of radius $|a|$ in the xy-plane. As expected, the curvature is then $1/|a|$ (from (3.26)), and the torsion is zero (from (3.27)).

EXAMPLE 7

As a second example we show that, for a plane curve in the xy-plane, if the unit tangent in the direction of s increasing makes an angle ψ with the positive x-axis then the radius of curvature ρ is given by

$$\rho = |ds/d\psi|.$$

(This conclusion agrees with the usual elementary definition of ρ for a plane curve.)

Solution

In terms of the angle ψ (Fig. 3.10)

$$\hat{\mathbf{T}} = (\cos\psi, \sin\psi, 0),$$

and so

$$\frac{d\hat{\mathbf{T}}}{ds} = \frac{d\psi}{ds}\frac{d\hat{\mathbf{T}}}{d\psi}$$

$$= \frac{d\psi}{ds}(-\sin\psi, \cos\psi, 0).$$

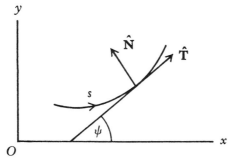

Fig. 3.10 The angle ψ is defined to be the angle between the unit tangent and the x-axis.

It follows by comparison with the formula

$$\frac{d\hat{\mathbf{T}}}{ds} = \kappa \hat{\mathbf{N}}$$

that

$$\kappa = \left| \frac{d\psi}{ds} \right|.$$

Thus the radius of curvature is

$$\rho = \kappa^{-1} = \left| \frac{ds}{d\psi} \right|.$$

Note that

$$\hat{\mathbf{N}} = \begin{cases} (-\sin\psi, \cos\psi, 0) & \text{when } ds/d\psi > 0 \\ (\sin\psi, -\cos\psi, 0) & \text{when } ds/d\psi < 0. \end{cases}$$

EXERCISES

3.19 Show that the unit tangent vector to the curve

$$\mathbf{r} = (4\cos t, \cos 2t, 2t + \sin 2t)$$

is

$$\hat{\mathbf{T}} = (-\sin t, -\sin t \cos t, \cos^2 t).$$

Show also that the curvature is $\frac{1}{4}(1 + \cos^2 t)^{\frac{1}{2}}$.

3.20 By finding $\hat{\mathbf{T}}, \hat{\mathbf{N}}, \hat{\mathbf{B}}$ in turn, verify that the plane parabolic curve

$$\mathbf{r} = (t, \tfrac{1}{2} t^2, 0)$$

has zero torsion.

3.21 With the notation used in the text, show that

$$\frac{d\hat{\mathbf{N}}}{ds} = -\kappa \hat{\mathbf{T}} + \tau \hat{\mathbf{B}}.$$

[*Hint*. Differentiate the relation $\hat{\mathbf{N}} = \hat{\mathbf{B}} \times \hat{\mathbf{T}}$.]

Note

This result, together with the relations $d\hat{\mathbf{T}}/ds = \kappa \hat{\mathbf{N}}$ and $d\hat{\mathbf{B}}/ds = -\tau \hat{\mathbf{N}}$, constitute the **Serret–Frenet formulae**. These formulae are fundamental to the differential geometry of curves.

3.7 APPLICATIONS IN KINEMATICS

The components of acceleration of a point moving along a curve

When a particle (or a point) P is in motion, its position relative to a given coordinate system will depend upon time t. If $\mathbf{r} = \mathbf{r}(t)$ is the position vector of P, then the **velocity** $\mathbf{v}$ and the **acceleration** $\mathbf{f}$ relative to the chosen coordinate system are defined as

$$\mathbf{v} = \dot{\mathbf{r}}, \quad \mathbf{f} = \ddot{\mathbf{r}} = \dot{\mathbf{v}}, \tag{3.28}$$

where $\dot{\mathbf{r}}$ denotes $d\mathbf{r}/dt$, $\ddot{\mathbf{r}}$ denotes $d^2\mathbf{r}/dt^2$, etc. We define $|\mathbf{v}|$ to be the **speed** of P.

Let $s = s(t)$ denote the arc length of the curve $\mathbf{r} = \mathbf{r}(t)$ that has been covered by the particle at time t. Then

$$\mathbf{v} = \dot{\mathbf{r}} = \dot{s}\,\frac{d\mathbf{r}}{ds} = \dot{s}\,\hat{\mathbf{T}}, \tag{3.29}$$

where $\hat{\mathbf{T}}$ is the unit tangent vector directed, at every instant, in the sense in which P is moving and $\dot{s} = |\mathbf{v}|$ is the speed of P. Also

$$\mathbf{f} = \frac{d}{dt}(\dot{s}\,\hat{\mathbf{T}}) = \ddot{s}\,\hat{\mathbf{T}} + \dot{s}^2\,\frac{d\hat{\mathbf{T}}}{ds}$$

$$= \ddot{s}\,\hat{\mathbf{T}} + \rho^{-1}\dot{s}^2\,\hat{\mathbf{N}}, \tag{3.30}$$

where $\hat{\mathbf{N}}$ is the principal unit normal of the curve described by P and ρ is the radius of curvature. The components of acceleration are therefore $\ddot{s}$ in the direction of the tangent vector and $\dot{s}^2/\rho$ along the **principal** normal. *This result is not restricted to plane motion.*

The components of acceleration in plane polar coordinates

Using the ideas introduced in this chapter we can derive well-known formulae for components of velocity and acceleration in terms of plane polar coordinates (r, θ).

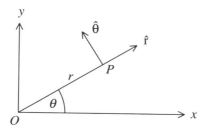

Fig. 3.11 The unit vectors $\hat{\mathbf{r}}$ and $\hat{\boldsymbol{\theta}}$ in the plane of polar coordinates r, θ.

Let $\hat{\mathbf{r}}, \hat{\boldsymbol{\theta}}$ denote unit vectors in the r, θ plane such that $\hat{\mathbf{r}}$ points away from the origin O and $\hat{\boldsymbol{\theta}}$ is normal to $\hat{\mathbf{r}}$ and in the direction θ increasing (Fig. 3.11). We first prove two important subsidiary results, namely

$$\frac{d\hat{\mathbf{r}}}{d\theta} = \hat{\boldsymbol{\theta}}, \quad \frac{d\hat{\boldsymbol{\theta}}}{d\theta} = -\hat{\mathbf{r}}. \tag{3.31}$$

Take rectangular cartesian axes Ox in the direction $\theta = 0$, Oy in the direction $\theta = \frac{1}{2}\pi$ and Oz to complete a right-handed triad. Then

$$\hat{\mathbf{r}} = (\cos\theta, \sin\theta, 0), \quad \hat{\boldsymbol{\theta}} = (-\sin\theta, \cos\theta, 0).$$

Hence

$$\frac{d\hat{\mathbf{r}}}{d\theta} = (-\sin\theta, \cos\theta, 0), \quad \frac{d\hat{\boldsymbol{\theta}}}{d\theta} = -(\cos\theta, \sin\theta, 0)$$

and the required formulae (3.31) follow at once.

Now the position vector of P is

$$\mathbf{r} = r\hat{\mathbf{r}}. \tag{3.32}$$

Hence

$$\dot{\mathbf{r}} = \dot{r}\hat{\mathbf{r}} + r\dot{\theta}\,\frac{d\hat{\mathbf{r}}}{d\theta}$$

$$= \dot{r}\hat{\mathbf{r}} + r\dot{\theta}\,\hat{\boldsymbol{\theta}}. \tag{3.33}$$

The radial and transverse components of velocity are therefore $\dot{r}, r\dot{\theta}$ respectively. Also

$$\ddot{\mathbf{r}} = \ddot{r}\hat{\mathbf{r}} + \dot{r}\dot{\theta}\,\frac{d\hat{\mathbf{r}}}{d\theta} + (\dot{r}\dot{\theta} + r\ddot{\theta})\hat{\boldsymbol{\theta}} + r\dot{\theta}^2\,\frac{d\hat{\boldsymbol{\theta}}}{d\theta}$$

$$= (\ddot{r} - r\dot{\theta}^2)\hat{\mathbf{r}} + (2\dot{r}\dot{\theta} + r\ddot{\theta})\hat{\boldsymbol{\theta}}. \tag{3.34}$$

Thus the radial and transverse components of acceleration are $\ddot{r} - r\dot{\theta}^2$ and $2\dot{r}\dot{\theta} + r\ddot{\theta}$.

EXAMPLE 8

When an electron moves in a magnetic field it experiences a force $e\mathbf{v} \times \mathbf{B}$, where e is the electronic charge, $\mathbf{v}$ is the velocity, and $\mathbf{B}$ is the magnetic induction vector. Hence if $\mathbf{f}$ is the acceleration and m is the mass of the electron, its equation of motion is

$$m\mathbf{f} = e\mathbf{v} \times \mathbf{B}. \tag{3.35}$$

If $\mathbf{B}$ is uniform and independent of time t, show that the path described by the electron is a circular helix.

Solution

Let the z-axis be chosen in the direction of $\mathbf{B}$, so that $\mathbf{B} = B\mathbf{k}$, and let $\mathbf{r}$ be the position vector of the electron at time t. Then (3.35) becomes

$$\ddot{\mathbf{r}} = p\dot{\mathbf{r}} \times \mathbf{k}, \tag{3.36}$$

where $p = eB/m$ and dots denote differentiation with respect to t. Now

$$\mathbf{r} = x\mathbf{i} + y\mathbf{j} + z\mathbf{k},$$

and hence

$$\ddot{x}\mathbf{i} + \ddot{y}\mathbf{j} + \ddot{z}\mathbf{k} = p(\dot{x}\mathbf{i} + \dot{y}\mathbf{j} + \dot{z}\mathbf{k}) \times \mathbf{k}.$$

Separating into components,

$$\ddot{x} = p\dot{y}, \quad \ddot{y} = -p\dot{x}, \quad \ddot{z} = 0. \tag{3.37}$$

Let the origin be chosen at the position of the particle when $t = 0$ and choose the direction of the x-axis so that the initial velocity of the particle is $u\mathbf{i} + w\mathbf{k}$ (that is, so that the initial velocity component in the y-direction is zero). Then equations (3.37) have to be solved subject to the initial conditions

$$\dot{x} = u, \quad \dot{y} = 0, \quad \dot{z} = w \tag{3.38}$$

and

$$x = y = z = 0. \tag{3.39}$$

The solution of the last of equations (3.37) is obtained at once as

$$z = wt. \tag{3.40}$$

Integrating the first two of equations (3.37) and using the initial conditions gives

$$\dot{x} = py + u, \quad \dot{y} = -px.$$

Substituting these results, the first two of equations (3.37) may now be rewritten as

$$\ddot{x} + p^2 x = 0 \quad \text{and} \quad \ddot{y} + p^2 y = -pu.$$

The general solutions for x, y are therefore

$$x = A \cos pt + B \sin pt,$$

$$y = C \cos pt + D \sin pt - u/p,$$

where A, B, C, D are arbitrary constants. Using the initial conditions (3.38), (3.39) it follows easily that $A = D = 0$, $B = C = u/p$. Thus

$$x = \frac{u}{p} \sin pt, \quad y = \frac{u}{p} (\cos pt - 1), \quad z = wt$$

and these are the parametric equations of a circular helix with axis $x = 0$, $y = -u/p$.

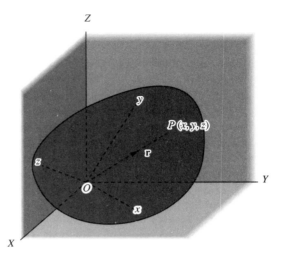

Fig. 3.12 A rigid body rotating relative to fixed axes.

Angular velocity

Consider a rigid body pivoted at an origin O, where O is both fixed in space and fixed in the body. Let $Oxyz$ be axes fixed in the body and let P (x, y, z) be a point fixed in the body and having position vector $\mathbf{r}$ relative to O. Then we shall show that there exists a vector $\boldsymbol{\omega}$ such that, relative to axes $OXYZ$ fixed in space, the velocity of P is

$$\mathbf{V} = \boldsymbol{\omega} \times \mathbf{r}. \tag{3.41}$$

We define $\boldsymbol{\omega}$ to be the **angular velocity** of the body.

Proof

Let $\mathbf{i}, \mathbf{j}, \mathbf{k}$ be the usual unit vectors associated with the coordinate system $Oxyz$. As the body rotates, the directions of $\mathbf{i}, \mathbf{j}, \mathbf{k}$ will, of course, change relative to the axes $OXYZ$, and hence $\mathbf{i}, \mathbf{j}, \mathbf{k}$ are functions of time t. Since

$$\mathbf{i} = \mathbf{j} \times \mathbf{k}, \quad \mathbf{j} = \mathbf{k} \times \mathbf{i}, \quad \mathbf{k} = \mathbf{i} \times \mathbf{j}, \tag{3.42}$$

$$\frac{d\mathbf{i}}{dt} = \frac{d\mathbf{j}}{dt} \times \mathbf{k} + \mathbf{j} \times \frac{d\mathbf{k}}{dt}$$

$$= \frac{d\mathbf{j}}{dt} \times (\mathbf{i} \times \mathbf{j}) + (\mathbf{k} \times \mathbf{i}) \times \frac{d\mathbf{k}}{dt}. \tag{3.43}$$

As $\mathbf{i} \cdot \mathbf{i} = 1$, we have $\mathbf{i} \cdot di/dt = 0$ with similar relationships for $\mathbf{j}$ and $\mathbf{k}$. Hence expanding the triple products in (3.43) gives

$$\frac{d\mathbf{i}}{dt} = -\mathbf{i} \cdot \frac{d\mathbf{j}}{dt}\mathbf{j} - \mathbf{i} \cdot \frac{d\mathbf{k}}{dt}\mathbf{k}. \tag{3.44}$$

But $\mathbf{j} = \mathbf{k} \times \mathbf{i}$ and $-\mathbf{k} = \mathbf{j} \times \mathbf{i}$. Hence (3.44) can be written as

$$\frac{d\mathbf{i}}{dt} = \left(-\mathbf{i} \cdot \frac{d\mathbf{j}}{dt}\mathbf{k} + \mathbf{i} \cdot \frac{d\mathbf{k}}{dt}\mathbf{j}\right) \times \mathbf{i}. \tag{3.45}$$

Furthermore, as $\mathbf{i} \cdot \mathbf{j} = 0$, we have $\mathbf{i} \cdot d\mathbf{j}/dt = -\mathbf{j} \cdot d\mathbf{i}/dt$. Thus (3.45) becomes

$$\frac{d\mathbf{i}}{dt} = \left(\mathbf{i} \cdot \frac{d\mathbf{k}}{dt}\mathbf{j} + \mathbf{j} \cdot \frac{d\mathbf{i}}{dt}\mathbf{k}\right) \times \mathbf{i}. \tag{3.46}$$

Since $\mathbf{i} \times \mathbf{i} = \mathbf{0}$, we may add a term $(\mathbf{k} \cdot d\mathbf{j}/dt)\mathbf{i}$ within the bracket of (3.46) to obtain the cyclically invariant form

$$\frac{d\mathbf{i}}{dt} = \boldsymbol{\omega} \times \mathbf{i} \tag{3.47}$$

where

$$\boldsymbol{\omega} = \mathbf{k} \cdot \frac{d\mathbf{j}}{dt}\mathbf{i} + \mathbf{i} \cdot \frac{d\mathbf{k}}{dt}\mathbf{j} + \mathbf{j} \cdot \frac{d\mathbf{i}}{dt}\mathbf{k}. \tag{3.48}$$

Likewise

$$\frac{d\mathbf{j}}{dt} = \boldsymbol{\omega} \times \mathbf{j} \quad \text{and} \quad \frac{d\mathbf{k}}{dt} = \boldsymbol{\omega} \times \mathbf{k} \tag{3.49}$$

where, by cyclic interchange, $\boldsymbol{\omega}$ is the same for each.
   Finally, as

$$\mathbf{r} = x\mathbf{i} + y\mathbf{j} + z\mathbf{k} \tag{3.50}$$

and as x, y, z are independent of time t then

$$\frac{d\mathbf{r}}{dt} = \mathbf{V} = \boldsymbol{\omega} \times \mathbf{r}. \tag{3.51}$$

Rotation about a fixed axis

Suppose that the rotation of the body is about the Z-axis and that the axes $Oxyz$ are initially coincident with the axes $OXYZ$. Let φ (radians) be the angle between Ox and OX at time t.
Relative to the axes $OXYZ$,

$$\mathbf{i} = (\cos \varphi, \ \sin \varphi, \ 0)$$
$$\mathbf{j} = (-\sin \varphi, \ \cos \varphi, \ 0)$$

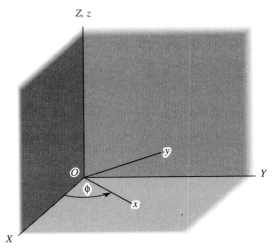

Fig. 3.13 Axes $Oxyz$ rotating about OZ.

and

$$\mathbf{k} = (0, 0, 1).$$

Thus

$$\frac{d\mathbf{i}}{dt} = (-\dot{\varphi} \sin\varphi, \; \dot{\varphi} \cos\varphi, \; 0) = \dot{\varphi}\,\mathbf{j}$$

and

$$\frac{d\mathbf{j}}{dt} = -\dot{\varphi}\,\mathbf{i}$$

where dots denote derivatives with respect to t. As $d\mathbf{k}/dt = \mathbf{0}$, equation (3.48) gives

$$\boldsymbol{\omega} = \dot{\varphi}\,\mathbf{k}. \qquad (3.52)$$

We define $|\boldsymbol{\omega}| = |\dot{\varphi}|$ to be the angular speed of the body. It measures the rate of rotation in radians per second.

General motion of a rigid body

Suppose that the axes $Oxyz$ move with the body, as before, but that the fixed axes are $O'XYZ$ where O and O' no longer coincide. Let the position vectors of the points O and P (both moving with the body) relative to O' be $\mathbf{R}$ and $\mathbf{r}'$ respectively. Then

$$\mathbf{r}' = \mathbf{R} + \mathbf{r} \qquad (3.53)$$

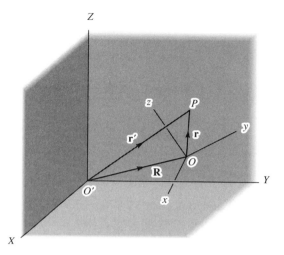

Fig. 3.14 Position vectors of O and P, moving with the body, relative to the fixed origin O'.

where $\mathbf{r}$ is given by (3.50). The components of $\mathbf{i}, \mathbf{j}, \mathbf{k}$ relative to the fixed axes $O'XYZ$ depend only on the orientation of the axes $Oxyz$ relative to the fixed system and not on the position of O relative to O'. Hence relations (3.48) and (3.51) follow as before. Differentiating (3.53) with respect to time t we therefore have

$$\mathbf{V}' = \mathbf{u} + \boldsymbol{\omega} \times \mathbf{r} \tag{3.54}$$

where $\mathbf{V}' = d\mathbf{r}'/dt$ is the velocity of P relative to O' and $\mathbf{u} = d\mathbf{R}/dt$ is the velocity of O relative to O'. This result is of importance for discussing the general motion of a rigid body relative to fixed axes.

The most general motion consists of a translation of the mass centre together with a superimposed rotational motion. If $\mathbf{u}$ denotes the velocity of the mass centre with respect to fixed axes and $\boldsymbol{\omega}$ is the angular velocity of the body, then equation (3.54) gives the velocity, relative to the fixed axes, of a point P having position vector $\mathbf{r}$ relative to the mass centre and moving with the body. If the velocities of three points of the body are known, the angular velocity $\boldsymbol{\omega}$ can be determined as follows.

Let A, B, C be points fixed in the body and having velocities $\mathbf{V}_A, \mathbf{V}_B, \mathbf{V}_C$, respectively, relative to the fixed axes $O'XYZ$ and position vectors $\mathbf{R}_A, \mathbf{R}_B, \mathbf{R}_C$ relative to O'. The position vectors of A, B, C relative to O are, respectively,

$$\mathbf{r}_A = \mathbf{R}_A - \mathbf{R}, \quad \mathbf{r}_B = \mathbf{R}_B - \mathbf{R}, \quad \mathbf{r}_C = \mathbf{R}_C - \mathbf{R}. \tag{3.55}$$

Using (3.54)

$$\mathbf{V}_A = \mathbf{u} + \boldsymbol{\omega} \times (\mathbf{R}_A - \mathbf{R})$$

and

$$\mathbf{V}_B = \mathbf{u} + \boldsymbol{\omega} \times (\mathbf{R}_B - \mathbf{R}).$$

Hence

$$\mathbf{V}_B - \mathbf{V}_A = \boldsymbol{\omega} \times (\mathbf{R}_B - \mathbf{R}_A). \tag{3.56}$$

Similarly

$$\mathbf{V}_C - \mathbf{V}_A = \boldsymbol{\omega} \times (\mathbf{R}_C - \mathbf{R}_A). \tag{3.57}$$

It follows from these equations that $\mathbf{V}_B - \mathbf{V}_A$ and $\mathbf{V}_C - \mathbf{V}_A$ are both perpendicular to $\boldsymbol{\omega}$. Hence there exists a scalar λ such that

$$\boldsymbol{\omega} = \lambda \, (\mathbf{V}_B - \mathbf{V}_A) \times (\mathbf{V}_C - \mathbf{V}_A). \tag{3.58}$$

Substituting this expression for $\boldsymbol{\omega}$ into (3.56) we have

$$\mathbf{V}_B - \mathbf{V}_A = \lambda \, [\, (\mathbf{V}_B - \mathbf{V}_A) \times (\mathbf{V}_C - \mathbf{V}_A) \,] \times (\mathbf{R}_B - \mathbf{R}_A)$$

$$= \lambda \, [(\mathbf{V}_B - \mathbf{V}_A) \cdot (\mathbf{R}_B - \mathbf{R}_A)] \, (\mathbf{V}_C - \mathbf{V}_A)$$

$$- \lambda \, [\mathbf{V}_C - \mathbf{V}_A) \cdot (\mathbf{R}_B - \mathbf{R}_A)] \, (\mathbf{V}_B - \mathbf{V}_A). \tag{3.59}$$

Now

$$(\mathbf{R}_B - \mathbf{R}_A) \cdot (\mathbf{R}_B - \mathbf{R}_A) = d^2 \tag{3.60}$$

where d is the distance between A and B. Since the body is rigid, the distance d remains constant as the body moves. Differentiating (3.60) with respect to t, we obtain

$$(\mathbf{V}_B - \mathbf{V}_A) \cdot (\mathbf{R}_B - \mathbf{R}_A) = 0. \tag{3.61}$$

The first term on the right hand side of (3.59) is therefore zero. Hence, comparing the coefficients of $\mathbf{V}_B - \mathbf{V}_A$ on each side of the equation, we find

$$\lambda = - \frac{1}{(\mathbf{V}_C - \mathbf{V}_A) \cdot (\mathbf{R}_B - \mathbf{R}_A)} \, .$$

Substituting this into equation (3.58) and simplifying the vector product gives

$$\boldsymbol{\omega} = - \frac{\mathbf{V}_A \times \mathbf{V}_B + \mathbf{V}_B \times \mathbf{V}_C + \mathbf{V}_C \times \mathbf{V}_A}{(\mathbf{V}_C - \mathbf{V}_A) \cdot (\mathbf{R}_B - \mathbf{R}_A)} \, . \tag{3.62}$$

Similar to (3.61) we have

$$(\mathbf{V}_B - \mathbf{V}_C) \cdot (\mathbf{R}_B - \mathbf{R}_C) = 0$$

which gives

$$\mathbf{V}_C \cdot \mathbf{R}_B = \mathbf{V}_B \cdot \mathbf{R}_B + \mathbf{V}_C \cdot \mathbf{R}_C - \mathbf{V}_B \cdot \mathbf{R}_C. \tag{3.63}$$

Expanding the denominator on the right-hand side of (3.62) and then using (3.63) to eliminate $\mathbf{V}_C \cdot \mathbf{R}_B$ we obtain

$$\boldsymbol{\omega} = \frac{\mathbf{V}_A \times \mathbf{V}_B + \mathbf{V}_B \times \mathbf{V}_C + \mathbf{V}_C \times \mathbf{V}_A}{\mathbf{V}_A \cdot \mathbf{R}_B + \mathbf{V}_B \cdot \mathbf{R}_C + \mathbf{V}_C \cdot \mathbf{R}_A - \mathbf{V}_A \cdot \mathbf{R}_A - \mathbf{V}_B \cdot \mathbf{R}_B - \mathbf{V}_C \cdot \mathbf{R}_C}. \tag{3.64}$$

This is the required expression for $\boldsymbol{\omega}$. It is cyclically symmetric in A, B, C.

Motion of a point relative to a rotating coordinate system

Let $Oxyz$ be a system of axes which are rotating with angular velocity $\boldsymbol{\omega}$ relative to a fixed coordinate system $OXYZ$ with the same origin O. Let P be a moving point which at time t has coordinates (x, y, z) relative to the axes $Oxyz$. The position vector of P relative to O is

$$\mathbf{r} = x\mathbf{i} + y\mathbf{j} + z\mathbf{k} \tag{3.65}$$

and the velocity of P relative to the moving axes is

$$\mathbf{v} = \dot{x}\mathbf{i} + \dot{y}\mathbf{j} + \dot{z}\mathbf{k} \tag{3.66}$$

where the dots denote first order derivatives with respect to t. Differentiating (3.65) with respect to t and using relations (3.47), (3.49) and (3.66), the velocity of P relative to the fixed axes is therefore

$$\frac{d\mathbf{r}}{dt} = \mathbf{V} = \mathbf{v} + \boldsymbol{\omega} \times \mathbf{r}. \tag{3.67}$$

This extends our earlier result (3.51) to the case when x, y and z vary with time t. The acceleration of P relative to the moving axes is

$$\mathbf{a} = \ddot{x}\mathbf{i} + \ddot{y}\mathbf{j} + \ddot{z}\mathbf{k} \tag{3.68}$$

where the dots denote second order derivatives with respect to t. Hence, using relations (3.47) and (3.49) again together with (3.68), we obtain from differentiation of (3.66)

$$\frac{d\mathbf{v}}{dt} = \mathbf{a} + \dot{x}\boldsymbol{\omega} \times \mathbf{i} + \dot{y}\boldsymbol{\omega} \times \mathbf{j} + \dot{z}\boldsymbol{\omega} \times \mathbf{k}$$

$$= \mathbf{a} + \boldsymbol{\omega} \times \mathbf{v}. \tag{3.69}$$

Differentiating (3.67) with respect to t and using (3.69) the acceleration of P relative to the fixed axes is therefore

$$\mathbf{A} = \frac{d\mathbf{V}}{dt} = \mathbf{a} + \boldsymbol{\omega} \times \mathbf{v} + \dot{\boldsymbol{\omega}} \times \mathbf{r} + \boldsymbol{\omega} \times \mathbf{V}.$$

Using (3.67) this gives

$$\mathbf{A} = \mathbf{a} + 2\boldsymbol{\omega} \times \mathbf{v} + \dot{\boldsymbol{\omega}} \times \mathbf{r} + \boldsymbol{\omega} \times (\boldsymbol{\omega} \times \mathbf{r}). \tag{3.70}$$

This result has an important application in the discussion of the motion of particles near the Earth's surface relative to an observer moving with the Earth.

Suppose that O is a point on the Earth's surface and that the coordinate system $Oxyz$ is fixed relative to the Earth. Let $O'XYZ$ be axes fixed in space with the origin O' at the Earth's centre and with $O'Z$ along the polar axis. The Earth is assumed to rotate with constant angular velocity $\boldsymbol{\omega}$ so that the axes $Oxyz$ are rotating in space with that angular velocity. Using the result of Exercise 3.25 at the end of this chapter, the acceleration of O relative to the axes $O'XYZ$ is $\boldsymbol{\omega} \times (\boldsymbol{\omega} \times \mathbf{R})$ where $\mathbf{R} = \overrightarrow{O'O}$. Hence the acceleration of a moving point P relative to the axes $O'XYZ$ is $\mathbf{A} + \boldsymbol{\omega} \times (\boldsymbol{\omega} \times \mathbf{R})$, where $\mathbf{A}$ is given by equation (3.70).

Suppose now that P is a particle of mass m which is in motion near the Earth's surface under the Earth's gravitational attractive force $\mathbf{F}$. By Newton's law of motion

$$m(\mathbf{A} + \boldsymbol{\omega} \times (\boldsymbol{\omega} \times \mathbf{R})) = \mathbf{F}.$$

Hence

$$\mathbf{A} = (\mathbf{F}/m) - \boldsymbol{\omega} \times (\boldsymbol{\omega} \times \mathbf{R}). \tag{3.71}$$

This is defined to be the acceleration due to gravity*, conventionally denoted by $\mathbf{g}$. Substituting $\mathbf{A}$ from equation (3.70) and noting that because the Earth's angular velocity is assumed constant $\dot{\boldsymbol{\omega}} = \mathbf{0}$, we have

$$\mathbf{a} = \mathbf{g} - 2\boldsymbol{\omega} \times \mathbf{v} - \boldsymbol{\omega} \times (\boldsymbol{\omega} \times \mathbf{r}). \tag{3.72}$$

The term $\boldsymbol{\omega} \times (\boldsymbol{\omega} \times \mathbf{r})$ is, however, small compared to $|\mathbf{g}|$ and $|\boldsymbol{\omega} \times \mathbf{v}|$ because $|\boldsymbol{\omega}|$ is very small ($|\boldsymbol{\omega}| \approx 7 \times 10^{-5}$ radians per second). Hence, to a good approximation, the acceleration of P relative to an observer at O is

$$\mathbf{a} = \mathbf{g} - 2\boldsymbol{\omega} \times \mathbf{v}. \tag{3.73}$$

To such an observer, the particle P appears to be acted upon by the force of gravity, $m\mathbf{g}$, and another force $-2m\boldsymbol{\omega} \times \mathbf{v}$; the latter is called the **Coriolis force**.

The Coriolis force makes a negligible contribution to calculations of the trajectory of a particle in flight near the Earth's surface provided that the time of flight is relatively small. However, for larger times, the contribution is significant. For example, a particle which is in flight for 16 seconds near the equator is deflected about 1m east by the Coriolis force. Furthermore, this deflection may be shown to be proportional to the cube of the time of flight, so large errors are possible when this term is neglected.

Another important consequence of the Coriolis force is indicated in Exercise 3.30.

---

* Note that $\mathbf{g}$ is actually the acceleration due to the Earth's gravitational attractive force minus the centripetal acceleration due to the Earth's rotation.

EXERCISES

3.22 At an origin O on the Earth's surface, the z-axis points vertically up-wards. A particle, moving under the influence of constant gravity only, has acceleration

$$\frac{d^2\mathbf{r}}{dt^2} = (0, 0, -g)$$

where $\mathbf{r}$ is the position vector and t denotes time. If the particle is projected from the origin when $t = 0$ with velocity $(u, 0, v)$, show by integrating the above differential equation that the locus of its path is

$$\mathbf{r} = (ut, 0, vt - \tfrac{1}{2}gt^2).$$

3.23 A particle moves with velocity $\mathbf{v}$ and acceleration $\mathbf{f}$. Show that the radius of curvature of its path is

$$\rho = v^3/|\mathbf{v} \times \mathbf{f}|.$$

Use this formula to determine the radius of curvature at the origin of the path of a particle whose position vector at time t is

$$\mathbf{r} = (t, t^2, t^3).$$

3.24 A point moves so that its position vector $\mathbf{r}$ satisfies the differential equation

$$\frac{d^2\mathbf{r}}{dt^2} = \mathbf{g} - \lambda \frac{d\mathbf{r}}{dt}$$

where t denotes time, $\mathbf{g}$ is a constant vector and λ is a constant scalar. If the point is at the origin at time $t = 0$ and is then moving with velocity $\mathbf{u}$, show that

$$\mathbf{r} = \frac{\mathbf{g}}{\lambda^2}(\lambda t + e^{-\lambda t} - 1) + \frac{\mathbf{u}}{\lambda}(1 - e^{-\lambda t}).$$

[*Hint.* Show first that

$$\frac{d\mathbf{r}}{dt} + \lambda \mathbf{r} = \mathbf{g}t + \mathbf{u}$$

then multiply this equation by $e^{\lambda t}$ and integrate.]

3.25 Let P be a point fixed on the Earth's surface and having position vector $\mathbf{R}$ relative to the Earth's centre O. Using equation (3.51) show that the acceleration of P relative to axes with origin at O and fixed in space is $\boldsymbol{\omega} \times (\boldsymbol{\omega} \times \mathbf{R})$ where $\boldsymbol{\omega}$ is the angular velocity of the Earth about the polar axis.

Deduce that, if the angle between OP and the polar axis is θ, the acceleration of P is directly towards the polar axis and of magnitude $\omega^2 R \sin \theta$, where R is the Earth's radius.

3.26 A rigid body is pivoted at a point O and is rotating at 90 radians per second about a fixed axis in the direction of the vector $(-1, 2, 2)$. To an observer looking in the direction of this vector the sense of rotation of the body is clockwise. Find the velocity of the point P with position vector $\left(\frac{1}{3}, \frac{2}{3}, -\frac{1}{3}\right)$ metres.

3.27 A rigid body is rotating about the fixed origin O. The points $A(0, -1, 2)$ and $B(2, 0, 0)$ are moving with velocities $(7, -2, -1)$ and $(0, 6, -4)$, respectively, the units being metres and seconds. Find the angular velocity of the body.

3.28 A rigid rod OA rotates in the XY-plane about the fixed vertical axis OZ as shown in Fig. 3.15. A small ring P is threaded on the rod and is free to slide along it. At time t, P is at a distance r from O and the rod makes an angle φ with OX. Use equation (3.70) to show that, relative to the axes $OXYZ$, the acceleration of P has a radial component $\ddot{r} - r\dot{\varphi}^2$ away from O and a transverse component $2\dot{r}\dot{\varphi} + r\ddot{\varphi}$ in the XY-plane perpendicular to the rod in the direction of φ increasing.

3.29 The point O lies on the Earth's surface at co-latitude λ (i.e. the angle between the Earth's diameter through O and the polar axis is λ). Axes $Oxyz$ fixed to the Earth are chosen such that Ox points east, Oy points north and Oz points vertically upwards. Show that, relative to the axes $Oxyz$, the Earth's angular velocity is

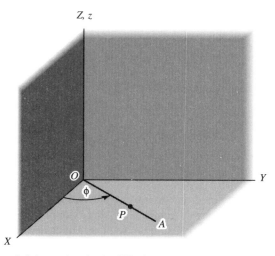

Fig. 3.15 Rigid rod OA rotating in the XY-plane.

$$\boldsymbol{\omega} = (0, \omega \sin \lambda, \omega \cos \lambda)$$

where ω is the angular speed of the Earth.

A particle P moves across an ice-rink which may be assumed to be frictionless and is in the xy-plane. The particle is projected from O eastward with initial speed u. If the coordinates of P at a later time t are $(x, y, 0)$ show that

$$\ddot{x} = 2\omega\dot{y} \cos \lambda, \quad \ddot{y} = -2\omega\dot{x} \cos \lambda.$$

Deduce that $\dot{x}^2 + \dot{y}^2 = u^2$.

Find the equation of the path of P and deduce that P moves in the xy-plane with constant speed u along a circular arc of radius $u/(2\omega \cos \lambda)$.

3.30 (This exercise requires the work of Chapter 4.) A simple model equation of the local fluid mechanics of the Earth's atmosphere is

$$2\rho\boldsymbol{\omega} \times \mathbf{v} = \mathbf{g} - \text{grad } p,$$

where ρ is the atmospheric density and p is the atmospheric pressure. Also, neglecting the Earth's curvature, we may take local axes $Oxyz$ with Ox and Oy fixed horizontally and Oz vertically upwards. In this simplified model we may then take $\mathbf{v} = (v_1, v_2, 0)$ and, at co-latitude θ, $\boldsymbol{\omega} = (0, 0, \omega \cos \theta)$ with $p = p(x, y, z)$.

(i) By writing out the x- and y- components of the model equation show that the flow is divergence free, that is div $\mathbf{v} = 0$.

(ii) Show that the wind blows along the isobars (lines of constant pressure).

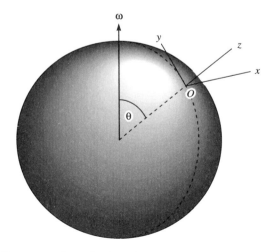

Fig. 3.16 Coordinate axes $Oxyz$ fixed to the rotating Earth.

Remark

The dominant forces, per unit volume, on a fluid element in the Earth's atmosphere are the Coriolis force $-2\rho\boldsymbol{\omega}\times\mathbf{v}$, the pressure gradient – grad p and gravity $\rho\mathbf{g}$. In steady atmospheric flow, the horizontal components of these forces balance, leading to the above model equation. The exercise shows the powerful influence of the Earth's rotation on the weather (through the agency of the Coriolis force).

3.31 An inextensible string has end A which is constrained to move with velocity $\mathbf{v}$, being attached to the end of a crank BA, as shown in Fig. 3.17. The string is threaded through a small fixed eye situated at O and the length OA is kept taut. If $\boldsymbol{\rho}$ is the vector from O to A at a particular instant

(i) show that the speed $\dot{s}$ with which the string slides through the eye is $\mathbf{v}.\boldsymbol{\rho}/\rho$,

(ii) assuming that there is no rotation of the string about its own axis, show that the angular velocity $\boldsymbol{\omega}$ of OA is $\boldsymbol{\rho}\times\mathbf{v}/\rho^2$.

If the crank BA turns with angular velocity $\boldsymbol{\Omega}$ so that $\mathbf{v}=\boldsymbol{\Omega}\times\mathbf{r}$, where $\mathbf{r}$ is the vector from B to A, what is the geometrical relationship between $\boldsymbol{\Omega}$, $\mathbf{r}$ and $\boldsymbol{\rho}$ if $\dot{s}=0$?

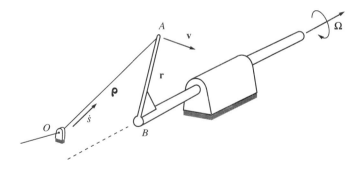

Fig. 3.17 String attached at A to a rotating crank and threaded through an eye O.

Scalar and vector fields | 4

4.1 REGIONS

In the further development of vector analysis we shall be concerned with functions defined on certain point sets. It is convenient to introduce here some associated definitions.*

Let U be a set of points in three-dimensional space. A point P belonging to U is called as **interior point** if there exists a sphere with centre at P such that every point within the sphere belongs to U.

A point P is called a **boundary point** of U if every sphere centred at P contains some interior points of U and some points which are not in U.

The set U is said to form a **region** $\mathcal{R}$ if each point is either an interior point or a boundary point and if every pair of points belonging to U can be joined by a continuous curve consisting entirely of points belonging to U. Further, $\mathcal{R}$ is said to be an **open region** if it has no boundary points; and it is said to be a **closed region** if all points not in R form one or more open regions.

The definitions above, applicable to point sets in three-dimensional space, can easily be modified to define interior points, boundary points, etc., for sets of points in a plane – the sphere is simply replaced by its two-dimensional counterpart, the circle.

A simple example of an open region is the set of points (x, y, z) such that

$$x^2 + y^2 + z^2 < 1,$$

that is, the set of all points lying inside – but not on the surface of – the sphere of unit radius centred at the origin. However, the set of points (x, y, z) such that

$$x^2 + y^2 + z^2 \leqslant 1$$

constitutes a closed region whose boundary is the sphere $x^2 + y^2 + z^2 = 1$.

---

* The definitions given here will be sufficient for our purposes. More formal definitions may be found in Baxandall, P. R. and Liebeck, H. (1981) *Differential Vector Calculus*, Longman Inc., New York.

EXERCISES

4.1 Does the set of all points in space constitute an open region or a closed region?

4.2 Consider the set of points (x, y) in the xy-plane such that

$$1 < x^2 + y^2 \leqslant 2.$$

Explain why this set constitutes neither an open region nor a closed region.

4.2 FUNCTIONS OF SEVERAL VARIABLES

For the reader whose knowledge of functions of several independent variables is slight, we give a brief account in this section of most of the concepts and results needed later.* As we shall be concerned mainly with functions of three independent real variables, attention is confined to this case. However, most results quoted extend in an obvious way to functions of more than three variables.

Throughout this section, the independent variables x, y, z are rectangular cartesian coordinates in three-dimensional space.

Continuity

Let the real valued function $f(x, y, z)$ be defined at all points in some region $\mathcal{R}$ containing the point $P(a, b, c)$. Let $Q(x, y, z)$ be any other point in the region. We say that $f(x, y, z)$ is *continuous* at the point $P(a, b, c)$ if the difference

$$f(x, y, z) - f(a, b, c)$$

between the values of f at Q and P tends to zero whenever Q approaches P *along any path in $\mathcal{R}$*.

Partial derivatives

The **first-order partial derivative** $\partial f/\partial x$ of the function $f(x, y, z)$ with respect to x is defined as

$$\lim_{h \to 0} \frac{f(x + h, y, z) - f(x, y, z)}{h} \tag{4.1}$$

whenever the limit exists. It follows that $\partial f/\partial x$ is obtained by differentiating f with respect to x, treating y, z as constants. The partial derivatives of f with

---

* Further information can be found in the following books: Hilton, P. J., *Partial Derivatives*, Routledge and Kegan Paul; and Courant, R. *Differential and Integral Calculus*, Blackie.

respect to y, z are defined similarly. There are thus three first-order partial derivatives of $f(x, y, z)$, namely

$$\frac{\partial f}{\partial x}, \quad \frac{\partial f}{\partial y}, \quad \frac{\partial f}{\partial z}.$$

They are also sometimes denoted by

$$f_x, \quad f_y, \quad f_z.$$

Note that at a boundary point of a region the derivatives may be one-sided (cf footnote in section 3.4).

EXAMPLE 1

Find f_x, f_y, f_z for the function

$$f = x^3 + x^2 y + xyz.$$

Solution

Treating y, z as constants and differentiating f with respect to x, we obtain immediately

$$f_x = 3x^2 + 2xy + yz.$$

Similarly, treating x, z as constants

$$f_y = x^2 + xz$$

and treating x, y as constants

$$f_z = xy.$$

Suppose that the value of a partial derivative, say f_x, is required at a particular point (x_0, y_0, z_0). Then, apart from the obvious procedure of evaluating f_x and then putting $x = x_0$, $y = y_0$, $z = z_0$, we can also put $y = y_0$, $z = z_0$ in the function $f(x, y, z)$ then find the derivative with respect to x and finally put $x = x_0$ in the result. Thus, in the example above, it is seen that

$$f_x(x_0, y_0, z_0) = 3x_0^2 + 2x_0 y_0 + y_0 z_0.$$

We also have

$$f(x, y_0, z_0) = x^3 + x^2 y_0 + xy_0 z_0.$$

Hence

$$f_x(x, y_0, z_0) = 3x^2 + 2x y_0 + y_0 z_0.$$

and so, as before,

$$f_x(x_0, y_0, z_0) = 3x_0^2 + 2x_0 y_0 + y_0 z_0.$$

Higher-order partial derivatives

It is clear that the first-order partial derivatives of $f(x, y, z)$ will themselves be functions of x, y, z. When these partial derivatives are differentiated partially with respect to x, y, or z we obtain **second-order partial derivatives**. The partial derivatives of $\partial f / \partial x$ with respect to x, y, z are denoted respectively by

$$\frac{\partial^2 f}{\partial x^2}, \quad \frac{\partial^2 f}{\partial y\, \partial x}, \quad \frac{\partial^2 f}{\partial z\, \partial x}$$

or alternatively by

$$f_{xx}, \quad f_{yx}, \quad f_{zx}.$$

A similar notation is used for the second-order derivatives arising from $\partial f / \partial y$ and $\partial f / \partial z$.

Derivatives of higher order are defined in a similar manner and an obvious extension of the above notation is used.

EXAMPLE 2

Find $\partial^3 f / \partial z^3$ for the function

$$f = (x + 2y + 3z)^4.$$

Solution

We find at once

$$\frac{\partial f}{\partial z} = 4(x + 2y + 3z)^3 \, \frac{\partial}{\partial z}(x + 2y + 3z)$$

$$= 12(x + 2y + 3z)^3.$$

Differentiating again with respect to z,

$$\frac{\partial^2 f}{\partial z^2} = 12 \times 3 \times 3(x + 2y + 3z)^2$$

hence

$$\frac{\partial^3 f}{\partial z^3} = 108 \times 2 \times 3(x + 2y + 3z)$$

$$= 648(x + 2y + 3z).$$

EXAMPLE 3

Find $\partial^2 f/\partial y\,\partial x$ and $\partial^2 f/\partial x\,\partial y$ for the function

$$f = \sin(ax + by + cz)$$

where a, b, c are constants.

Solution

Differentiating f partially with respect to x,

$$\frac{\partial f}{\partial x} = a\cos(ax + by + cz).$$

Hence, differentiating this with respect to y,

$$\frac{\partial^2 f}{\partial y\,\partial x} = -ab\sin(ax + by + cz).$$

Also, differentiating f partially with respect to y,

$$\frac{\partial f}{\partial y} = b\cos(ax + by + cz).$$

Hence

$$\frac{\partial^2 f}{\partial x\,\partial y} = -ab\sin(ax + by + cz).$$

In the example above, it is observed that $f_{yx} = f_{xy}$ so the order in which the x-, y-differentiations are performed is in this case immaterial. The same situation arises for nearly all commonly occurring functions; in fact if the reader were to write down a function $f(x, y, z)$ 'at random' it is most unlikely that pairs of mixed derivatives f_{xy}, f_{yx} would not be identical. For reference, we quote the following theorem which sets out sufficient conditions for the order of partial differentiation to be immaterial.

Theorem

If all the mixed second-order derivatives (f_{xy}, f_{xz}, etc.) of the function $f(x, y, z)$ exist and are continuous at a given point, then, at that point,

$$f_{xy} = f_{yx}, \quad f_{yz} = f_{zy}, \quad f_{zx} = f_{xz}. \tag{4.2}$$

Continuously differentiable functions

The function $f(x, y, z)$ is said to be **continuously differentiable** in a region $\mathcal{R}$ if its first-order partial derivatives f_x, f_y, f_z exist and are continuous at every point of $\mathcal{R}$.

Functions which are not continuously differentiable require special treatment and will not be discussed in this book. For emphasis we shall occasionally remind the reader of this when stating important results, *but otherwise it will be assumed without comment that the functions discussed are properly defined and continuously differentiable in the region under consideration. Furthermore, whenever we have occasion to introduce second- or higher-order derivatives of a function, we shall assume that they too exist and are continuous throughout the region considered.*

The chain rule

Let F be a continuously differentiable function of f, g, h and suppose that each of f, g, h is a continuously differentiable function of x, y, z. Then it may be shown that F is a continuously differentiable composite function of x, y, z and that

$$\frac{\partial F}{\partial x} = \frac{\partial F}{\partial f}\frac{\partial f}{\partial x} + \frac{\partial F}{\partial g}\frac{\partial g}{\partial x} + \frac{\partial F}{\partial h}\frac{\partial h}{\partial x}$$

$$\frac{\partial F}{\partial y} = \frac{\partial F}{\partial f}\frac{\partial f}{\partial y} + \frac{\partial F}{\partial g}\frac{\partial g}{\partial y} + \frac{\partial F}{\partial h}\frac{\partial h}{\partial y} \qquad (4.3)$$

$$\frac{\partial F}{\partial z} = \frac{\partial F}{\partial f}\frac{\partial f}{\partial z} + \frac{\partial F}{\partial g}\frac{\partial g}{\partial z} + \frac{\partial F}{\partial h}\frac{\partial h}{\partial z} . \quad *$$

This is the rule for differentiation of a function of functions, usually called the chain rule.[†]

To illustrate this result let

$$F = f - 4g + h,$$

where

$$f = 3x^2 + 2y^2, \quad g = (x - z)^2, \quad h = y + 1.$$

Then

$$\frac{\partial F}{\partial x} = \frac{\partial F}{\partial f}\frac{\partial f}{\partial x} + \frac{\partial F}{\partial g}\frac{\partial g}{\partial x} + \frac{\partial F}{\partial h}\frac{\partial h}{\partial x}$$

$$= 1 \times 3 \times 2x + (-4) \times 2(x - z) + 1 \times 0$$

$$= 8z - 2x.$$

Similarly $\partial F/\partial y$ and $\partial F/\partial z$ may also be evaluated using the chain rule and without writing down F explicitly as a function of x, y, z.

---

* A proof, together with a full statement of the conditions under which the result holds, is given in Baxandall, P.R. and Liebeck, H. (1981) *Differential Vector Calculus*, Longman Inc., New York.
† It is convenient to use the symbol F to denote both the function of f, g and h and the composite function. Hence we denote

$$F[f(x, y, z), \ g(x, y, z), h(x, y, z)]$$

by $F(x, y, z)$ rather than, say, $G(x, y, z)$. We note that this common practice has some advantage and it will be used throughout the book whenever composite functions occur. In practice this should not cause any difficulty for the reader.

EXERCISES

4.3 If

$$f(x, y) = ax^2 + 2hxy + by^2,$$

where a, b, h are constants, find f_x, f_y, f_{xx}, f_{yy}, f_{xy} and f_{yx}.

4.4 If

$$f(x, y) = \sin xy$$

verify that $f_{xy} = f_{yx}$.

4.5 If

$$r^2 = x^2 + y^2 + z^2$$

find $\partial r/\partial x$, $\partial r/\partial y$ and $\partial r/\partial z$. Hence show that

$$\frac{\partial^2}{\partial x^2}\left(\frac{1}{r}\right) + \frac{\partial^2}{\partial y^2}\left(\frac{1}{r}\right) + \frac{\partial^2}{\partial z^2}\left(\frac{1}{r}\right) = 0$$

except when $r = 0$.

4.6 If

$$u = (Ar^n + Br^{-n})\cos n\theta,$$

where A, B and n are constants, verify that

$$\frac{\partial^2 u}{\partial r^2} + \frac{1}{r}\frac{\partial u}{\partial r} + \frac{1}{r^2}\frac{\partial^2 u}{\partial \theta^2} = 0.$$

4.7 Let u be a function of x and y which satisfies the differential equation

$$\frac{\partial u}{\partial x} + u = \frac{\partial^2 u}{\partial y^2}.$$

If $v = ue^x$, prove that

$$\frac{\partial v}{\partial x} = \frac{\partial^2 v}{\partial y^2}.$$

4.8 The function $u(x, t)$ satisfies the differential equation

$$\frac{\partial^2 u}{\partial x^2} = \frac{1}{c^2}\frac{\partial^2 u}{\partial t^2}$$

where c is a constant. If the independent variables are changed to

$$\xi = x + ct \quad \text{and} \quad \eta = x - ct,$$

prove that

$$\frac{\partial^2 u}{\partial \xi \partial \eta} = 0.$$

[*Hint*. Show first that

$$\partial u/\partial x = \partial u/\partial \xi + \partial u/\partial \eta$$

and

$$\partial u/\partial t = c\ \partial u/\partial \xi - c\ \partial u/\partial \eta\ .]$$

4.3 DEFINITIONS OF SCALAR AND VECTOR FIELDS

Suppose that a scalar $\Omega(x, y, z)$ is defined on a point set U in three-dimensional space, that is, to each point $P(x, y, z)$ in U there corresponds a value of $\Omega(x, y, z)$. Then Ω is called a **scalar function of position** or a **scalar field**. Likewise, if a vector $\mathbf{F}(x, y, z)$ is defined on the set U, then $\mathbf{F}$ is called a **vector function of position** or a **vector field**. In practice the set U will nearly always constitute a region, as defined in section 4.1, and accordingly we shall henceforth restrict our attention to this case. Since the point $P(x, y, z)$ is completely specified by its position vector $\mathbf{r} = (x, y, z)$ relative to the origin, the notation

$$\Omega = \Omega(\mathbf{r}), \quad \mathbf{F} = \mathbf{F}(\mathbf{r}) \tag{4.4}$$

may also be used to signify that $\Omega, \mathbf{F}$ are functions of position.

Simple examples of scalar fields are

$$\Omega = x^2 + y^2 + z^2 \tag{4.5}$$

and

$$\Omega = 1/x. \tag{4.6}$$

Examples of vector fields are

$$\mathbf{F} = x\mathbf{i} + y\mathbf{j} + z\mathbf{k} \tag{4.7}$$

and

$$\mathbf{F} = (1 - x^2 - y^2 - z^2)^{\frac{1}{2}}(\mathbf{i} + \mathbf{j}). \tag{4.8}$$

The scalar field (4.5) and the vector field (4.7) are defined over the whole of space. The scalar field (4.6) is defined at all points except those lying on the plane $x = 0$, and the vector field (4.8) is defined only at points lying inside or on the sphere

$$x^2 + y^2 + z^2 = 1.$$

Scalar and vector fields arise naturally in a variety of physical situations. For example, when a gas flows along a pipe there are associated with any point in the pipe the gas pressure p, the density ρ, and the velocity $\mathbf{v}$ at that point. Thus p and ρ are scalar fields and $\mathbf{v}$ is a vector field associated with the motion.

4.4 GRADIENT OF A SCALAR FIELD

If the scalar field $\Omega(x, y, z)$ is defined and continuously differentiable in some open region $\mathcal{R}$, then the **gradient** of Ω is defined as

$$\text{grad}\,\Omega = \frac{\partial \Omega}{\partial x}\mathbf{i} + \frac{\partial \Omega}{\partial y}\mathbf{j} + \frac{\partial \Omega}{\partial z}\mathbf{k} = \left(\frac{\partial \Omega}{\partial x}, \frac{\partial \Omega}{\partial y}, \frac{\partial \Omega}{\partial z}\right). \tag{4.9}$$

The gradient of Ω is a vector field on $\mathcal{R}$

Proof

At every point of $\mathcal{R}$, grad Ω clearly satisfies condition 1. of the definition of a vector given in section 2.2. To complete the proof we must establish (a) that the components are invariant under a translation of the coordinate axes and (b) that the components transform according to the vector law (see equations (2.1)) under a rotation of the axes.

(a) Consider a translation to new coordinate axes $O'XYZ$ such that the coordinates (X, Y, Z) are related to the original coordinates (x, y, z) by the equations

$$x = X + a, \quad y = Y + b, \quad z = Z + c \tag{4.10}$$

where a, b, c are constants. By the chain rule (equations (4.3))

$$\frac{\partial \Omega}{\partial X} = \frac{\partial x}{\partial X}\frac{\partial \Omega}{\partial x} + \frac{\partial y}{\partial X}\frac{\partial \Omega}{\partial y} + \frac{\partial z}{\partial X}\frac{\partial \Omega}{\partial z} = \frac{\partial \Omega}{\partial x}.$$

Similarly

$$\frac{\partial \Omega}{\partial Y} = \frac{\partial \Omega}{\partial y}, \quad \frac{\partial \Omega}{\partial Z} = \frac{\partial \Omega}{\partial z}.$$

The components of grad Ω are therefore invariant under a translation of the coordinate axes.

(b) Consider a rotation of the axes defined by equations (1.19), namely

$$x = l_{11}x' + l_{21}y' + l_{31}z',$$

$$y = l_{12}x' + l_{22}y' + l_{32}z', \tag{4.11}$$

$$z = l_{13}x' + l_{23}y' + l_{33}z'.$$

Using the chain rule,

$$\frac{\partial \Omega}{\partial x'} = \frac{\partial x}{\partial x'}\frac{\partial \Omega}{\partial x} + \frac{\partial y}{\partial x'}\frac{\partial \Omega}{\partial y} + \frac{\partial z}{\partial x'}\frac{\partial \Omega}{\partial z}$$

$$= l_{11}\frac{\partial \Omega}{\partial x} + l_{12}\frac{\partial \Omega}{\partial y} + l_{13}\frac{\partial \Omega}{\partial z}.$$

Similarly

$$\frac{\partial \Omega}{\partial y'} = l_{21} \frac{\partial \Omega}{\partial x} + l_{22} \frac{\partial \Omega}{\partial y} + l_{23} \frac{\partial \Omega}{\partial z}$$

and

$$\frac{\partial \Omega}{\partial z'} = l_{31} \frac{\partial \Omega}{\partial x} + l_{32} \frac{\partial \Omega}{\partial y} + l_{33} \frac{\partial \Omega}{\partial z}.$$

These equations show that under a rotation of the axes the components of grad Ω transform as components of a vector (cf. equations (2.1)), and so the verification that grad Ω is a vector is complete.

Remark

Mere existence of the first-order partial derivatives Ω_x, Ω_y, Ω_z is not sufficient to ensure that grad Ω is a vector field. For the chain rule must be used during the course of the proof and this demands a stronger condition on Ω. We have assumed that Ω is continuously differentiable in an open region and this (together with continuous differentiability of x, y, z as functions of the new variables which is obviously the case by inspection of equations (4.10) and (4.11)) is certainly sufficient.

EXAMPLE 4

Find grad Ω when

(i) $\Omega = x^2 + xy + y^2$,
(ii) $\Omega = r$

where r denotes distance from the origin.

Solution

(i) We have

$$\frac{\partial \Omega}{\partial x} = 2x + y, \quad \frac{\partial \Omega}{\partial y} = x + 2y, \quad \frac{\partial \Omega}{\partial z} = 0.$$

Hence, from the definition (4.9),

$$\text{grad } \Omega = (2x + y)\mathbf{i} + (x + 2y)\mathbf{j}.$$

(ii) We have

$$r^2 = x^2 + y^2 + z^2.$$

Hence

$$2r \frac{\partial r}{\partial x} = 2x$$

giving

$$\frac{\partial r}{\partial x} = \frac{x}{r}.$$

Similarly

$$\frac{\partial r}{\partial y} = \frac{y}{r} \quad \text{and} \quad \frac{\partial r}{\partial z} = \frac{z}{r}.$$

Thus

$$\text{grad } r = \frac{x}{r}\mathbf{i} + \frac{y}{r}\mathbf{j} + \frac{z}{r}\mathbf{k}$$

$$= (x\mathbf{i} + y\mathbf{j} + z\mathbf{k})/r$$

$$= \mathbf{r}/r = \hat{\mathbf{r}}, \tag{4.12}$$

where $\hat{\mathbf{r}}$ is the radial unit vector.

EXERCISES

4.9 If

$$\Omega = xy^2z^3 - x^3y^2z,$$

find grad Ω at the point $(1, -1, 1)$.

4.10 If

$$\Omega = x^n + y^n + z^n,$$

show that $\mathbf{r}.\,\text{grad } \Omega = n\Omega$.

4.11 Show that grad $r^n = nr^{n-2}\mathbf{r}$.

4.12 If $\mathbf{a}$ is a constant vector field (i.e. a vector field of constant magnitude and direction) and $\mathbf{r}$ is the position vector, prove that

$$\text{grad }(\mathbf{a}.\,\mathbf{r}) = \mathbf{a}.$$

4.5 PROPERTIES OF GRADIENT

The directional derivative $\partial\Omega/\partial n$

Let P be a fixed point and let P' be another point which varies in such a way that the vector $\overrightarrow{PP'}$ is always parallel to a fixed unit vector $\hat{\mathbf{n}}$. Assume that a scalar field Ω takes values $\Omega(P)$, $\Omega(P')$ at P, P' respectively. The derivative of Ω at P in the direction of $\hat{\mathbf{n}}$ is defined as

$$\frac{\partial\Omega}{\partial n} = \lim_{PP' \to 0} \frac{\Omega(P') - \Omega(P)}{PP'} \tag{4.13}$$

whenever the limit exists. It is clear that, in general, Ω will vary at different rates as we move away from P in different directions; the directional derivative $\partial\Omega/\partial n$ measures the rate of variation in the direction of $\hat{\mathbf{n}}$.

An important property of grad Ω is that

$$\hat{\mathbf{n}} \cdot \operatorname{grad} \Omega = \frac{\partial\Omega}{\partial n}. \tag{4.14}$$

The proof is quite simple. Choose axes $Oxyz$ such that the directions of the unit vectors $\mathbf{i}$ and $\hat{\mathbf{n}}$ coincide. Then

$$\hat{\mathbf{n}} \cdot \operatorname{grad} \Omega = \mathbf{i} \cdot \operatorname{grad} \Omega = \frac{\partial\Omega}{\partial x}$$

which is the rate of change of Ω in the x-direction, that is to say the direction of $\hat{\mathbf{n}}$.

Geometrical interpretation of grad Ω

Let θ denote the angle between the vector grad Ω and the unit vector $\hat{\mathbf{n}}$. Then

$$\frac{\partial\Omega}{\partial n} = \hat{\mathbf{n}} \cdot \operatorname{grad} \Omega = |\operatorname{grad}\Omega| \cos\theta. \tag{4.15}$$

Suppose that θ is allowed to vary and consider a point where grad $\Omega \neq \mathbf{0}$. At this point, (4.15) shows that the greatest value of $\partial\Omega/\partial n$ will occur when $\theta = 0$, i.e. when $\hat{\mathbf{n}}$ and grad Ω have the same direction. It follows that, *at a point where grad $\Omega \neq \mathbf{0}$, the vector grad Ω points in the direction in which Ω increases most rapidly and $|\operatorname{grad}\Omega|$ is the rate of change of Ω in this direction. The gradient of a scalar field thus describes completely the manner in which the field varies.*

This property is extremely useful in the geometry of surfaces and the well known Method of Steepest Descent (section 4.15).

A point where grad $\Omega = \mathbf{0}$ is called a **stationary point**.

Level surfaces

Consider the equation

$$\Omega(x, y, z) = \lambda \tag{4.16}$$

where λ is a parameter. For each fixed value of λ the equation represents a surface and if λ is allowed to take a variety of different values we obtain a family of surfaces.

For example, for any fixed positive value of λ the equation

$$x^2 + y^2 + z^2 = \lambda \tag{4.17}$$

represents a sphere with centre at the origin and of radius $\sqrt{\lambda}$. If λ varies, the equation represents a family of concentric spheres.

The family of surfaces (4.16) are called the **iso-surfaces** or **level surfaces** of Ω. On each level surface Ω is constant. A particular example of a set of level surfaces are the surfaces of constant atmospheric pressure. Their intersections with the Earth's surface are shown on weather maps (and usually referred to as 'isobars', i.e. lines of constant barometric pressure).

An important connection between the gradient of a scalar field and the level surfaces of the field is the following. Let P be any point on the level surface

$$\Omega(x, y, z) = c \tag{4.18}$$

where c is a constant. Suppose that $\mathscr{C}_1$ and $\mathscr{C}_2$ are two curves through P which lie on the surface and that $\hat{\mathbf{T}}_1$ and $\hat{\mathbf{T}}_2$ are their unit tangents at P. Assume that grad Ω does not vanish at P and denote its value at this point by $(\text{grad } \Omega)_P$. Then the vector $(\text{grad } \Omega)_P$ is directed along the normal to the surface at P (Fig. 4.1).

Proof

Let the intrinsic equation of $\mathscr{C}_1$ be

$$\mathbf{r} = \mathbf{r}(s) = (x(s), y(s), z(s)). \tag{4.19}$$

Since the curve lies on the surface (4.18) the coordinates of any point on the curve must satisfy (4.18) so

$$\Omega(x(s), y(s), z(s)) = c.$$

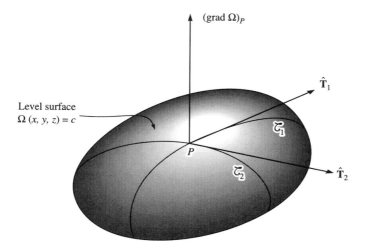

Fig. 4.1 The vector $(\text{grad } \Omega)_P$ is normal to the level surface of Ω which passes through P.

Differentiating this equation with respect to s, using the chain rule, we have

$$\frac{\partial \Omega}{\partial x}\frac{dx}{ds} + \frac{\partial \Omega}{\partial y}\frac{dy}{ds} + \frac{\partial \Omega}{\partial z}\frac{dz}{ds} = 0. \tag{4.20}$$

Now

$$\operatorname{grad} \Omega = \left(\frac{\partial \Omega}{\partial x}, \frac{\partial \Omega}{\partial y}, \frac{\partial \Omega}{\partial z} \right)$$

and the unit tangent to the curve (4.19) is

$$\hat{\mathbf{T}} = \left(\frac{dx}{ds}, \frac{dy}{ds}, \frac{dz}{ds} \right).$$

Thus (4.20) reduces to

$$\hat{\mathbf{T}} . \operatorname{grad} \Omega = 0,$$

and hence, at P,

$$\hat{\mathbf{T}}_1 . (\operatorname{grad} \Omega)_P = 0.$$

Since $(\operatorname{grad} \Omega)_P \neq \mathbf{0}$, the unit vector $\hat{\mathbf{T}}_1$ and the vector $(\operatorname{grad} \Omega)_P$ are perpendicular. Also, by repeating the above argument, $(\operatorname{grad} \Omega)_P$ is perpendicular to the unit tangent $\hat{\mathbf{T}}_2$ to $\mathscr{C}_2$. Now the unit tangents to $\mathscr{C}_1$, $\mathscr{C}_2$ are also tangents to the surface (since the curves lie on the surface). Hence, since the vector $(\operatorname{grad} \Omega)_P$ is perpendicular to both these unit tangents, it is directed along the normal to the surface.

EXAMPLE 5

Let

$$\Omega(x, y, z) = x^2 + y^2 + z^2 = a^2.$$

Then

$$\operatorname{grad} \Omega = (2x, 2y, 2z).$$

Thus the vector

$$(2x_0, 2y_0, 2z_0)$$

is normal to the level surface through the point $P(x_0, y_0, z_0)$. Reference to Fig. 4.2, and the observation that

$$(x_0, y_0, z_0) = \overrightarrow{OP}$$

confirm this conclusion.

Taylor's expansion

Let $P(x, y, z)$ and $Q(x + \delta x, y + \delta y, z + \delta z)$ be neighbouring points in an open region $\mathscr{R}$ and put

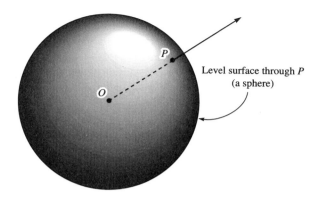

Fig. 4.2

$$\overrightarrow{PQ} = \delta\mathbf{r} = (\delta x, \ \delta y, \ \delta z). \tag{4.21}$$

If the scalar field $\Omega(x, y, z)$ is continuously differentiable in $\mathcal{R}$, it can be shown (see for example the references given in section 4.2) that

$$\Omega(x + \delta x, \ y + \delta y, \ z + \delta z) = \ \Omega(x, y, z) + \delta x \ \frac{\partial\Omega}{\partial x} + \delta y \ \frac{\partial\Omega}{\partial y} + \delta z \ \frac{\partial\Omega}{\partial z} + \varepsilon |\delta\mathbf{r}|$$

where $\varepsilon \to 0$ as $|\delta\mathbf{r}| \to 0$, and $\partial\Omega/\partial x$, $\partial\Omega/\partial y$, $\partial\Omega/\partial z$ are evaluated at $P(x, y, z)$. The notation introduced in this section enables us to express this in a more concise form, namely

$$\Omega_Q = \Omega_P + \delta\mathbf{r} \cdot (\operatorname{grad}\Omega)_P + \varepsilon |\delta\mathbf{r}| \tag{4.22}$$

where the suffixes P, Q indicate that the quantity concerned is to be evaluated at P, Q respectively.

An explicit expression for ε can be given if it is assumed that all second order derivatives of Ω exist and are continuous; the expression for Ω_Q is then the Taylor expansion of Ω to second order terms. However, for many purposes it is sufficient to know that $\varepsilon \to 0$ as $|\delta\mathbf{r}| \to 0$.

We may note that, to a first order approximation, (4.22) implies that the difference $\delta\Omega$ in the values of Ω at P, Q is given by the formula

$$\Omega_Q - \Omega_P = \delta\Omega \approx \delta\mathbf{r} \cdot (\operatorname{grad}\Omega)_P. \tag{4.23}$$

EXERCISES

4.13 Let a_1, a_2 be constants and let Ω_1, Ω_2 be continuously differentiable scalar fields. Prove from the definition of gradient that

$$\operatorname{grad}(a_1\Omega_1 + a_1\Omega_2) \equiv a_1 \operatorname{grad}\Omega_1 + a_2 \operatorname{grad}\Omega_2.$$

[N.B. This is an important property of grad, showing that it is a **linear operator** – cf. section 4.7.]

4.14 Find the derivative of the scalar field $\Omega = x^2yz + 4xz^2$ in the direction of the vector $(2, -1, -1)$ at the point $P(1, -2, -1)$. [*Hint.* Write down the unit vector $\hat{\mathbf{n}}$ which has the same direction as the given vector and evaluate $\hat{\mathbf{n}} \cdot \operatorname{grad} \Omega$ at the point P.]

4.15 Relative to axes $Oxyz$, the temperature in a certain medium is given by

$$T = T_0(1 + ax + by)e^{cz},$$

where a, b, c, and $T_0 \, (> 0)$ are constants. At the origin O, find the direction in which the temperature changes most rapidly.

4.16 Find direction ratios of the normal at any point to the ellipsoid

$$x^2/a^2 + y^2/b^2 + z^2/c^2 = \text{constant}$$

where a, b, c are constants.

4.6 THE DIVERGENCE AND CURL OF A VECTOR FIELD

Suppose that the components of the vector field $\mathbf{F} = (F_1, F_2, F_3)$ are continuously differentiable functions of the coordinates x, y, z in some open region $\mathcal{R}$. Then, in $\mathcal{R}$, the **divergence** of the vector field is defined as

$$\operatorname{div} \mathbf{F} = \frac{\partial F_1}{\partial x} + \frac{\partial F_2}{\partial y} + \frac{\partial F_3}{\partial z} \tag{4.24}$$

and the **curl** of the vector field is defined as

$$\operatorname{curl} \mathbf{F} = \left(\frac{\partial F_3}{\partial y} - \frac{\partial F_2}{\partial z}\right)\mathbf{i} + \left(\frac{\partial F_1}{\partial z} - \frac{\partial F_3}{\partial x}\right)\mathbf{j} + \left(\frac{\partial F_2}{\partial x} - \frac{\partial F_1}{\partial y}\right)\mathbf{k}. \tag{4.25}$$

The formula for curl $\mathbf{F}$ can also be conveniently expressed in the symbolic determinantal form

$$\operatorname{curl} \mathbf{F} = \begin{vmatrix} \mathbf{i} & \mathbf{j} & \mathbf{k} \\ \dfrac{\partial}{\partial x} & \dfrac{\partial}{\partial y} & \dfrac{\partial}{\partial z} \\ F_1 & F_2 & F_3 \end{vmatrix}. \tag{4.26}$$

It is clear that div $\mathbf{F}$ *is a scalar field; it also has the fundamental property that it is invariant under a translation or a rotation of the coordinate axes.* On the other hand, curl $\mathbf{F}$ *is a vector field* as the notation used anticipates. The ground covered in Chapter 2 enables these statements to be proved easily, but it is convenient to defer the proofs until operator notation is introduced in the next section.

It was shown in section 4.5 that the geometrical interpretation of grad Ω is that it measures the variation of the scalar field Ω. The divergence and curl of a vector field cannot be interpreted in such simple terms and discussion of the significance of these concepts is left for a later stage (Chapter 6). However, we may note that divergence and curl play an important role in several branches of applied mathematics, e.g. hydrodynamics, elasticity and electromagnetism.

EXAMPLE 6

Find the divergence of the following vector fields.

(i) $\mathbf{F} = (x^2, 3y, x^3)$,

(ii) $\mathbf{F} = \mathbf{r}$, where $\mathbf{r}$ is the position vector.

Solution

(i) Using (4.24)

$$\operatorname{div} \mathbf{F} = \frac{\partial(x^2)}{\partial x} + \frac{\partial(3y)}{\partial y} + \frac{\partial(x^3)}{\partial z}$$

$$= 2x + 3.$$

(ii) Since $\mathbf{r} = (x, y, z)$

$$\operatorname{div} \mathbf{r} = 3. \tag{4.27}$$

EXAMPLE 7

Find curl $\mathbf{F}$ for the vector field

$$\mathbf{F} = (z, x, y)$$

and show that curl curl $\mathbf{F} = \mathbf{0}$.

Solution

We have

$$\operatorname{curl} \mathbf{F} = \begin{vmatrix} \mathbf{i} & \mathbf{j} & \mathbf{k} \\ \dfrac{\partial}{\partial x} & \dfrac{\partial}{\partial y} & \dfrac{\partial}{\partial z} \\ z & x & y \end{vmatrix}$$

$$= (1 - 0)\mathbf{i} + (1 - 0)\mathbf{j} + (1 - 0)\mathbf{k}$$

$$= \mathbf{i} + \mathbf{j} + \mathbf{k}$$

$$= (1, 1, 1).$$

Also

$$\text{curl curl } \mathbf{F} = \begin{vmatrix} \mathbf{i} & \mathbf{j} & \mathbf{k} \\ \dfrac{\partial}{\partial x} & \dfrac{\partial}{\partial y} & \dfrac{\partial}{\partial z} \\ 1 & 1 & 1 \end{vmatrix}$$

$$= \mathbf{0}.$$

EXERCISES

4.17 If a_1, a_2 are constants and $\mathbf{F}_1, \mathbf{F}_2$ are continuously differentiable vector fields, prove from the definitions of divergence and curl that

$$\text{div}(a_1\mathbf{F}_1 + a_2\mathbf{F}_2) \equiv a_1 \text{ div } \mathbf{F}_1 + a_2 \text{ div } \mathbf{F}_2$$

and

$$\text{curl}(a_1\mathbf{F}_1 + a_2\mathbf{F}_2) \equiv a_1 \text{ curl } \mathbf{F}_1 + a_2 \text{ curl } \mathbf{F}_2.$$

[N.B. These are important properties showing that, like grad (see section 4.5, Exercise 4.13), div and curl are linear operators (cf. section 4.7 which follows).]

4.18 Find the divergence and curl of the vector field $\mathbf{F} = (xy, yz, 0)$ at the point $(1, 1, 1)$. Also evaluate grad (div $\mathbf{F}$).

4.19 If $\Omega = x + y^2 + z^3$, find div (grad Ω) and curl (grad Ω).

4.20 If $\mathbf{a}$ is a constant vector field and $\mathbf{r}$ is the position vector, show that

$$\text{curl}(\mathbf{a} \times \mathbf{r}) = 2\mathbf{a}.$$

4.21 The point $A(a, b, c)$ is fixed and $P(x, y, z)$ is a variable point. Show that

$$\text{div } \overrightarrow{AP} = 3, \quad \text{curl } \overrightarrow{AP} = \mathbf{0}.$$

4.22 At all points a vector field $\mathbf{F}$ is parallel to the xy-plane. If the components of $\mathbf{F}$ are functions of x and y only, prove that curl curl $\mathbf{F}$ is parallel to the xy-plane.

4.7 THE DEL–OPERATOR

The reader may be familiar with the concept of operators in connection with ordinary differential equations. Thus the differential equation

$$\frac{d^2y}{dx^2} + 2\frac{dy}{dx} + 3y = 0$$

is sometimes written in the form

$$\left(\frac{d^2}{dx^2} + 2\frac{d}{dx} + 3\right)y = 0$$

and expressed as

$$(D^2 + 2D + 3)y = 0$$

where $D \equiv d/dx$. In this example D is an **operator**. When it acts upon the function $y(x)$ it gives the derivative dy/dx and when D^2 acts upon $y(x)$ it gives the second derivative d^2y/dx^2.

The D-operator obeys some, but not all, of the rules of ordinary algebra. For example

$$D(Dy) \equiv D^2y$$

and

$$(D^2 + 2D + 3)y \equiv D^2y + 2Dy + 3y,$$

but

$$D(xy) \neq xDy \neq xyD.$$

Motivated by these ideas, we now turn to operators in vector analysis.

The del-operator

The expression

$$\nabla \equiv \mathbf{i}\frac{\partial}{\partial x} + \mathbf{j}\frac{\partial}{\partial y} + \mathbf{k}\frac{\partial}{\partial z} \equiv \left(\frac{\partial}{\partial x}, \frac{\partial}{\partial y}, \frac{\partial}{\partial z}\right) \qquad (4.28)$$

is called the **del-operator** or, more briefly, **del** or **nabla**.

Under a translation or rotation of the axes, the components $\partial/\partial x$, $\partial/\partial y$, $\partial/\partial z$ of the del-operator transform in the same way as the components of a vector (this explains the notation used in (4.28)). The proof of this can be obtained at once from the proof in section 4.4 that grad Ω is a vector field. For, if in parts (a) and (b) of the proof, Ω is omitted from every line, we obtain

$$\frac{\partial}{\partial X} = \frac{\partial}{\partial x}, \quad \frac{\partial}{\partial Y} = \frac{\partial}{\partial y}, \quad \frac{\partial}{\partial Z} = \frac{\partial}{\partial z}$$

and

$$\frac{\partial}{\partial x'} = l_{11}\frac{\partial}{\partial x} + l_{12}\frac{\partial}{\partial y} + l_{13}\frac{\partial}{\partial z}$$

together with similar expressions for $\partial/\partial y'$ and $\partial/\partial z'$. In this respect then, the del-operator behaves as a vector and it is often called a vector operator. The ideas associated with the D-operator, discussed above, can now be extended to the del-operator.

We define

$$\nabla\Omega = \operatorname{grad}\Omega. \qquad (4.29)$$

The left-hand side is

$$\left(\mathbf{i}\frac{\partial}{\partial x} + \mathbf{j}\frac{\partial}{\partial y} + \mathbf{k}\frac{\partial}{\partial z}\right)\Omega$$

and it is natural to interpret this as

$$\mathbf{i}\frac{\partial\Omega}{\partial x} + \mathbf{j}\frac{\partial\Omega}{\partial y} + \mathbf{k}\frac{\partial\Omega}{\partial z} = \operatorname{grad}\Omega.$$

It is also natural to define

$$\nabla.\,\mathbf{F} = \operatorname{div}\mathbf{F} \qquad (4.30)$$

and

$$\nabla \times \mathbf{F} = \operatorname{curl}\mathbf{F}. \qquad (4.31)$$

For, expanding the scalar product formally

$$\nabla.\,\mathbf{F} = \left(\frac{\partial}{\partial x},\ \frac{\partial}{\partial y},\ \frac{\partial}{\partial z}\right).\,(F_1, F_2, F_3)$$

$$= \frac{\partial F_1}{\partial x} + \frac{\partial F_2}{\partial y} + \frac{\partial F_3}{\partial z}$$

$$= \operatorname{div}\mathbf{F};$$

and likewise we have

$$\nabla \times \mathbf{F} = \left(\frac{\partial}{\partial x},\ \frac{\partial}{\partial y},\ \frac{\partial}{\partial z}\right) \times (F_1, F_2, F_3)$$

$$= \begin{vmatrix} \mathbf{i} & \mathbf{j} & \mathbf{k} \\ \dfrac{\partial}{\partial x} & \dfrac{\partial}{\partial y} & \dfrac{\partial}{\partial z} \\ F_1 & F_2 & F_3 \end{vmatrix}$$

$$= \operatorname{curl}\mathbf{F}.$$

Observe that, as with the D-operator, the components of ∇ act only upon functions on their right.

We are now in a position to prove two results left over from the previous section, namely

(a) *The scalar field* div $\mathbf{F}$ *is invariant under a translation or rotation of the coordinate axes,*

(b) curl $\mathbf{F}$ *is a vector field.*

Proofs

(a) Writing

$$\text{div } \mathbf{F} = \nabla . \mathbf{F}$$

the invariance under a translation of the axes follows easily. For, it has been seen that the components of ∇ are invariant under a translation of the coordinate axes, and since the components of the vector $\mathbf{F}$ are also invariant under such a translation, $\nabla . \mathbf{F}$ is invariant.

To establish the invariance of $\nabla . \mathbf{F}$ under a rotation of the axes it is convenient to denote the coordinates by x_1, x_2, x_3 instead of x, y, z. With this notation

$$\nabla . \mathbf{F} = \frac{\partial F_1}{\partial x_1} + \frac{\partial F_2}{\partial x_2} + \frac{\partial F_3}{\partial x_3} = \frac{\partial F_i}{\partial x_i},$$

using the summation convention.

Consider the rotation to new axes $Ox'_1 x'_2 x'_3$ defined by equations (1.25). Using primes to denote components relative to the new axes, the vector transformation law (equation (2.2)) gives

$$\frac{\partial}{\partial x'_j} = l_{ji} \frac{\partial}{\partial x_i}$$

and

$$F'_j = l_{jk} F_k.$$

Hence, since the coefficients l_{jk} are independent of x_1, x_2, x_3,

$$\frac{\partial F'_j}{\partial x'_j} = l_{ji} l_{jk} \frac{\partial F_k}{\partial x_i}.$$

Using the orthonormality conditions

$$l_{ji} l_{jk} = \delta_{ik}$$

(equation (1.29)), it follows that

$$\frac{\partial F'_j}{\partial x'_j} = \delta_{ik} \frac{\partial F_k}{\partial x_i} = \frac{\partial F_i}{\partial x_i},$$

i.e.

$$\frac{\partial F'_1}{\partial x'_1} + \frac{\partial F'_2}{\partial x'_2} + \frac{\partial F'_3}{\partial x'_3} = \frac{\partial F_1}{\partial x_1} + \frac{\partial F_2}{\partial x_2} + \frac{\partial F_3}{\partial x_3}.$$

The left-hand side is div $\mathbf{F}$ referred to the new axes and the right-hand side is div $\mathbf{F}$ referred to the original axes; the invariance of div $\mathbf{F}$ under a rotation of the axes is thus established.

(b) The proof of the invariance of $\nabla \cdot \mathbf{F}$ closely follows the proof in section 2.6 that the scalar product $\mathbf{a} \cdot \mathbf{b}$ is a scalar invariant. Similarly, the proof that $\nabla \times \mathbf{F}$ is a vector field follows closely the proof in section 2.7 that the vector product $\mathbf{a} \times \mathbf{b}$ is a vector. The essential features are similar because the operator ∇ transforms under a translation or rotation of the axes as a vector. It is left as an exercise for the reader to verify this.

We have already remarked, in Exercises 4.13 and 4.17, that grad, div and curl are all linear operators and we can now express these same facts concisely in the single statement that del is a linear operator.

The differential operator $D \equiv d/dx$, to which we have already referred, possesses the characteristic property of linearity:

$$D(a_1 y_1 + a_2 y_2) = a_1 Dy_1 + a_2 Dy_2$$

for any constants a_1, a_2; and, in the same way, the partial differential operator $\partial/\partial x$ may be said to be linear, since

$$\frac{\partial}{\partial x}(a_1 f_1 + a_2 f_2) = a_1 \frac{\partial f_1}{\partial x} + a_2 \frac{\partial f_2}{\partial x}$$

for any functions f_1, f_2 that can be differentiated partially with respect to x and any constants a_1, a_2.

Now

$$\nabla \equiv \mathbf{i}\frac{\partial}{\partial x} + \mathbf{j}\frac{\partial}{\partial y} + \mathbf{k}\frac{\partial}{\partial z}$$

is a linear combination of such partial differential operators and we may, therefore, expect it also to behave as a linear operator. As the reader can easily verify, ∇ does indeed behave in this way in all three of its possible applications: $\nabla\Omega$, $\nabla \cdot \mathbf{F}$ and $\nabla \times \mathbf{F}$, as is asserted in Exercises 4.13 and 4.17.

EXERCISES

4.23 Express in operator notation, (i) grad (div $\mathbf{F}$), (ii) div (grad Ω), (iii) div (curl $\mathbf{F}$), (iv) curl (grad Ω), (v) curl curl $\mathbf{F}$.

4.24 Prove that $(\mathbf{F} \times \nabla) \times \mathbf{G}$ is a vector field.

4.8 SCALAR INVARIANT OPERATORS

Let $\mathcal{D}$ denote a linear partial differential operator which involves only the rectangular cartesian coordinates x, y, z as independent variables. For example, we might have

$$\mathcal{D} = \frac{\partial}{\partial x} + \frac{\partial}{\partial y} + 2\frac{\partial}{\partial z} \quad \text{or} \quad \mathcal{D} = \frac{\partial^2}{\partial x^2} + \frac{\partial}{\partial y} + \frac{\partial^2}{\partial y\,\partial z}.$$

The operator $\mathscr{D}$ is called a **scalar invariant operator** if its form is unchanged under a translation or rotation of the coordinate axes. Thus, for example, if the first of the operators above were invariant (it is not, in fact) then upon changing to new axes $O'x'y'z'$ it would become

$$\mathscr{D}' = \frac{\partial}{\partial x'} + \frac{\partial}{\partial y'} + 2\frac{\partial}{\partial z'}.$$

The following result concerning scalar invariant operators will be required.

Let $\mathscr{D}$ be a scalar invariant operator and define its operation on a vector field **F** by

$$\mathscr{D}\mathbf{F} = \mathscr{D}(F_1, F_2, F_3) = (\mathscr{D}F_1, \mathscr{D}F_2, \mathscr{D}F_3). \tag{4.32}$$

Then $\mathscr{D}\mathbf{F}$ is a vector field.

Proof

Under a translation of the axes the components of $\mathscr{D}\mathbf{F}$ are unchanged because both $\mathscr{D}$ and the components of **F** are unchanged.

Under a rotation to new axes $Ox'y'z'$, whose positions relative to the original axes $Oxyz$ are defined by the transformation matrix (1.12), we have

$$F'_j = l_{jk} F_k,$$

where, as usual, primes denote components relative to the new axes. Thus, using the property of invariance of the form of $\mathscr{D}$, and also the linearity of $\mathscr{D}$,

$$\mathscr{D}' F'_j = \mathscr{D}(l_{jk} F_k) = l_{jk} \mathscr{D} F_k,$$

showing that the components of $\mathscr{D}\mathbf{F}$ transform according to the vector law under a rotation of the axes. It follows that $\mathscr{D}\mathbf{F}$ is a vector field

The Laplacian operator ∇^2

The most important of the scalar invariant operators is the **Laplacian**, defined as

$$\nabla^2 \equiv \frac{\partial^2}{\partial x^2} + \frac{\partial^2}{\partial y^2} + \frac{\partial^2}{\partial z^2}. \tag{4.33}$$

Alternatively, using rectangular coordinates x_1, x_2, x_3 and the summation convention,

$$\nabla^2 \equiv \frac{\partial^2}{\partial x_1^2} + \frac{\partial^2}{\partial x_2^2} + \frac{\partial^2}{\partial x_3^2} \equiv \frac{\partial}{\partial x_i} \frac{\partial}{\partial x_i}. \tag{4.34}$$

Formally, the Laplacian operator is the 'square' of the del-operator. This connection is easily understood by observing that, formally,

$$\mathbf{\nabla} \cdot \mathbf{\nabla} = \left(\frac{\partial}{\partial x}, \frac{\partial}{\partial y}, \frac{\partial}{\partial z}\right) \cdot \left(\frac{\partial}{\partial x}, \frac{\partial}{\partial y}, \frac{\partial}{\partial z}\right) = \frac{\partial^2}{\partial x^2} + \frac{\partial^2}{\partial y^2} + \frac{\partial^2}{\partial z^2}.$$

By analogy with the relation $\mathbf{a} \cdot \mathbf{a} = a^2$ it is thus natural to use the notation $\mathbf{\nabla} \cdot \mathbf{\nabla} = \nabla^2$.

The invariance of the Laplacian operator follows from the fact that the del-operator transforms as a vector. Thus, under a translation of the axes, $\partial/\partial x_i$ is invariant and hence so also is ∇^2. Under a rotation to coordinate axes $Ox_1' \, x_2' \, x_3'$, we have

$$\frac{\partial}{\partial x_j'} = l_{ji} \frac{\partial}{\partial x_i}$$

(cf. the proof of the invariance of div $\mathbf{F}$ in section 4.7). Thus

$$\frac{\partial}{\partial x_j'} \frac{\partial}{\partial x_j'} = \left(l_{ji} \frac{\partial}{\partial x_i}\right)\left(l_{jk} \frac{\partial}{\partial x_k}\right)$$

$$= \delta_{ik} \frac{\partial}{\partial x_i} \frac{\partial}{\partial x_k}$$

$$= \frac{\partial}{\partial x_i} \frac{\partial}{\partial x_i}$$

which shows that ∇^2 is also invariant under a rotation of the axes.

The Laplacian operator can act upon a scalar field Ω or on a vector field $\mathbf{F}$. The following important identities hold:

$$\nabla^2 \Omega \equiv \operatorname{div}(\operatorname{grad} \Omega); \tag{4.35}$$

$$\nabla^2 \mathbf{F} \equiv \operatorname{grad}(\operatorname{div} \mathbf{F}) - \operatorname{curl} \operatorname{curl} \mathbf{F}. \tag{4.36}$$

These results are proved by expanding the right-hand sides. Thus

$$\operatorname{div}(\operatorname{grad} \Omega) \equiv \operatorname{div}\left(\mathbf{i} \frac{\partial \Omega}{\partial x} + \mathbf{j} \frac{\partial \Omega}{\partial y} + \mathbf{k} \frac{\partial \Omega}{\partial z}\right)$$

$$\equiv \frac{\partial}{\partial x}\left(\frac{\partial \Omega}{\partial x}\right) + \frac{\partial}{\partial y}\left(\frac{\partial \Omega}{\partial y}\right) + \frac{\partial}{\partial z}\left(\frac{\partial \Omega}{\partial z}\right)$$

$$\equiv \nabla^2 \Omega,$$

which proves (4.35). If $\mathbf{F} = (F_1, F_2, F_3)$, the x-component of grad(div $\mathbf{F}$) is

$$\frac{\partial}{\partial x}\left(\frac{\partial F_1}{\partial x} + \frac{\partial F_2}{\partial y} + \frac{\partial F_3}{\partial z}\right) \equiv \frac{\partial^2 F_1}{\partial x^2} + \frac{\partial^2 F_2}{\partial x \, \partial y} + \frac{\partial^2 F_3}{\partial x \, \partial z}.$$

Also

$$\text{curl curl } \mathbf{F} \equiv \begin{vmatrix} \mathbf{i} & \mathbf{j} & \mathbf{k} \\ \dfrac{\partial}{\partial x} & \dfrac{\partial}{\partial y} & \dfrac{\partial}{\partial z} \\ \dfrac{\partial F_3}{\partial y} - \dfrac{\partial F_2}{\partial z} & \dfrac{\partial F_1}{\partial z} - \dfrac{\partial F_3}{\partial x} & \dfrac{\partial F_2}{\partial x} - \dfrac{\partial F_1}{\partial y} \end{vmatrix}.$$

Thus the x-component of curl curl $\mathbf{F}$ is

$$\frac{\partial}{\partial y}\left(\frac{\partial F_2}{\partial x} - \frac{\partial F_1}{\partial y}\right) - \frac{\partial}{\partial z}\left(\frac{\partial F_1}{\partial z} - \frac{\partial F_3}{\partial x}\right) \equiv \frac{\partial^2 F_2}{\partial y\,\partial x} - \frac{\partial^2 F_1}{\partial y^2} - \frac{\partial^2 F_1}{\partial z^2} + \frac{\partial^2 F_3}{\partial z\,\partial x}.$$

Hence, assuming that the order of differentiation in the mixed derivatives can be inverted (see the theorem in section 4.2), the x-component of the right-hand side of (4.36) is

$$\frac{\partial^2 F_1}{\partial x^2} + \frac{\partial^2 F_1}{\partial y^2} + \frac{\partial^2 F_1}{\partial z^2} \equiv \nabla^2 F_1.$$

The y- and z-components of the right-hand side of (4.36) can be similarly evaluated, giving

$$\text{grad}(\text{div } \mathbf{F}) - \text{curl curl } \mathbf{F} \equiv (\nabla^2 F_1, \nabla^2 F_2, \nabla^2 F_3) \equiv \nabla^2 \mathbf{F}$$

as required.

The Laplacian operator occurs in several fundamental differential equations governing physical phenomena. The simplest of these differential equations is **Laplace's equation**, namely

$$\nabla^2 \Omega = 0. \tag{4.37}$$

The operator $\mathbf{F} \cdot \nabla$

This is a useful scalar invariant operator, defined as

$$\mathbf{F} \cdot \nabla \equiv F_1 \frac{\partial}{\partial x} + F_2 \frac{\partial}{\partial y} + F_3 \frac{\partial}{\partial z}. \tag{4.38}$$

The proof of invariance is straightforward and is left to the reader.

The operator can act upon a scalar field Ω giving

$$(\mathbf{F} \cdot \nabla)\Omega = F_1 \frac{\partial \Omega}{\partial x} + F_2 \frac{\partial \Omega}{\partial y} + F_3 \frac{\partial \Omega}{\partial z}. \tag{4.39}$$

Since the right-hand side of (4.39) is $\mathbf{F} \cdot (\nabla \Omega)$, no ambiguity can arise by writing

$$(\mathbf{F} \cdot \nabla)\Omega = \mathbf{F} \cdot \nabla \Omega = \mathbf{F} \cdot \text{grad } \Omega. \tag{4.40}$$

The operator $\mathbf{F} \cdot \nabla$ can also act upon a vector field $\mathbf{G} = (G_1, G_2, G_3)$ giving, in rectangular cartesian coordinates,

$$(\mathbf{F} \cdot \nabla)\mathbf{G} = (\mathbf{F} \cdot \nabla G_1, \mathbf{F} \cdot \nabla G_2, \mathbf{F} \cdot \nabla G_3). \tag{4.41}$$

The result proved at the beginning of this section shows that $(\mathbf{F} \cdot \nabla)\mathbf{G}$ is a vector field.

N.B. Interpretation of the expressions $\nabla^2\mathbf{F}$ and $(\mathbf{F} \cdot \nabla)\,\mathbf{G}$ requires a little care when orthogonal coordinates other than the rectangular cartesian system are used (see section 4.13, Example 14 and Exercise 4.52).

EXERCISES

4.25 If $\mathbf{F}_1 = (x, y, z)$, $\mathbf{F}_2 = (1, 2, 3)$, $\mathbf{G} = (x^2, y^2, z^2)$ and $\Omega = xyz$, evaluate

(i) $(\mathbf{F}_1 \cdot \nabla)\mathbf{G}$,   (ii) $(\mathbf{F}_2 \cdot \nabla)\,\mathbf{G}$,   (iii) $(\mathbf{F}_1 \cdot \nabla)\,\mathbf{F}_2$

(iv) $\mathbf{F}_1 \cdot \nabla\Omega$,   (v) $\nabla^2\mathbf{G}$,   (vi) $\nabla \times \{(\mathbf{F}_2 \cdot \nabla)\,\mathbf{G}\}$.

4.26 If $\mathbf{F} = (xy, yz, zx)$ and $\mathbf{a} = (1, 2, 3)$ verify that

(i) $\nabla \times (\nabla \times \mathbf{F}) \equiv \nabla(\nabla \cdot \mathbf{F}) - \nabla^2\mathbf{F}$,

(ii) $\nabla \times (\mathbf{F} \times \mathbf{a}) \equiv \mathbf{F}(\nabla \cdot \mathbf{a}) - \mathbf{a}(\nabla \cdot \mathbf{F}) + (\mathbf{a} \cdot \nabla)\mathbf{F} - (\mathbf{F} \cdot \nabla)\mathbf{a}$.

4.9 USEFUL IDENTITIES

Alternative expressions for grad φ, div $\mathbf{F}$ and curl $\mathbf{F}$

It is convenient to introduce here new expressions for grad φ, div $\mathbf{F}$ and curl $\mathbf{F}$ which make the proofs of some of the identities given later less laborious.

Take rectangular cartesian axes Ox_1, Ox_2, Ox_3 and let $\mathbf{e}_1, \mathbf{e}_2, \mathbf{e}_3$ denote the unit vectors along the respective axes. Then, using the summation convention,

(a)
$$\operatorname{grad} \Omega = \mathbf{e}_i \frac{\partial \Omega}{\partial x_i}, \tag{4.42}$$

(b)
$$\operatorname{div} \mathbf{F} = \mathbf{e}_i \cdot \frac{\partial \mathbf{F}}{\partial x_i}, \tag{4.43}$$

(c)
$$\operatorname{curl} \mathbf{F} = \mathbf{e}_i \times \frac{\partial \mathbf{F}}{\partial x_i} \tag{4.44}$$

where

$$\frac{\partial \mathbf{F}}{\partial x_i} = \left(\frac{\partial F_1}{\partial x_i}, \frac{\partial F_2}{\partial x_i}, \frac{\partial F_3}{\partial x_i} \right), \quad i = 1, 2, 3. \tag{4.45}$$

Proof

(a) Expression (4.42) follows at once from the definition of grad Ω.

(b) Since $\mathbf{F} = F_1\mathbf{e}_1 + F_2\mathbf{e}_2 + F_3\mathbf{e}_3 = F_j\mathbf{e}_j$ and the vectors $\mathbf{e}_j$ are constant,

$$\mathbf{e}_i \cdot \frac{\partial \mathbf{F}}{\partial x_i} = \mathbf{e}_i \cdot \mathbf{e}_j \frac{\partial F_j}{\partial x_i}$$

$$= \delta_{ij} \frac{\partial F_j}{\partial x_i},$$

using the fact that $\mathbf{e}_1, \mathbf{e}_2, \mathbf{e}_3$ form an orthonormal triad. Thus

$$\mathbf{e}_i \cdot \frac{\partial \mathbf{F}}{\partial x_i} = \frac{\partial F_i}{\partial x_i} = \operatorname{div} \mathbf{F}.$$

(c) We have

$$\mathbf{e}_i \times \frac{\partial \mathbf{F}}{\partial x_i} = \mathbf{e}_i \times \mathbf{e}_j \frac{\partial F_j}{\partial x_i}.$$

The terms obtained by allowing each of i, j to take the values 1, 2 are

$$\mathbf{e}_1 \times \mathbf{e}_1 \frac{\partial F_1}{\partial x_1} + \mathbf{e}_1 \times \mathbf{e}_2 \frac{\partial F_2}{\partial x_1} + \mathbf{e}_2 \times \mathbf{e}_1 \frac{\partial F_1}{\partial x_2} + \mathbf{e}_2 \times \mathbf{e}_2 \frac{\partial F_2}{\partial x_2} = \left(\frac{\partial F_2}{\partial x_1} - \frac{\partial F_1}{\partial x_2} \right) \mathbf{e}_3$$

and this is the x_3-component of curl $\mathbf{F}$. Similarly the other two possible pairs of values of i, j give the x_1- and x_2-components of curl $\mathbf{F}$. Hence (4.44) follows.

Identities

If it is assumed (in accordance with our usual practice) that all necessary derivatives exist and are continuous, then the following identities hold:

$\qquad$ (i) $\quad \operatorname{div}(\operatorname{curl} \mathbf{F}) \equiv 0;$ $\qquad$ (4.46)

$\qquad$ (ii) $\quad \operatorname{curl}(\operatorname{grad} \Omega) \equiv \mathbf{0};$ $\qquad$ (4.47)

$\qquad$ (iii) $\quad \operatorname{grad}(\Omega_1 \Omega_2) \equiv \Omega_1 \operatorname{grad} \Omega_2 + \Omega_2 \operatorname{grad} \Omega_1;$ $\qquad$ (4.48)

$\qquad$ (iv) $\quad \operatorname{div}(\Omega \mathbf{F}) \equiv \Omega \operatorname{div} \mathbf{F} + \mathbf{F} \cdot \operatorname{grad} \Omega;$ $\qquad$ (4.49)

$\qquad$ (v) $\quad \operatorname{curl}(\Omega \mathbf{F}) \equiv \Omega \operatorname{curl} \mathbf{F} - \mathbf{F} \times \operatorname{grad} \Omega;$ $\qquad$ (4.50)

$\qquad$ (vi) $\quad \operatorname{grad}(\mathbf{F} \cdot \mathbf{G}) \equiv \mathbf{F} \times \operatorname{curl} \mathbf{G} + \mathbf{G} \times \operatorname{curl} \mathbf{F} + (\mathbf{F} \cdot \nabla)\mathbf{G} + (\mathbf{G} \cdot \nabla)\mathbf{F};$ $\quad$ (4.51)

$\qquad$ (vii) $\quad \operatorname{div}(\mathbf{F} \times \mathbf{G}) \equiv \mathbf{G} \cdot \operatorname{curl} \mathbf{F} - \mathbf{F} \cdot \operatorname{curl} \mathbf{G};$ $\qquad$ (4.52)

$\qquad$ (viii) $\quad \operatorname{curl}(\mathbf{F} \times \mathbf{G}) \equiv \mathbf{F} \operatorname{div} \mathbf{G} - \mathbf{G} \operatorname{div} \mathbf{F} + (\mathbf{G} \cdot \nabla)\mathbf{F} - (\mathbf{F} \cdot \nabla)\mathbf{G}.$ $\qquad$ (4.53)

In operator notation:

$\qquad$ (i)′ $\quad \nabla \cdot (\nabla \times \mathbf{F}) \equiv 0;$ $\qquad$ (4.54)

$\qquad$ (ii)′ $\quad \nabla \times (\nabla \Omega) \equiv \mathbf{0};$ $\qquad$ (4.55)

$\qquad$ (iii)′ $\quad \nabla(\Omega_1 \Omega_2) \equiv \Omega_1 \nabla \Omega_2 + \Omega_2 \nabla \Omega_1;$ $\qquad$ (4.56) (iv)' $\mathbf{\nabla} . (\Omega\mathbf{F}) \equiv \Omega\mathbf{\nabla} . \mathbf{F} + \mathbf{F} . \mathbf{\nabla}\Omega;$ (4.57)

(v)' $\mathbf{\nabla} \times (\Omega\mathbf{F}) \equiv \Omega\mathbf{\nabla} \times \mathbf{F} - \mathbf{F} \times \mathbf{\nabla}\Omega;$ (4.58)

(vi)' $\mathbf{\nabla}(\mathbf{F} . \mathbf{G}) \equiv \mathbf{F} \times (\mathbf{\nabla} \times \mathbf{G}) + \mathbf{G} \times (\mathbf{\nabla} \times \mathbf{F}) + (\mathbf{F} . \mathbf{\nabla})\,\mathbf{G} + (\mathbf{G} . \mathbf{\nabla})\mathbf{F};$ (4.59)

(vii)' $\mathbf{\nabla} . (\mathbf{F} \times \mathbf{G}) \equiv \mathbf{G} . (\mathbf{\nabla} \times \mathbf{F}) - \mathbf{F} . (\mathbf{\nabla} \times \mathbf{G});$ (4.60)

(viii)' $\mathbf{\nabla} \times (\mathbf{F} \times \mathbf{G}) \equiv \mathbf{F}(\mathbf{\nabla} . \mathbf{G}) - \mathbf{G}(\mathbf{\nabla} . \mathbf{F}) + (\mathbf{G} . \mathbf{\nabla})\mathbf{F} - (\mathbf{F} . \mathbf{\nabla})\mathbf{G}.$ (4.61)

Proofs

(i) $\operatorname{div} (\operatorname{curl} \mathbf{F}) \equiv \dfrac{\partial}{\partial x}\left(\dfrac{\partial F_3}{\partial y} - \dfrac{\partial F_2}{\partial z}\right) + \dfrac{\partial}{\partial y}\left(\dfrac{\partial F_1}{\partial z} - \dfrac{\partial F_3}{\partial x}\right) + \dfrac{\partial}{\partial z}\left(\dfrac{\partial F_2}{\partial x} - \dfrac{\partial F_1}{\partial y}\right)$

$$\equiv 0$$

using the assumption that the derivatives are continous so that the order of differentiation can be inverted.

(ii)

$$\operatorname{curl} (\operatorname{grad} \Omega) \equiv \begin{vmatrix} \mathbf{i} & \mathbf{j} & \mathbf{k} \\ \dfrac{\partial}{\partial x} & \dfrac{\partial}{\partial y} & \dfrac{\partial}{\partial z} \\ \dfrac{\partial\Omega}{\partial x} & \dfrac{\partial\Omega}{\partial y} & \dfrac{\partial\Omega}{\partial z} \end{vmatrix}$$

$$\equiv \mathbf{0}$$

because $\partial^2\Omega/\partial y\,\partial z - \partial^2\,\Omega/\partial z\,\partial y = 0$ with two other similar relations.

(iii) Using (4.42)

$$\operatorname{grad} (\Omega_1\,\Omega_2) \equiv \mathbf{e}_i\,\frac{\partial}{\partial x_i}\,(\Omega_1\,\Omega_2)$$

$$\equiv \Omega_1\mathbf{e}_i\,\frac{\partial\Omega_2}{\partial x_i} + \Omega_2\mathbf{e}_i\,\frac{\partial\Omega_1}{\partial x_i}$$

$$\equiv \Omega_1\operatorname{grad}\Omega_2 + \Omega_2\operatorname{grad}\Omega_1.$$

(iv) Using (4.43)

$$\operatorname{div} (\Omega\mathbf{F}) \equiv \mathbf{e}_i . \frac{\partial}{\partial x_i}\,(\Omega\mathbf{F})$$

$$\equiv \Omega\mathbf{e}_i . \frac{\partial\mathbf{F}}{\partial x_i} + \left(\mathbf{e}_i\,\frac{\partial\Omega}{\partial x_i}\right). \mathbf{F}$$

$$\equiv \Omega\operatorname{div}\mathbf{F} + \mathbf{F} . \operatorname{grad}\Omega.$$

(v) The proof is similar to the proof of (iv) and is left to the reader as an exercise.

(vi) Using (4.44) we have

$$\mathbf{F} \times \operatorname{curl}\mathbf{G} + \mathbf{G} \times \operatorname{curl}\mathbf{F} \equiv \mathbf{F} \times \left(\mathbf{e}_i \times \frac{\partial\mathbf{G}}{\partial x_i}\right) + \mathbf{G} \times \left(\mathbf{e}_i \times \frac{\partial\mathbf{F}}{\partial x_i}\right)$$

$$\equiv \left(\mathbf{F} \cdot \frac{\partial \mathbf{G}}{\partial x_i}\right)\mathbf{e}_i - (\mathbf{F} \cdot \mathbf{e}_i)\frac{\partial \mathbf{G}}{\partial x_i} + \left(\mathbf{G} \cdot \frac{\partial \mathbf{F}}{\partial x_i}\right)\mathbf{e}_i - (\mathbf{G} \cdot \mathbf{e}_i)\frac{\partial \mathbf{F}}{\partial x_i}$$

$$\equiv \mathbf{e}_i\frac{\partial}{\partial x_i}(\mathbf{F} \cdot \mathbf{G}) - \left(\mathbf{F} \cdot \mathbf{e}_i\frac{\partial}{\partial x_i}\right)\mathbf{G} - \left(\mathbf{G} \cdot \mathbf{e}_i\frac{\partial}{\partial x_i}\right)\mathbf{F}$$

$$\equiv \operatorname{grad}(\mathbf{F} \cdot \mathbf{G}) - (\mathbf{F} \cdot \boldsymbol{\nabla})\mathbf{G} - (\mathbf{G} \cdot \boldsymbol{\nabla})\mathbf{F}$$

which proves identity (vi).

(vii) Using (4.43)

$$\operatorname{div}(\mathbf{F} \times \mathbf{G}) \equiv \mathbf{e}_i \cdot \frac{\partial}{\partial x_i}(\mathbf{F} \times \mathbf{G})$$

$$\equiv \mathbf{e}_i \cdot \left(\frac{\partial \mathbf{F}}{\partial x_i} \times \mathbf{G}\right) + \mathbf{e}_i \cdot \left(\mathbf{F} \times \frac{\partial \mathbf{G}}{\partial x_i}\right)$$

$$\equiv \left(\mathbf{e}_i \times \frac{\partial \mathbf{F}}{\partial x_i}\right) \cdot \mathbf{G} - \left(\mathbf{e}_i \times \frac{\partial \mathbf{G}}{\partial x_i}\right) \cdot \mathbf{F}$$

$$\equiv \mathbf{G} \cdot \operatorname{curl}\mathbf{F} - \mathbf{F} \cdot \operatorname{curl}\mathbf{G},$$

using (4.44).

(viii) The proof of this identity is left to the reader as an exercise.

EXAMPLE 8

Using the formula $\operatorname{grad} r^n = nr^{n-2}\,\mathbf{r}$ (see Exercise 4.11, section 4.4) prove that

$$\nabla^2 r^n = n(n+1)\,r^{n-2} \quad (r \neq 0 \text{ if } n \leqslant 2).$$

Solution

We have

$$\nabla^2 r^n = \operatorname{div}\operatorname{grad} r^n$$
$$= \operatorname{div}(nr^{n-2}\,\mathbf{r})$$
$$= nr^{n-2}\operatorname{div}\mathbf{r} + n\mathbf{r} \cdot \operatorname{grad} r^{n-2}$$

using identity (4.49). Hence

$$\nabla^2 r^n = 3nr^{n-2} + n(n-2)\,r^{n-4}\,\mathbf{r} \cdot \mathbf{r}$$
$$= n(n+1)\,r^{n-2}.$$

EXAMPLE 9

Show that the vector field $\mathbf{H} = \Phi\boldsymbol{\nabla}\Psi$ is perpendicular to curl $\mathbf{H}$ at all points where neither vector field vanishes.

Solution

Using identity (4.58),

$$\nabla \times (\Phi \nabla \Psi) = \Phi \nabla \times \nabla \Psi - \nabla \Psi \times \nabla \Phi$$

$$= \nabla \Phi \times \nabla \Psi \quad \text{by (4.55).}$$

Thus

$$\mathbf{H} . \operatorname{curl} \mathbf{H} = \Phi \nabla \Psi . (\nabla \Phi \times \nabla \Psi)$$

$$= 0$$

which shows that $\mathbf{H}$ and curl $\mathbf{H}$ are perpendicular at points where $\mathbf{H} \neq \mathbf{0}$ and curl $\mathbf{H} \neq \mathbf{0}$.

EXERCISES

4.27 Prove the identities (4.50) and (4.53).

4.28 Prove that curl $(\mathbf{r}/r^2) = \mathbf{0}$. Also, show that div $(\mathbf{r}/r^2) = 1/r^2$.

4.29 Using the notation introduced at the beginning of this section, show that

$$\nabla^2 \equiv \mathbf{e}_i . \mathbf{e}_j \, \partial^2/\partial x_i \partial x_j.$$

Using expression (4.44) for curl, prove the identity

$$\operatorname{curl} \operatorname{curl} \mathbf{F} \equiv \operatorname{grad} (\operatorname{div} \mathbf{F}) - \nabla^2 \mathbf{F}.$$

4.30 Show that

$$\operatorname{curl} (\mathbf{r} \times \operatorname{curl} \mathbf{F}) + (\mathbf{r} . \nabla)\operatorname{curl} \mathbf{F} + 2 \operatorname{curl} \mathbf{F} \equiv \mathbf{0}.$$

4.31 Prove that

$$\nabla^2 (\Phi \, \Psi) \equiv \Phi \, \nabla^2 \Psi + 2 \nabla \Phi . \nabla \Psi + \Psi \nabla^2 \Phi.$$

4.32 If Φ satisfies Laplace's equation $\nabla^2 \Phi = 0$ and is such that $\mathbf{r} . \nabla \Phi = m\Phi$, where m is constant, prove that Φ/r^{2m+1} is also a solution of Laplace's equation. [*Hint.* Put $\Psi = 1/r^{2m+1}$ in Exercise 31, and use the formula of Exercise 11, section 4.4. See also Example 8.]

4.10 CYLINDRICAL AND SPHERICAL POLAR COORDINATES

In the theory developed so far we have worked entirely with rectangular cartesian coordinates. However, in practice other coordinate systems are often more convenient. Cylindrical polar coordinates and spherical polar coordinates are especially important.

Cylindrical polar coordinates

Let a point P have rectangular cartesian coordinates (x, y, z). Denote by R the perpendicular distance from P to the z-axis and let φ be the angle between the zx-plane and the plane containing P and the axis Oz. We count φ positive in the sense indicated by the arrow in Fig. 4.3 and restrict its values to the range $-\pi < \varphi \leqslant \pi$, or sometimes $0 \leqslant \varphi < 2\pi$. Then R, φ, z are the **cylindrical polar coordinates** of P.

From the geometry of Fig. 4.3 the rectangular coordinates (x, y, z) of P are related to the cylindrical polar coordinates (R, φ, z) by the equations

$$x = R \cos \varphi, \quad y = R \sin \varphi, \quad z = z. \tag{4.62}$$

The equations can be solved for R, φ giving, in the case when $-\pi < \varphi \leqslant \pi$,

$$\left. \begin{aligned} R &= (x^2 + y^2)^{\frac{1}{2}}, \\ \varphi &= \tan^{-1}(y/x) + n\pi, \end{aligned} \right\} \tag{4.63}$$

where $n = 0$ if $x > 0$, $n = 1$ if $x < 0$ and $y \geqslant 0$, and $n = -1$ if $x < 0$ and $y < 0$.

Spherical polar coordinates

As before, let P have rectangular cartesian coordinates (x, y, z), let r denote the distance of P from the origin O, let θ be the angle that OP makes with the z-axis and φ the angle between the zx-plane and the plane containing P and the z-axis. The angles θ, φ are measured positive in the senses shown by the arrows in Fig. 4.4 and their values are restricted to the ranges $0 \leqslant \theta \leqslant \pi$, $-\pi < \varphi \leqslant \pi$ (or sometimes $0 \leqslant \varphi < 2\pi$). Then r, θ, φ are the **spherical polar coordinates** of P.

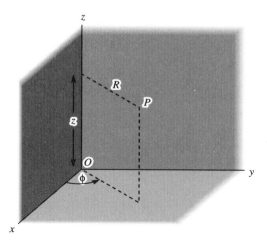

Fig. 4.3 Cylindrical polar coordinates (R, φ, z).

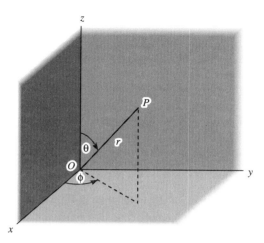

Fig. 4.4 Spherical polar coordinates (r, θ, φ).

From the geometry of Fig. 4.4 the rectangular cartesian coordinates (x, y, z) of P are related to the spherical polar coordinates (r, θ, φ) by the equations

$$x = r \sin \theta \, \cos \varphi, \quad y = r \sin \theta \, \sin \varphi, \quad z = r \cos \theta. \tag{4.64}$$

Solving for r, θ, φ in terms of x, y, z, we have in the case when $-\pi < \varphi \leqslant \pi$,

$$r = (x^2 + y^2 + z^2)^{\frac{1}{2}},$$

$$\theta = \cos^{-1} \frac{z}{(x^2 + y^2 + z^2)^{\frac{1}{2}}}, \tag{4.65}$$

$$\varphi = \tan^{-1}(y/x) + n\pi,$$

where n is chosen as before [equation (4.63)].

Coordinate lines

If the coordinates y, z of the point $P(x, y, z)$ are fixed, and x is allowed to range over its permissible values, the locus of P is a straight line parallel to the x-axis – it is called an **x-coordinate line**. Similarly if x, z are fixed and y varies, the locus is a **y-coordinate line**, and if x, y are fixed and z varies, the locus is a **z-coordinate line** (Fig. 4.5).

In the case of cylindrical polar coordinates, the locus of $P(R, \varphi, z)$ when φ, z are fixed and R varies is called an R-coordinate line; the locus when R, z are fixed and φ varies is a φ-coordinate line and the locus when R, φ are fixed and z varies is a z-coordinate line. The R-coordinate lines are straight lines radiating from, and normal to, the z-axis, the φ-coordinate lines are circles centred on the z-axis and parallel to the xy-plane and the z-coordinate lines are straight lines parallel to the z-axis (Fig. 4.6a). The coordinate lines of a spherical polar coordinate system are defined similarly (Fig. 4.6b).

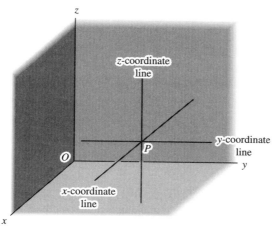

Fig. 4.5 Coordinate lines for rectangular cartesian coordinates.

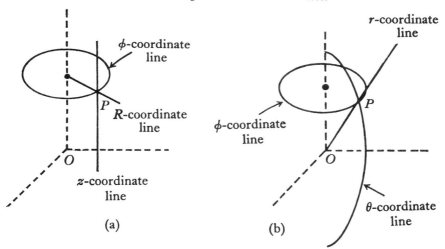

Fig. 4.6 Coordinate lines for (a) cylindrical and (b) spherical polar coordinates.

In each of the above coordinate systems the tangents at a point P to the three coordinate lines through P are mutually perpendicular. Such coordinate systems are said to be **orthogonal**. Because of this feature it is easy to discuss them all within the framework of the theory of rectangular cartesian systems.

EXERCISE

4.33 In rectangular cartesian coordinates the locus $x = $ constant is a plane parallel to the yz-plane. It is called a **coordinate surface**. Similarly, the coordinate surfaces $y = $ constant and $z = $ constant are also planes.

What geometrical shapes are the coordinate surfaces in (i) cylindrical polar coordinates, (ii) spherical polar coordinates?

4.11 GENERAL ORTHOGONAL CURVILINEAR COORDINATES

With cylindrical and spherical polar coordinates to refer to as examples, we now consider a transformation to general orthogonal curvilinear coordinates (u, v, w), defined by the equations

$$x = x(u, v, w), \quad y = y(u, v, w), \quad z = z(u, v, w). \tag{4.66}$$

We might have written $x = f(u, v, w)$, $y = g(u, v, w)$, $z = h(u, v, w)$. However, the common practice of using x, y, z to symbolize functions is more convenient and causes no confusion in this context. Writing $\mathbf{r} = x\mathbf{i} + y\mathbf{j} + z\mathbf{k}$ (as usual), equations (4.66) may be expressed more briefly as

$$\mathbf{r} = \mathbf{r}(u, v, w). \tag{4.67}$$

In order to have a coordinate system of practical value, certain restrictions are placed upon the functions x, y, z as follows.

(i) Throughout any region under consideration, we suppose that there is a one-to-one correspondence between the triads (x, y, z) and the triads (u, v, w). Since each point has a unique set of rectangular coordinates (x, y, z), each point will then have a unique set of curvilinear coordinates (u, v, w). Furthermore, equations (4.66) can be solved for u, v, w, giving

$$u = u(x, y, z), \quad v = v(x, y, z), \quad w = w(x, y, z). \tag{4.68}$$

(ii) It is assumed that x, y, z are continuously differentiable functions of u, v, w and that the determinant

$$J = \begin{vmatrix} \dfrac{\partial x}{\partial u} & \dfrac{\partial y}{\partial u} & \dfrac{\partial z}{\partial u} \\ \dfrac{\partial x}{\partial v} & \dfrac{\partial y}{\partial v} & \dfrac{\partial z}{\partial v} \\ \dfrac{\partial x}{\partial w} & \dfrac{\partial y}{\partial w} & \dfrac{\partial z}{\partial w} \end{vmatrix} \tag{4.69}$$

does not vanish at any point. This determinant is called the **Jacobian** of the transformation.

In addition to these conditions, the tangents at a point P to the u-, v-, w-coordinate lines through this point are assumed to be mutually perpendicular so that the coordinate system is *orthogonal*. The parametric equations of the coordinate lines are obtained by assigning constant values to pairs of u, v, w in equation (4.67). Thus, if u_0, v_0, w_0 are constants, the equations

$$\mathbf{r} = \mathbf{r}(u, v_0, w_0), \quad \mathbf{r} = \mathbf{r}(u_0, v, w_0), \quad \mathbf{r} = \mathbf{r}(u_0, v_0, w) \qquad (4.70)$$

represent u-, v-, w-coordinate lines, respectively.

Transformations encountered in practice are usually chosen to satisfy the above conditions everywhere except possibly at isolated points or on certain curves or surfaces. Thus, in the case of cylindrical polar coordinates, the one-to-one correspondence between the triads (x, y, z) and the triads (R, φ, z) fails on the z-axis. The values $R = 0$, $z = z_0$ correspond to the fixed point $x = y = 0$, $z = z_0$ on the z-axis regardless of the value of φ. However, such exceptions are easily recognized and will usually cause no serious difficulty. They will not be considered further.

Consider the coordinate transformation defined by equations (4.67) and let

$$h_1 = \left| \frac{\partial \mathbf{r}}{\partial u} \right|, \quad h_2 = \left| \frac{\partial \mathbf{r}}{\partial v} \right|, \quad h_3 = \left| \frac{\partial \mathbf{r}}{\partial w} \right|. \qquad (4.71)$$

The elements of the first, second and third rows of the Jacobian (4.69) are the components of $\partial \mathbf{r}/\partial u$, $\partial \mathbf{r}/\partial v$, $\partial \mathbf{r}/\partial w$ respectively. As the Jacobian is assumed to be non-vanishing at least one element of each row must be non-zero and so it follows that h_1, h_2, h_3 do not vanish.

Let $\mathbf{e}_u$, $\mathbf{e}_v$, $\mathbf{e}_w$ respectively denote the unit tangents at a point P to the u-, v-, w-coordinate lines through P. Then (cf. section 3.4)

$$\mathbf{e}_u = \frac{1}{h_1} \frac{\partial \mathbf{r}}{\partial u}, \quad \mathbf{e}_v = \frac{1}{h_2} \frac{\partial \mathbf{r}}{\partial v}, \quad \mathbf{e}_w = \frac{1}{h_3} \frac{\partial \mathbf{r}}{\partial w}, \qquad (4.72)$$

where the right-hand sides are, of course, evaluated at P. These unit vectors will be mutually perpendicular since the coordinate system is orthogonal. For convenience, the parameters u, v, w of the coordinate lines are always chosen so that $\mathbf{e}_u$, $\mathbf{e}_v$, $\mathbf{e}_w$ (in that order) form a right-handed triad. This can always be achieved because if u is replaced by $-u$ in (4.67) the direction of $\partial \mathbf{r}/\partial u$ is then reversed.

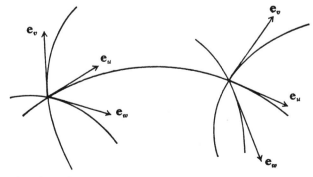

Fig. 4.7 Showing the unit vectors $\mathbf{e}_u$, $\mathbf{e}_v$, $\mathbf{e}_w$ at two different points on a u-coordinate line.

The vectors $\mathbf{e}_u$, $\mathbf{e}_v$, $\mathbf{e}_w$ form an orthonormal triad, just as do the vectors $\mathbf{i}$, $\mathbf{j}$, $\mathbf{k}$. There is, however, one fundamental difference between the two triads: whereas $\mathbf{i}$, $\mathbf{j}$, $\mathbf{k}$ point in fixed directions, *the directions of $\mathbf{e}_u$, $\mathbf{e}_v$, $\mathbf{e}_w$ will in general vary from point to point* because the coordinate lines will be curved (Fig. 4.7).

EXAMPLE 10

Spherical polar coordinates

In the case of spherical polar coordinates defined by equations (4.64) we have

$$\mathbf{r} = r(\sin \theta \cos \varphi, \ \sin \theta \sin \varphi, \ \cos \theta). \tag{4.73}$$

Therefore

$$\frac{\partial \mathbf{r}}{\partial r} = (\sin \theta \cos \varphi, \ \sin \theta \sin \varphi, \ \cos \theta),$$

$$\frac{\partial \mathbf{r}}{\partial \theta} = r(\cos \theta \cos \varphi, \ \cos \theta \sin \varphi, \ -\sin \theta)$$

and

$$\frac{\partial \mathbf{r}}{\partial \varphi} = r(-\sin \theta \sin \varphi, \ \sin \theta \cos \varphi, \ 0).$$

It follows that

$$h_1^2 = \sin^2 \theta \cos^2 \varphi + \sin^2 \theta \sin^2 \varphi + \cos^2 \theta$$

$$= \sin^2 \theta (\cos^2 \varphi + \sin^2 \varphi) + \cos^2 \theta$$

$$= 1.$$

Similarly

$$h_2^2 = r^2 \quad \text{and} \quad h_3^2 = r^2 \sin^2 \theta.$$

Since $0 \leqslant \theta \leqslant \pi$, we have $\sin \theta \geqslant 0$. Thus

$$h_1 = 1, \quad h_2 = r, \quad h_3 = r \sin \theta. \tag{4.74}$$

From the definition (4.71) it follows also that

$$\mathbf{e}_r = (\sin \theta \cos \varphi, \ \sin \theta \sin \varphi, \ \cos \theta),$$

$$\mathbf{e}_\theta = (\cos \theta \cos \varphi, \ \cos \theta \sin \varphi, \ -\sin \theta), \tag{4.75}$$

$$\mathbf{e}_\varphi = (-\sin \varphi, \ \cos \varphi, \ 0).$$

The directions of these unit vectors are shown in Fig. 4.8(a)

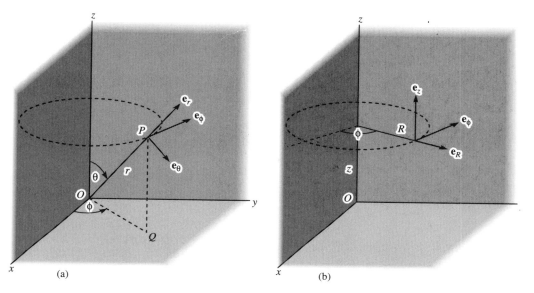

Fig. 4.8 (a) Unit vectors $\mathbf{e}_r$, $\mathbf{e}_\theta$, $\mathbf{e}_\varphi$ for spherical polar coordinates (r, θ, φ). $\mathbf{e}_\theta$ lies in the plane OPQ and $\mathbf{e}_\varphi$ is perpendicular to OPQ. (b) Unit vectors $\mathbf{e}_R$, $\mathbf{e}_\varphi$, $\mathbf{e}_z$ for cylindrical polar coordinates (R, φ, z). $\mathbf{e}_R$ and $\mathbf{e}_\varphi$ are parallel to the xy-plane.

EXAMPLE 11

Cylindrical polar coordinates

In the case of cylindrical polar coordinates (R, φ, z), defined by equations (4.62), it is found that

$$h_1 = 1, \quad h_2 = R, \quad h_3 = 1. \tag{4.76}$$

Also

$$\mathbf{e}_R = (\cos \varphi, \ \sin \varphi, \ 0),$$
$$\mathbf{e}_\varphi = (-\sin \varphi, \ \cos \varphi, \ 0), \tag{4.77}$$
$$\mathbf{e}_z = (0, 0, 1).$$

It is left as an exercise for the reader to establish these results. Fig. 4.8(b) shows the directions of the unit vectors $\mathbf{e}_R$, $\mathbf{e}_\varphi$, $\mathbf{e}_z$.

The formulae for h_1, h_2, h_3 can sometimes be obtained by the following geometrical argument. Suppose that u is given an infinitesimal positive increment du and that v, w are held constant. Then, from (4.67), the corresponding increment in $\mathbf{r}$ is

$$d\mathbf{r} = \frac{\partial \mathbf{r}}{\partial u} du$$

so

$$|d\mathbf{r}| = \left| \frac{\partial \mathbf{r}}{\partial u} \right| du = h_1 \, du.$$

Since **r** is the position vector of a point P relative to the origin, $|d\mathbf{r}| = h_1\,du$ is the displacement of P corresponding to the increment du in u. Similarly,

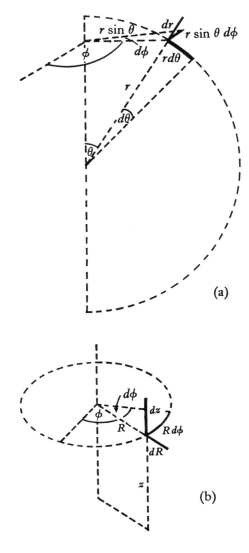

Fig. 4.9 (a) In spherical polar coordinates (r, θ, φ), $d\mathbf{r} = (dr, r\,d\theta, r\sin\theta\,d\varphi)$; thus $h_1 = 1$, $h_2 = r$, $h_3 = r\sin\theta$. (b) In cylindrical polar coordinates (R, θ, z), $d\mathbf{r} = (dR, R\,d\varphi, dz)$; thus $h_1 = 1$, $h_2 = R$, $h_3 = 1$.

$h_2 \, dv$ and $h_3 \, dw$ are the displacements corresponding to increments dv in v and dw in w respectively. Thus h_1, h_2, h_3 may be found by giving small increments to u, v, w and considering the geometrical effect. Fig. 4.9 indicates how this method may be applied in the case of spherical polar coordinates and cylindrical polar coordinates.

EXERCISES

4.34 Prove the results stated in equations (4.76) and (4.77).

4.35 *Toroidal curvilinear coordinates R, θ, φ are defined in terms of rectangular cartesian coordinates x, y, z by*

$$x = (a - R \cos \theta) \cos \varphi, \quad y = (a - R \cos \theta) \sin \varphi, \quad z = R \sin \theta$$

where a is a constant and $R < a$. Show that this is an orthogonal coordinate system and that, in the usual notation,

$$h_1 = 1, \quad h_2 = R, \quad h_3 = a - R \cos \theta.$$

4.36 *Elliptical coordinates ξ, η, z are such that the position vector is given by*

$$\mathbf{r} = (\cosh \xi \cos \eta, \; \sinh \xi \sin \eta, \; z)$$

where $0 \leqslant \xi < \infty$, $-\pi < \eta \leqslant \pi$ and $-\infty < z < \infty$. Show that this curvilinear system of coordinates is orthogonal and find h_1, h_2, h_3. What are the geometrical shapes of the ξ- and η–coordinate lines?

4.37 *Parabolic coordinates u, v, w are defined so that the position vector relative to rectangular cartesian axes is*

$$\mathbf{r} = (\tfrac{1}{2}(u^2 - v^2), uv, -w)$$

where $-\infty < u < \infty, 0 \leqslant v < \infty$ and $-\infty < w < \infty$. Show that, for any fixed value of w, the u- and v-coordinate curves are confocal parabolas. Verify that the system is orthogonal and show that the basic unit vectors at the point $u = v = w = 1$ are $(1, 1, 0)/\sqrt{2}, (-1, 1, 0)/\sqrt{2}$ and $(0, 0, -1)$. Show that cylindrical polar coordinates R, φ, z may be chosen so that

$$u = (2R)^{\frac{1}{2}} \cos \tfrac{1}{2} \varphi, \quad v = (2R)^{\frac{1}{2}} \sin \tfrac{1}{2} \varphi, \quad w = -z.$$

4.38 In spherical polar coordinates show that

$$\partial \mathbf{e}_r / \partial \theta = \mathbf{e}_\theta, \quad \partial \mathbf{e}_\theta / \partial \theta = -\mathbf{e}_r, \quad \partial \mathbf{e}_\varphi / \partial \theta = \mathbf{0},$$

$$\partial \mathbf{e}_r / \partial \varphi = \sin \theta \, \mathbf{e}_\varphi, \quad \partial \mathbf{e}_\theta / \partial \varphi = \cos \theta \, \mathbf{e}_\varphi,$$

$$\partial \mathbf{e}_\varphi / \partial \varphi = -\sin \theta \, \mathbf{e}_r - \cos \theta \, \mathbf{e}_\theta.$$

4.12 VECTOR COMPONENTS IN ORTHOGONAL CURVILINEAR COORDINATES

Suppose that a vector field has the value $\mathbf{F}$ at the point P with curvilinear coordinates (u, v, w). Since the unit vectors $\mathbf{e}_u, \mathbf{e}_v, \mathbf{e}_w$ are not coplanar, it follows from the worked Example 5 in section 2.8 that $\mathbf{F}$ can be expressed in the form

$$\mathbf{F} = F_u\mathbf{e}_u + F_v\mathbf{e}_v + F_w\mathbf{e}_w. \tag{4.78}$$

In other words, the vectors $\mathbf{e}_u, \mathbf{e}_v, \mathbf{e}_w$ associated with P can be taken as an orthonormal basis for the representation of $\mathbf{F}$. We call F_u, F_v, F_w **the components of F along the coordinate lines**.

When rectangular cartesian coordinates were used we wrote

$$\mathbf{F} = (F_1, F_2, F_3)$$

and it may seem natural now to use the notation

$$\mathbf{F} = (F_u, F_v, F_w). \tag{4.79}$$

However, in this case the variation of $\mathbf{F}$ with position is not represented completely by the variation of F_u, F_v, F_w because the base vectors $\mathbf{e}_u, \mathbf{e}_v, \mathbf{e}_w$ are themselves functions of u, v, w in general. Thus

$$\frac{\partial \mathbf{F}}{\partial u} = \frac{\partial F_u}{\partial u}\mathbf{e}_u + \frac{\partial F_v}{\partial u}\mathbf{e}_v + \frac{\partial F_w}{\partial u}\mathbf{e}_w + F_u\frac{\partial \mathbf{e}_u}{\partial u} + F_v\frac{\partial \mathbf{e}_v}{\partial u} + F_w\frac{\partial \mathbf{e}_w}{\partial u}$$

whence, in general,

$$\frac{\partial \mathbf{F}}{\partial u} \neq \left(\frac{\partial F_u}{\partial u}, \frac{\partial F_v}{\partial u}, \frac{\partial F_w}{\partial u}\right).$$

In order to avoid possible misunderstanding, the notation (4.78) will always be used in preference to (4.79) when the coordinate system is curvilinear.

The procedure for expressing a vector field

$$\mathbf{F}(x, y, z) = F_1(x, y, z)\mathbf{i} + F_2(x, y, z)\mathbf{j} + F_3(x, y, z)\mathbf{k} \tag{4.80}$$

in the form

$$\mathbf{F}(u, v, w) = F_u(u, v, w)\mathbf{e}_u + F_v(u, v, w)\mathbf{e}_v + F_w(u, v, w)\mathbf{e}_w \tag{4.81}$$

is straightforward in principle. First, F_1, F_2, F_3 can be expressed as functions of u, v, w by substitution from equations (4.66). Also, since the triad $\mathbf{e}_u, \mathbf{e}_v, \mathbf{e}_w$ is orthonormal,

$$F_u = \mathbf{F} \cdot \mathbf{e}_u = F_1\mathbf{i} \cdot \mathbf{e}_u + F_2\mathbf{j} \cdot \mathbf{e}_u + F_3\mathbf{k} \cdot \mathbf{e}_u.$$

But from the defining equations (4.71) and (4.72)

$$\mathbf{e}_u = \frac{1}{h_1}\frac{\partial \mathbf{r}}{\partial u} = \frac{1}{h_1}\left(\frac{\partial x}{\partial u}\mathbf{i} + \frac{\partial y}{\partial u}\mathbf{j} + \frac{\partial z}{\partial u}\mathbf{k}\right)$$

where

$$h_1 = \left\{ \left(\frac{\partial x}{\partial u} \right)^2 + \left(\frac{\partial y}{\partial u} \right)^2 + \left(\frac{\partial z}{\partial u} \right)^2 \right\}^{\frac{1}{2}}.$$

Hence

$$F_u = \frac{1}{h_1} \left(F_1 \frac{\partial x}{\partial u} + F_2 \frac{\partial y}{\partial u} + F_3 \frac{\partial z}{\partial u} \right).$$

Similarly, F_v and F_w can be found in terms of u, v, w and $\mathbf{F}$ is thus represented in the form (4.81).

EXAMPLE 12

Express the vector field

$$\mathbf{F} = z\mathbf{i} = (z, 0, 0)$$

in the form

$$\mathbf{F} = F_r \mathbf{e}_r + F_\theta \mathbf{e}_\theta + F_\varphi \mathbf{e}_\varphi$$

where r, θ, φ are spherical polar coordinates.

Solution

The unit vectors $\mathbf{e}_r, \mathbf{e}_\theta, \mathbf{e}_\varphi$ have already been obtained and are given by equations (4.75). Thus, using those equations, and also the relation $z = r \cos \theta$,

$$F_r = \mathbf{F} \cdot \mathbf{e}_r = z \sin \theta \cos \varphi = r \cos \theta \sin \theta \cos \varphi,$$

$$F_\theta = \mathbf{F} \cdot \mathbf{e}_\theta = z \cos \theta \cos \varphi = r \cos^2 \theta \cos \varphi,$$

$$F_\varphi = \mathbf{F} \cdot \mathbf{e}_\varphi = -z \sin \varphi = -r \cos \theta \sin \varphi.$$

Hence

$$\mathbf{F} = r \cos \theta \sin \theta \cos \varphi \, \mathbf{e}_r + r \cos^2 \theta \cos \varphi \, \mathbf{e}_\theta - r \cos \theta \sin \varphi \, \mathbf{e}_\varphi.$$

EXERCISES

4.39 Express the vector field $\mathbf{F} = (-y, x, 0)$ in cylindrical polar and in spherical polar component form.

4.40 Express the position vector in cylindrical polar coordinate form.

4.41 In spherical polar coordinates

$$\mathbf{F} = r \, \mathbf{e}_\theta + r\mathbf{e}_\varphi.$$

Find the spherical polar components of $\partial \mathbf{F}/\partial r$, $\partial \mathbf{F}/\partial \theta$ and $\partial \mathbf{F}/\partial \varphi$.

4.13 EXPRESSIONS FOR GRAD Ω, DIV F, CURL F, AND ∇^2 IN ORTHOGONAL CURVILINEAR COORDINATES

Using the notation of sections 4.11 and 4.12, the following formulae will now be established:

(i)
$$\operatorname{grad} \Omega = \frac{\mathbf{e}_u}{h_1} \frac{\partial \Omega}{\partial u} + \frac{\mathbf{e}_v}{h_2} \frac{\partial \Omega}{\partial v} + \frac{\mathbf{e}_w}{h_3} \frac{\partial \Omega}{\partial w}; \tag{4.82}$$

or alternatively, in operator notation,

$$\mathbf{\nabla} = \frac{\mathbf{e}_u}{h_1} \frac{\partial}{\partial u} + \frac{\mathbf{e}_v}{h_2} \frac{\partial}{\partial v} + \frac{\mathbf{e}_w}{h_3} \frac{\partial}{\partial w} \cdot \tag{4.83}$$

(ii)
$$\operatorname{div} \mathbf{F} = \frac{1}{h_1 h_2 h_3} \left\{ \frac{\partial}{\partial u} (h_2 h_3 F_u) + \frac{\partial}{\partial v} (h_3 h_1 F_v) + \frac{\partial}{\partial w} (h_1 h_2 F_w) \right\}. \tag{4.84}$$

(iii)
$$\operatorname{curl} \mathbf{F} = \frac{1}{h_1 h_2 h_3} \begin{vmatrix} h_1 \mathbf{e}_u & h_2 \mathbf{e}_v & h_3 \mathbf{e}_w \\ \dfrac{\partial}{\partial u} & \dfrac{\partial}{\partial v} & \dfrac{\partial}{\partial w} \\ h_1 F_u & h_2 F_v & h_3 F_w \end{vmatrix}. \tag{4.85}$$

(iv)
$$\nabla^2 = \frac{1}{h_1 h_2 h_3} \left\{ \frac{\partial}{\partial u} \left(\frac{h_2 h_3}{h_1} \frac{\partial}{\partial u} \right) + \frac{\partial}{\partial v} \left(\frac{h_3 h_1}{h_2} \frac{\partial}{\partial v} \right) + \frac{\partial}{\partial w} \left(\frac{h_1 h_2}{h_3} \frac{\partial}{\partial w} \right) \right\}. \tag{4.86}$$

N.B. The formulae in the particular cases of cylindrical and spherical polar coordinates are listed in Appendix B for easy reference.

Proofs

(i) We have

$$\mathbf{e}_u = \frac{1}{h_1} \frac{\partial \mathbf{r}}{\partial u} = \frac{1}{h_1} \left(\frac{\partial x}{\partial u}, \frac{\partial y}{\partial u}, \frac{\partial z}{\partial u} \right)$$

and

$$\operatorname{grad} \Omega = \left(\frac{\partial \Omega}{\partial x}, \frac{\partial \Omega}{\partial y}, \frac{\partial \Omega}{\partial z} \right).$$

Hence the component of grad Ω in the direction of $\mathbf{e}_u$ is

$$\mathbf{e}_u \cdot \operatorname{grad} \Omega = \frac{1}{h_1} \left(\frac{\partial \Omega}{\partial x} \frac{\partial x}{\partial u} + \frac{\partial \Omega}{\partial y} \frac{\partial y}{\partial u} + \frac{\partial \Omega}{\partial z} \frac{\partial z}{\partial u} \right)$$

$$= \frac{1}{h_1} \frac{\partial \Omega}{\partial u} \quad \text{(by the chain rule (4.3)).} \tag{4.87}$$

The components of grad Ω in the directions $\mathbf{e}_v$, $\mathbf{e}_w$ follow similarly, and hence

$$\text{grad } \Omega = \frac{\mathbf{e}_u}{h_1} \frac{\partial \Omega}{\partial u} + \frac{\mathbf{e}_v}{h_2} \frac{\partial \Omega}{\partial v} + \frac{\mathbf{e}_w}{h_3} \frac{\partial \Omega}{\partial w}.$$

(ii) It is convenient to use operator notation in this proof.
We note first that taking Ω to be u, v, w in turn in (4.82) gives

$$\mathbf{e}_u = h_1 \, \nabla u, \quad \mathbf{e}_v = h_2 \, \nabla v, \quad \mathbf{e}_w = h_3 \, \nabla w. \tag{4.88}$$

Thus, since $\mathbf{e}_u$, $\mathbf{e}_v$, $\mathbf{e}_w$ are a right-handed orthonormal triad,

$$\mathbf{e}_u = \mathbf{e}_v \times \mathbf{e}_w = h_2 \, h_3 \, \nabla v \times \nabla w.$$

Now

$$\nabla . (F_u \, \mathbf{e}_u) = \nabla . (h_2 \, h_3 \, F_u \, \nabla v \times \nabla w)$$

$$= h_2 \, h_3 \, F_u \, \nabla . (\nabla v \times \nabla w) + (\nabla v \times \nabla w) . \nabla (h_2 \, h_3 \, F_u),$$

using identity (4.57). But using (4.60) followed by (4.55) gives

$$\nabla . (\nabla v \times \nabla w) = \nabla w . (\nabla \times \nabla v) - \nabla v . (\nabla \times \nabla w)$$

$$= 0.$$

Hence

$$\nabla . (F_u \mathbf{e}_u) = (\nabla v \times \nabla w) . \nabla (h_2 \, h_3 \, F_u)$$

$$= \frac{\mathbf{e}_u}{h_2 \, h_3} . \nabla (h_2 \, h_3 \, F_u)$$

$$= \frac{1}{h_1 \, h_2 \, h_3} \frac{\partial}{\partial u} (h_2 \, h_3 \, F_u), \quad \text{using (4.87)}.$$

Similar results can be obtained for $\nabla . (F_v \, \mathbf{e}_v)$ and $\nabla . (F_w \, \mathbf{e}_w)$. Thus, since

$$\nabla . (F_u \mathbf{e}_u) + \nabla . (F_v \, \mathbf{e}_v) + \nabla . (F_w \mathbf{e}_w) = \nabla . (F_u \mathbf{e}_u + F_v \, \mathbf{e}_v + F_w \mathbf{e}_w)$$

$$= \text{div } \mathbf{F},$$

formula (4.84) follows.

(iii) We have

$$\nabla \times (F_u \mathbf{e}_u) = \nabla \times (h_1 F_u \, \nabla_u) \quad \text{(using (4.88))}$$

$$= h_1 F_u \nabla \times (\nabla u) - \nabla u \times \nabla (h_1 F_u) \quad \text{(using identity(4.58))}$$

$$= \nabla (h_1 F_u) \times \nabla u \quad \text{(using identity(4.55))}$$

$$= \left[\frac{\mathbf{e}_u}{h_1} \frac{\partial}{\partial u} (h_1 F_u) + \frac{\mathbf{e}_v}{h_2} \frac{\partial}{\partial v} (h_1 F_u) + \frac{\mathbf{e}_w}{h_3} \frac{\partial}{\partial w} (h_1 F_u) \right] \times \frac{\mathbf{e}_u}{h_1}$$

$$= \frac{\mathbf{e}_v}{h_1 h_3} \frac{\partial}{\partial w} (h_1 F_u) - \frac{\mathbf{e}_w}{h_1 h_2} \frac{\partial}{\partial v} (h_1 F_u)$$

$$= \frac{1}{h_1 h_2 h_3} \begin{vmatrix} h_1 \mathbf{e}_u & h_2 \mathbf{e}_v & h_3 \mathbf{e}_w \\ \dfrac{\partial}{\partial u} & \dfrac{\partial}{\partial v} & \dfrac{\partial}{\partial w} \\ h_1 F_u & 0 & 0 \end{vmatrix}.$$

Similarly

$$\nabla \times (F_v \, \mathbf{e}_v) = \frac{1}{h_1 h_2 h_3} \begin{vmatrix} h_1 \mathbf{e}_u & h_2 \mathbf{e}_v & h_3 \mathbf{e}_w \\ \dfrac{\partial}{\partial u} & \dfrac{\partial}{\partial v} & \dfrac{\partial}{\partial w} \\ 0 & h_2 F_v & 0 \end{vmatrix}$$

and

$$\nabla \times (F_w \, \mathbf{e}_w) = \frac{1}{h_1 h_2 h_3} \begin{vmatrix} h_1 \mathbf{e}_u & h_2 \mathbf{e}_v & h_3 \mathbf{e}_w \\ \dfrac{\partial}{\partial u} & \dfrac{\partial}{\partial v} & \dfrac{\partial}{\partial w} \\ 0 & 0 & h_3 F_w \end{vmatrix}.$$

Addition of the three results gives formula (4.85).
   (iv) We have

$$\nabla^2 = \nabla \cdot \nabla = \nabla \cdot \left[\frac{\mathbf{e}_u}{h_1} \frac{\partial}{\partial u} + \frac{\mathbf{e}_v}{h_2} \frac{\partial}{\partial v} + \frac{\mathbf{e}_w}{h_3} \frac{\partial}{\partial w} \right].$$

Hence, applying (4.84) with

$$F_u, F_v, F_w \quad \text{replaced by} \quad \frac{1}{h_1} \frac{\partial}{\partial u}, \; \frac{1}{h_2} \frac{\partial}{\partial v}, \; \frac{1}{h_3} \frac{\partial}{\partial w}$$

respectively, formula (4.86) follows.

EXAMPLE 13

Find the curl and divergence of the vector field

$$\mathbf{H} = r^2 \cos \theta \, \mathbf{e}_r + (\mathbf{e}_\theta \sin \theta + \mathbf{e}_\varphi) / r \sin \theta$$

in spherical polar coordinates r, θ, φ.

Solution

In spherical polar coordinates

$$h_1 = 1, \quad h_2 = r, \quad h_3 = r \sin \theta.$$

Thus

$$\text{curl } \mathbf{H} = \frac{1}{r^2 \sin \theta} \begin{vmatrix} \mathbf{e}_r & r\mathbf{e}_\theta & r \sin \theta \, \mathbf{e}_\varphi \\ \partial/\partial r & \partial/\partial\theta & \partial/\partial\varphi \\ r^2 \cos \theta & 1 & 1 \end{vmatrix}$$

$$= r \sin \theta \, \mathbf{e}_\varphi.$$

Also

$$\text{div } \mathbf{H} = \frac{1}{r^2} \frac{\partial}{\partial r} (r^2 H_r) + \frac{1}{r \sin \theta} \frac{\partial}{\partial\theta} (H_\theta \sin \theta) + \frac{1}{r \sin \theta} \frac{\partial H_\varphi}{\partial \varphi}$$

$$= 4r \cos \theta + r^{-2} \cot \theta.$$

EXAMPLE 14

If $\psi(R, z)$ is a function of R and z only, in cylindrical polar coordinates R, φ, z, show that the vector field

$$\mathbf{H} = -\frac{1}{R} \frac{\partial \psi}{\partial z} \mathbf{e}_R + \frac{1}{R} \frac{\partial \psi}{\partial R} \mathbf{e}_z$$

satisfies the equations

$$\text{div } \mathbf{H} = 0$$

and

$$\text{curl } \mathbf{H} = \left(\frac{1}{R^2} \frac{\partial \psi}{\partial R} - \frac{1}{R} \frac{\partial^2 \psi}{\partial R^2} - \frac{1}{R} \frac{\partial^2 \psi}{\partial z^2} \right) \mathbf{e}_\varphi.$$

Solution

In cylindrical polar coordinates

$$h_1 = 1, \quad h_2 = R, \quad h_3 = 1.$$

Thus

$$\text{div } \mathbf{H} = \frac{1}{R} \frac{\partial}{\partial R} (RH_R) + \frac{1}{R} \frac{\partial H_\varphi}{\partial \varphi} + \frac{\partial H_z}{\partial z}$$

$$= -\frac{1}{R} \frac{\partial^2 \psi}{\partial R \, \partial z} + \frac{1}{R} \frac{\partial^2 \psi}{\partial z \, \partial R}$$

$$= 0$$

provided that the order of differentiation is immaterial. Further,

$$\text{curl } \mathbf{H} = \frac{1}{R} \begin{vmatrix} \mathbf{e}_R & R\,\mathbf{e}_\varphi & \mathbf{e}_z \\ \partial/\partial R & \partial/\partial \varphi & \partial/\partial z \\ -\dfrac{1}{R}\dfrac{\partial \psi}{\partial z} & 0 & \dfrac{1}{R}\dfrac{\partial \psi}{\partial R} \end{vmatrix}$$

$$= -\left[\frac{\partial}{\partial R}\left(\frac{1}{R}\frac{\partial \psi}{\partial R} \right) + \frac{1}{R}\frac{\partial^2 \psi}{\partial z^2} \right] \mathbf{e}_\varphi$$

giving the required result.

EXAMPLE 15

A vector field $\mathbf{F}$ is such that, in terms of cylindrical polar coordinates, R, φ, z,

$$\mathbf{F} = F_R\,\mathbf{e}_R + F_\varphi\,\mathbf{e}_\varphi$$

where F_R, F_φ are independent of z. Express $(\mathbf{F}\,.\boldsymbol{\nabla})\mathbf{F}$, $\frac{1}{2}\operatorname{grad} F^2$ and $\mathbf{F} \times \operatorname{curl}\mathbf{F}$ in terms of cylindrical polar components and verify that the identity

$$(\mathbf{F}\,.\boldsymbol{\nabla})\mathbf{F} = \tfrac{1}{2}\operatorname{grad} F^2 - \mathbf{F} \times \operatorname{curl}\mathbf{F} \qquad (4.89)$$

is satisfied.

Solution

In the case of cylindrical polar coordinates, $h_1 = 1$, $h_2 = R$, $h_3 = 1$ (equations (4.76)) so

$$\boldsymbol{\nabla} = \mathbf{e}_R \frac{\partial}{\partial R} + \mathbf{e}_\varphi \frac{\partial}{R\,\partial \varphi} + \mathbf{e}_z \frac{\partial}{\partial z}\,.$$

Hence

$$(\mathbf{F}\,.\boldsymbol{\nabla})\mathbf{F} = \left(F_R \frac{\partial}{\partial R} + F_\varphi \frac{\partial}{R\,\partial \varphi} \right)(F_R\,\mathbf{e}_R + F_\varphi\,\mathbf{e}_\varphi)$$

$$= F_R \frac{\partial F_R}{\partial R}\,\mathbf{e}_R + F_R^2 \frac{\partial \mathbf{e}_R}{\partial R} + F_R \frac{\partial F_\varphi}{\partial R}\,\mathbf{e}_\varphi + F_R F_\varphi \frac{\partial \mathbf{e}_\varphi}{\partial R}$$

$$+ F_\varphi \frac{\partial F_R}{R\,\partial \varphi}\,\mathbf{e}_R + F_\varphi F_R \frac{\partial \mathbf{e}_R}{R\,\partial \varphi} + F_\varphi \frac{\partial F_\varphi}{R\,\partial \varphi}\,\mathbf{e}_\varphi + F_\varphi^2 \frac{\partial \mathbf{e}_\varphi}{R\,\partial \varphi}\,.$$

From equations (4.77)

$$\mathbf{e}_R = (\cos \varphi,\ \sin \varphi,\ 0), \quad \mathbf{e}_\varphi = (-\sin \varphi,\ \cos \varphi,\ 0).$$

Hence

$$\frac{\partial \mathbf{e}_R}{\partial \varphi} = (-\sin \varphi, \; \cos \varphi, \; 0) = \mathbf{e}_\varphi,$$

$$\frac{\partial \mathbf{e}_\varphi}{\partial \varphi} = (-\cos \varphi, \; -\sin \varphi, \; 0) = -\mathbf{e}_R,$$

and the derivatives of $\mathbf{e}_R$ and $\mathbf{e}_\varphi$ with respect to R are zero. Thus

$$(\mathbf{F} \cdot \boldsymbol{\nabla})\mathbf{F} = \left(F_R \frac{\partial F_R}{\partial R} + F_\varphi \frac{\partial F_R}{R \, \partial \varphi} - \frac{F_\varphi^2}{R} \right) \mathbf{e}_R$$

$$+ \left(F_R \frac{\partial F_\varphi}{\partial R} + \frac{F_R F_\varphi}{R} + F_\varphi \frac{\partial F_\varphi}{R \, \partial \varphi} \right) \mathbf{e}_\varphi. \tag{4.90}$$

Also

$$\tfrac{1}{2} \, \text{grad} \, F^2 = \tfrac{1}{2} \left(\mathbf{e}_R \frac{\partial}{\partial R} + \mathbf{e}_\varphi \frac{\partial}{R \, \partial \varphi} + \mathbf{e}_z \frac{\partial}{\partial z} \right) (F_R^2 + F_\varphi^2)$$

$$= \left(F_R \frac{\partial F_R}{\partial R} + F_\varphi \frac{\partial F_\varphi}{\partial R} \right) \mathbf{e}_R + \left(F_R \frac{\partial F_R}{R \, \partial \varphi} + F_\varphi \frac{\partial F_\varphi}{R \, \partial \varphi} \right) \mathbf{e}_\varphi \tag{4.91}$$

since F_R, F_φ are independent of z.

Using formula (4.85)

$$\mathbf{F} \times \text{curl} \, \mathbf{F} = \mathbf{F} \times \frac{1}{R} \begin{vmatrix} \mathbf{e}_R & R\mathbf{e}_\varphi & \mathbf{e}_z \\ \dfrac{\partial}{\partial R} & \dfrac{\partial}{\partial \varphi} & \dfrac{\partial}{\partial z} \\ F_R & RF_\varphi & 0 \end{vmatrix}$$

$$= \mathbf{F} \times \frac{1}{R} \left[\frac{\partial}{\partial R} (RF_\varphi) - \frac{\partial F_R}{\partial \varphi} \right] \mathbf{e}_z$$

$$= (F_R \, \mathbf{e}_R + F_\varphi \, \mathbf{e}_\varphi) \times \left(\frac{\partial F_\varphi}{\partial R} + \frac{F_\varphi}{R} - \frac{\partial F_R}{R \, \partial \varphi} \right) \mathbf{e}_z$$

$$= \left(F_\varphi \frac{\partial F_\varphi}{\partial R} + \frac{F_\varphi^2}{R} - F_\varphi \frac{\partial F_R}{R \, \partial \varphi} \right) \mathbf{e}_R$$

$$+ \left(F_R \frac{\partial F_R}{R \, \partial \varphi} - \frac{F_R F_\varphi}{R} - F_R \frac{\partial F_\varphi}{\partial R} \right) \mathbf{e}_\varphi. \tag{4.92}$$

By substituting (4.90), (4.91) and (4.92) into (4.89) it is seen at once that the identity is satisfied. We note that (4.89) is a particular case of identity (4.51), namely the case when $\mathbf{F} \equiv \mathbf{G}$.

EXERCISES

4.42 When u, v, w are identical to the rectangular cartesian coordinates x, y, z, show that $h_1 = h_2 = h_3 = 1$. Verify that, in this case, expressions (4.82) – (4.86) reduce to the original expressions whereby grad Ω, div $\mathbf{F}$, curl $\mathbf{F}$ and ∇^2 were defined.

4.43 Find $\operatorname{grad}(R^2 z \sin \varphi \cos \varphi)$ at the point $R = 1$, $\varphi = \frac{1}{4}\pi$, $z = 2$.

4.44 The scalar field $\Omega(r, \theta, \varphi)$ is such that

$$\mathbf{r} \cdot \operatorname{grad} \Omega = n \Omega$$

where n is a constant. Show that Ω is of the form

$$\Omega = r^n f(\theta, \varphi).$$

4.45 Find the divergence of the vector field $R \cos \varphi \, \mathbf{e}_R + R \sin \varphi \, \mathbf{e}_\varphi$ in cylindrical polar coordinates R, φ, z.

4.46 In spherical polar coordinates find the components of $\operatorname{grad}(\operatorname{div} \mathbf{e}_\theta)$.

4.47 Working entirely in cylindrical polar coordinates verify that the vector field

$$\mathbf{F} = (R \sin \varphi \cos \varphi + z \cos \varphi) \, \mathbf{e}_R + (R \cos^2 \varphi - z \sin \varphi) \, \mathbf{e}_\varphi + R \sin \varphi \, \mathbf{e}_z$$

is such that curl curl $\mathbf{F} \equiv \mathbf{0}$.

4.48 In spherical polar coordinates, $\mathbf{F} = F_r \mathbf{e}_r + F_\theta \mathbf{e}_\theta$ where F_r, F_θ are independent of φ. Prove that curl curl $\mathbf{F}$ is of a similar form to $\mathbf{F}$.

4.49 Using spherical polar coordinates show that if n is constant

$$\nabla^2 r^n = n(n + 1) \, r^{n-2} \quad (r \neq 0 \text{ if } n \leq 2).$$

4.50 Show that, in cylindrical polar coordinates, Laplace's equation (4.37) is

$$\frac{\partial^2 \Omega}{\partial R^2} + \frac{1}{R} \frac{\partial \Omega}{\partial R} + \frac{1}{R^2} \frac{\partial^2 \Omega}{\partial \varphi^2} + \frac{\partial^2 \Omega}{\partial z^2} = 0.$$

4.51 If $\mathbf{F} = R \cos \varphi \, \mathbf{e}_R + \sin \varphi \, \mathbf{e}_\varphi$ evaluate $(\mathbf{F} \cdot \nabla)\mathbf{F}$.

4.52 Using cylindrical polar coordinates throughout, verify the identity

$$\operatorname{curl} \operatorname{curl} \mathbf{F} \equiv \operatorname{grad} (\operatorname{div} \mathbf{F}) - \nabla^2 \mathbf{F}$$

for the vector field $\mathbf{F} = \mathbf{e}_\varphi$.

4.14 VECTOR ANALYSIS IN n-DIMENSIONAL SPACE

The reader is probably aware of the possibility of generalizing certain concepts associated with three-dimensional space to spaces of higher dimensions. Such generalizations are not, by any means, valueless. For example, in relativity theory, distance and time are found to be inextricably linked and it is necessary to work within the framework of a four-dimensional coordinate system.

This is a convenient point in our exposition of vector analysis to explain how some of the ideas with which we have been concerned so far may be extended to spaces of higher dimensions. The digression will be brief and must necessarily, therefore, be somewhat superficial.

In an n-dimensional space, **points** are defined as ordered n-tuples of real numbers of the form $(x_1, x_2, \ldots, x_n)$. If the distance between two points $P(x_1, x_2, \ldots, x_n)$ and $Q(y_1, y_2, \ldots, y_n)$ is defined as

$$d = [(x_1 - y_1)^2 + (x_2 - y_2)^2 + \ldots + (x_n - y_n)^2]^{\frac{1}{2}}, \qquad (4.93)$$

the space is said to be a **Euclidean metric space of n-dimensions**; $(x_1, x_2, \ldots, x_n)$ are called the **rectangular cartesian coordinates** of P.

An orthogonal transformation of coordinates to a new system

$$(x_1', \ x_2', \ \ldots, \ x_n')$$

with the same origin as the original system may be defined in much the same way as was done for a three-dimensional space in Chapter 1. The transformation matrix becomes

$$\begin{pmatrix} l_{11} & l_{12} & \ldots & l_{1n} \\ l_{21} & l_{22} & \ldots & l_{2n} \\ \vdots & & & \vdots \\ l_{n1} & l_{n2} & \ldots & l_{nn} \end{pmatrix} \qquad (4.94)$$

and is such that

$$l_{ki}\, l_{kj} = \delta_{ij} \qquad (4.95)$$

and

$$x_i' = l_{ij}\, x_j. \qquad (4.96)$$

Observe that here we are using the summation convention and that each repeated suffix ranges from 1 to n.

The definition of a vector given in section 2.2 and the definitions of addition, subtraction, etc., extend in a natural way. The n-dimensional analogue of (2.3) is

$$\mathbf{a} = (a_1, a_2, \ldots, a_n). \qquad (4.97)$$

The basic unit vectors of the n-dimensional system are

$$\mathbf{e}_1 = (1, 0, 0, \ldots, 0)$$

$$\mathbf{e}_2 = (0, 1, 0, \ldots, 0)$$

$$\vdots \qquad\qquad (4.98)$$

$$\mathbf{e}_n = (0, 0, 0, \ldots, 1).$$

The scalar product of two vectors $\mathbf{a} = (a_1, a_2, \ldots, a_n)$ and $\mathbf{b} = (b_1, b_2, \ldots, b_n)$ is defined as

$$\mathbf{a} \cdot \mathbf{b} = a_1 b_1 + a_2 b_2 + \ldots + a_n b_n. \tag{4.99}$$

This may be shown to be invariant under a translation or rotation of the axes, as in the three-dimensional case.

There is, however, one concept which does not extend to spaces of dimensionality greater than three, namely that of the vector product. In three-dimensions there is just one unit vector normal to two given vectors (and associated with the given vectors by a right-hand screw rule) and it is essentially for this reason that it is possible to define a product of two vectors which is again a vector. When an attempt is made to generalise this to an n-dimensional space ($n > 3$) it is found that there are $n - 2$ unit vectors normal to two given vectors and no analogue of the vector product exists.

Some of the definitions introduced in the present chapter extend quite readily to spaces of higher dimensions. For example, the gradient of a scalar field Ω is defined as

$$\operatorname{grad} \Omega = \left(\frac{\partial \Omega}{\partial x_1}, \frac{\partial \Omega}{\partial x_2}, \ldots, \frac{\partial \Omega}{\partial x_n} \right). \tag{4.100}$$

Definition (4.13) of the directional derivative may also be extended in an obvious way, and it is easily shown that the relation (4.14) between the directional derivative and the gradient of a scalar field still holds. Further, the divergence of a vector field $\mathbf{F} = (F_1, F_2, \ldots, F_n)$ may be defined as

$$\operatorname{div} \mathbf{F} = \frac{\partial F_1}{\partial x_1} + \frac{\partial F_2}{\partial x_2} + \ldots + \frac{\partial F_n}{\partial x_n}. \tag{4.101}$$

However, the definition of curl $\mathbf{F}$ does not generalize as may be expected in view of its close relation to the vector product.

Enough has perhaps now been said to indicate the lines upon which some parts of vector analysis may be generalized to spaces of n-dimensions. There are many other extensions less obvious than those mentioned here but a full discussion lies well outside the scope of this book.

EXERCISES

4.53 Define the Laplacian in an n-dimensional rectangular cartesian coordinate system and verify that

$$\operatorname{div}(\operatorname{grad} \Omega) \equiv \nabla^2 \Omega.$$

4.54 Prove the identity

$$\operatorname{div}(\varphi \mathbf{F}) \equiv \varphi \operatorname{div} \mathbf{F} + \mathbf{F} \cdot \operatorname{grad} \varphi$$

for an n-dimensional space.

4.15 METHOD OF STEEPEST DESCENT

We now apply the results of this chapter in a simple way, introducing an important application of vector analysis in the form of the Method of Steepest Descent. Although this is introduced here as a practical example of the use of vector analysis in geometry, the basic idea is very powerful and will be found useful at all levels, including research. Volumes on operational research, optimization or the conjugate gradient method will give more advanced information.

Following equation (4.18) in section 4.5 we discussed the family of surfaces $\Omega(x, y, z) = c$, where c is a constant. It was shown that the gradient, grad Ω, is at right angles to the level surface through the field point. This is a particularly useful concept with wide applications. Here we discuss one of the most important of these applications, namely the Method of Steepest Descent, which is used for locating the places where a function has a minimum value. Since one of the purposes is to link vector analysis with geometry, the method will be formulated in n-dimensions. Then the more general form of the equation of the family of surfaces $\Omega(x, y, z) = c$ is

$$f(x_1, x_2, \ldots, x_n) = c \qquad (4.102)$$

where $(x_1, x_2, \ldots, x_n)$ are the n coordinates.

The name of the method springs from the property, demonstrated earlier, that a function f varies most rapidly in the direction of grad f. Thus, from any starting point within the valley which contains the minimum, the direction of steepest descent is known and can in principle be followed stepwise until the lowest point of the n-dimensional valley is reached. At that point f will have a minimum. The method can also be used to find a maximum value of f by replacing it by $-f$ and finding its minimum.

The Method of Steepest Descent works very well in practice and no scientist or engineer should be without some experience of it. The method can also be exploited in other ways. It is one of the few totally reliable techniques available, for example, in numerically determining the roots of complex analytic functions.

Consider first the case of two dimensions. Close to a minimum of $f(x, y)$ the level contours $f(x, y) = c$ consist of closed loops surrounding the minimum point P_{MIN} which has coordinates x_{MIN}, y_{MIN}. These will be ovals and may be sketched as ellipses (Fig 4.10). The surface $z = f(x, y)$ is shown in Fig.4.11 with the z-axis conventionally vertical but not shown. The minimum point P_{MIN} lies at the lowest point of the valley, as shown.

To find the point P_{MIN} (x_{MIN}, y_{MIN}) a search may be carried out numerically or otherwise. Many variations may be envisaged to suit the user's individual requirements. The simplest and most time-consuming method is to proceed stepwise along the path of steepest descent, shown as a curved line joining the starting point A to the finishing point P_{MIN} on Fig.4.10. A much shorter method, in practice, is as follows.

By some means obtain a first approximation $A(x_A, y_A)$ and determine the line l of steepest descent to the level contour through A. This is shown in Fig. 4.10. The line will be in one or other of the directions $\pm \operatorname{grad} f$ and along it f will have a minimum at the point B at which l is tangential to another level contour lying inside the first. This point (x_B, y_B) is the next approximation to (x_{MIN}, y_{MIN}). Repeat the process until sufficient accuracy is achieved. A few steps $A, B, C,...$ of the process are shown in Fig.4.10.

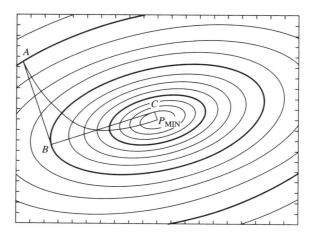

Fig. 4.10 Level surfaces $f(x, y) = c$ near a minimum.

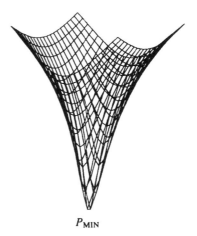

P_{MIN}

Fig. 4.11 The minimum point P_{MIN} of f shown on the surface $z = f(x, y)$.

Steepest Descent solution of linear equations

Since many students will have some experience of matrix notation, from methods and linear algebra courses or from sixth form studies, it is convenient here to mix the notations of vector analysis and linear algebra in an uncritical way. This procedure is used only for the purposes of the present section. Thus, depending on the context, we may freely use either of the vector notations

$$\mathbf{x} = (x_1, \ x_2, \ldots, \ x_n) \ \text{ or } \ \mathbf{x} = \begin{pmatrix} x_1 \\ x_2 \\ x_n \end{pmatrix}. \tag{4.103}$$

In the latter case we would write $\mathbf{x}^T = (x_1, \ x_2, \ldots, \ x_n)$. The meanings will be clear from the context. Note that

$$\mathbf{a}^T \mathbf{b} = a_1 \, b_1 + a_2 \, b_2 + \ldots + a_n \, b_n \quad \text{and} \quad \mathbf{x}^T \mathbf{x} = x_1^2 + x_2^2 + \ldots + x_n^2 .$$

Consider any *symmetric* set of linear equations *with positive diagonal* such as

$$A\mathbf{x} = \mathbf{b}. \tag{4.104}$$

Then if

$$c = \tfrac{1}{2} \mathbf{x}^T A \mathbf{x} - \mathbf{x}^T \mathbf{b} \tag{4.105}$$

we see that in n-dimensions the gradient is

$$\text{grad } c = A\mathbf{x} - \mathbf{b}. \tag{4.106}$$

However, it is usual to call this vector quantity the **residue vector R** rather than the gradient. So at this point we define the residue vector by

$$\mathbf{R} = A\mathbf{x} - \mathbf{b}. \tag{4.107}$$

The form of (4.106) shows that the solution of the linear equations (4.104) lies at the minimum value of c. The surfaces given by (4.105), corresponding to constant values of c, form a family of nested quadrics similar to the ellipses of Fig. 4.10 with their common centre at the solution of (4.104). This may be determined numerically by using the Method of Steepest Descent along with expression (4.107), the residue vector, for the direction of steepest descent.

Linear algebra tends to be dominated by notation, and the meaning of equations (4.104) to (4.106) will become clearer when particular numerical examples are considered. Often the geometrical reasoning is clearer in a coordinate system Oxy, or $Oxyz$.

EXAMPLE 16

By way of illustration, show that the linear equations (whose solution is obviously $x = 1$, $y = 1$)

$$2x - y = 1$$

$$-x + y = 0 \qquad (4.108)$$

can be solved by the Method of Steepest Descent.

Solution

The equations correspond to

$$A = \begin{pmatrix} 2 & -1 \\ -1 & 1 \end{pmatrix}, \quad \mathbf{b} = \begin{pmatrix} 1 \\ 0 \end{pmatrix},$$

but the positive diagonal and the symmetry are obvious without writing out A and $\mathbf{b}$ explicitly. Using (4.105) the set of nested quadrics is the family of conic sections given by

$$\tfrac{1}{2} \{ x(2x - y) + y(-x + y) \} - x = c,$$

that is

$$\tfrac{1}{2} (2x^2 - 2xy + y^2) - x = c.$$

By partial differentiation

$$\operatorname{grad} c = (2x - y - 1, \; -x + y)$$

which, as may be readily verified, agrees with the general expression (4.106). The minimization could be carried out directly from here but some useful general formulae are given below.

Practical details

It is convenient to work in terms of the residue vector defined by (4.107). Then, after some working which will not be given here, it is found that the parametric vector equation of the normal at the point $\mathbf{x} = \mathbf{p}$ (say) to the quadric surface in n-dimensions passing through the point $\mathbf{p} = (p_1, p_2, \ldots, p_n)$ is

$$\mathbf{x} = \mathbf{p} + s\,\mathbf{R}_p, \qquad (4.109)$$

where s is a parameter related to the arc length along the normal and where $\mathbf{R}_p = A\mathbf{p} - \mathbf{b}$. Moreover, it can be shown that the minimum value of c, from (4.105), along the normal (4.109), is achived when s assumes the value

$$s = -\mathbf{R}_p^T \mathbf{R}_p / (\mathbf{R}_p^T A \mathbf{R}_p). \tag{4.110}$$

The normal is tangential to another of the quadric surfaces at the value of s given by (4.110). For the example given above the reader should verify using (4.107) that if $\mathbf{R} = (R_1, R_2)$ and $\mathbf{x} = (p_1, p_2)$ then

$$R_1 = 2p_1 - p_2 - 1, \quad R_2 = -p_1 + p_2 \tag{4.111}$$

and

$$s = -\frac{R_1^2 + R_2^2}{2R_1^2 - 2R_1 R_2 + R_2^2}. \tag{4.112}$$

Hence if $\mathbf{x} = (p_1, p_2)$ is an approximation to the solution of the equation $A\mathbf{x} = \mathbf{b}$ then a better approximation is $\mathbf{x} = (x_1, x_2)$ where

$$x_1 = p_1 + sR_1, \quad x_2 = p_2 + sR_2 \tag{4.113}$$

and R_1, R_2 and s are given by equations (4.111) and (4.112). We see that we have here the basis of a purely algebraic method of obtaining successive approximations to the solution of the original set of linear equations.

To begin the iterative procedure we have to make an initial guess. Since convergence is here guaranteed by the nature of the method itself, $(0, 0)$ is chosen. Starting from that point the successive approximations are shown as $A, B, C, D, E, F, G, \ldots$ in Fig. 4.12 and given numerically in Table 4.1.

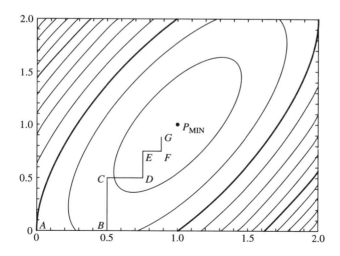

Fig. 4.12

Table 4.1 Giving the details of successive numerical steps starting from $(0, 0)$ and converging towards $(1, 1)$

Guess	R_1	R_2	s	New
$(0, 0)$	-1	0	$-\frac{1}{2}$	$(\frac{1}{2}, 0)$
$(\frac{1}{2}, 0)$	0	$-\frac{1}{2}$	-1	$(\frac{1}{2}, \frac{1}{2})$
$(\frac{1}{2}, \frac{1}{2})$	$-\frac{1}{2}$	0	$-\frac{1}{2}$	$(\frac{3}{4}, \frac{1}{2})$
$(\frac{3}{4}, \frac{1}{2})$	0	$-\frac{1}{4}$	-1	$(\frac{3}{4}, \frac{3}{4})$
$(\frac{3}{4}, \frac{3}{4})$	$-\frac{1}{4}$	0	$-\frac{1}{2}$	$(\frac{7}{8}, \frac{3}{4})$
$(\frac{7}{8}, \frac{3}{4})$	0	$-\frac{1}{8}$	-1	$(\frac{7}{8}, \frac{7}{8})$

The method slows down a great deal in the case of a long, curved, narrow valley – successive approximations will zigzag and converge more slowly. Signs of this can be seen in the diagram. In spite of this the method will often be found to be excellent both analytically and numerically for many problems in science and engineering. Enhanced convergence may be obtained by using the method of conjugate gradients; however, this lies beyond our present aims.

EXERCISES

4.55 Consider any member $\mathscr{E}$ of the family of ellipses

$$ax^2 + by^2 = d$$

where a and b are fixed positive numbers and d is the family parameter. Show vectorially that at (x_1, y_1) the vector parametric equations of the normal to $\mathscr{E}$ are

$$x = (1 + as)x_1, \quad y = (1 + bs)y_1$$

where s is a parameter related to distance along the normal. Show that this normal cuts the ellipse $ax^2 + by^2 = e$ at the two values of s which satisfy

$$ax_1^2(1 + as)^2 + by_1^2(1 + bs)^2 = e$$

and that when these two values coincide

$$e = d - \frac{(a^2x_1^2 + b^2y_1^2)^2}{a^3x_1^2 + b^3y_1^2}.$$

Use these results to discuss the convergence of the Method of Steepest Descent. Show that in the case $a = b$ convergence is achieved in a single step and explain geometrically why this happens.

4.56 By deriving the components of the vector parametric equation of the normal in the form

$$x = (1 + as)x_1, \quad y = (1 + bs)y_1, \quad z = (1 + cs)z_1,$$

generalize Exercise 4.55 to the family of ellipsoids

$$ax^2 + by^2 + cz^2 = d.$$

Derive further the expression for e in the form

$$e = d - \frac{(a^2x_1^2 + b^2y_1^2 + c^2z_1^2)^2}{a^3x_1^2 + b^3y_1^2 + c^3z_1^2}.$$

Show that in the case $a = b = c$, convergence is achieved in a single step and explain why.

4.57 Show that the linear equations

$$4x + y = 5, \quad x + y = 2$$

can be written in the form $A\mathbf{x} = \mathbf{b}$ where A is a symmetric 2×2 matrix with positive diagonal. By finding where $\text{grad}\, c = \mathbf{0}$ for the quadratic form c defined by

$$c = \tfrac{1}{2}(4x^2 + 2xy + y^2) - 5x - 2y,$$

show that the solution of the linear equations lies at the minimum of c. Use Fig. 4.13 to trace a few lines of steepest descent.

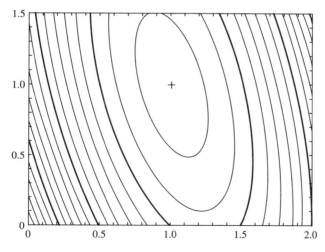

Fig. 4.13 Showing the level curves $c = $ constant for Exercise 4.57.

4.58 Verify that if $c = x^2 - y^2$, then $\operatorname{grad} c = \mathbf{0}$ gives the (simple) linear equations $x = 0$, $y = 0$. Show that the curve of steepest descent through the point $(-1, y_0)$ does not pass through the solution of these equations unless $y_0 = 0$. Alter c in such a way that convergence to $(0, 0)$ of the method of steepest descent is assured.

4.59 Verify by computer that for the quadratic form $0.001 x^2 + y^2$ the convergence of the Method of Steepest Descent, starting from the point $(1, 0.001)$, is very slow (Hestes, Magnus R. (1980) *Conjugate Gradient Methods in Optimization*, Springer-Verlag, Berlin, p. 1).

Line, surface and volume integrals

In this chapter the concept of the Riemann integral of a function of a single real variable is extended to functions of three real variables.

5.1 LINE INTEGRAL OF A SCALAR FIELD

Let $\mathscr{C}$ be a piecewise smooth curve with intrinsic equation

$$\mathbf{r} = \mathbf{r}(s) = (x(s), y(s), z(s)), \ 0 \leq s \leq l, \tag{5.1}$$

and let $\Omega(x, y, z)$ be a scalar field defined at all points on $\mathscr{C}$. Then, on $\mathscr{C}$, Ω is a function of arc length s only. We define

$$I = \int_{s_1}^{s_2} \Omega(x(s), y(s), z(s)) \, ds, \tag{5.2}$$

where $0 \leq s_1 \leq s_2 \leq l$, to be the **line integral** of Ω along the curve $\mathscr{C}$ from $s = s_1$ to $s = s_2$ provided, of course, that the integral exists. An alternative notation is

$$I = \int_{P_1}^{P_2} \Omega(s) \, ds, \tag{5.3}$$

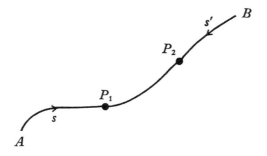

Fig. 5.1

where P_1, P_2, are the points on $\mathscr{C}$ corresponding to s_1, s_2.

The line integral of a scalar field along a curve $\mathscr{C}$ is independent of the sense of description (or orientation) of $\mathscr{C}$.

Proof

Let A, B be the end points of $\mathscr{C}$ corresponding to $s = 0$, $s = l$ respectively. Put
$$s' = l - s. \tag{5.4}$$
Then s' is arc length measured from B (Fig. 5.1). When s increases from 0 to l the curve is described in the sense A to B, and when s' increases from 0 to l the sense of description is from B to A.

Taking points P_1, P_2 on $\mathscr{C}$ corresponding to $s = s_1, s = s_2$
$$\int_{P_1}^{P_2} \Omega \, ds = \int_{s_1}^{s_2} \Omega \, ds$$
$$= \int_{l - s_1}^{l - s_2} - \Omega \, ds', \quad \text{using (5.4)}$$
$$= \int_{l - s_2}^{l - s_1} \Omega \, ds'.$$

But $l - s_2, l - s_1$ are the values of s' at P_2, P_1, respectively. Hence
$$\int_{P_1}^{P_2} \Omega \, ds = \int_{P_2}^{P_1} \Omega \, ds'$$

showing that the line integral does not depend on the sense of description of $\mathscr{C}$.

EXAMPLE 1

The curve $\mathscr{C}$ is defined by the intrinsic equation
$$\mathbf{r} = (1, \ \sinh^{-1}s, \ (1 + s^2)^{\frac{1}{2}}), \ \ 0 \leqslant s \leqslant 1.$$
Evaluate the line integral of
$$\Omega = z^2 - x^2$$
along $\mathscr{C}$ from $s = 0$ to $s = 1$.

Solution

On $\mathscr{C}$, $\Omega = (1 + s^2) - 1 = s^2$. Thus the required line integral is
$$\int_0^1 s^2 \, ds = \tfrac{1}{3}.$$

Integral around a closed curve

Let $\mathscr{C}$ be a simple closed curve of total length l. The line integral

$$\int_0^l \Omega \, ds,$$

taken all the way round $\mathscr{C}$, is usually denoted by

$$\oint_{\mathscr{C}} \Omega \, ds. \qquad (5.5)$$

We have already shown that the line integral of a scalar field along a curve is independent of the sense of description of the curve. *When the curve is closed, the line integral is also independent of the point from which arc length is measured.*

Proof

Let P, Q be different points on a closed curve $\mathscr{C}$. Let s denote arc length measured from P and let s' denote arc length measured from Q (Fig. 5.2). Then at any point

$$s = s' + \text{constant}. \qquad (5.6)$$

Hence

$$\oint_{\mathscr{C}} \Omega \, ds = \int_P^Q \Omega \, ds + \int_Q^P \Omega \, ds$$

$$= \int_P^Q \Omega \, ds' + \int_Q^P \Omega \, ds', \quad \text{using (5.6)}$$

$$= \oint_{\mathscr{C}} \Omega \, ds'$$

which proves the result stated.

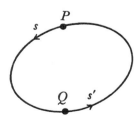

Fig. 5.2

EXAMPLE 2

Evaluate the line integral of

$$\Omega = (a^2 y^2/b^2 + b^2 x^2/a^2)^{\frac{1}{2}}$$

around the ellipse $\mathscr{C}$ with equation

$$x^2/a^2 + y^2/b^2 = 1, \quad z = 0.$$

Solution

The parametric equations of the ellipse can be taken as

$$x = a \cos \theta, \quad y = b \sin \theta, \quad z = 0, \quad 0 \leqslant \theta \leqslant 2\pi.$$

If s denotes arc length, we have

$$\frac{ds}{d\theta} = \left\{ \left(\frac{dx}{d\theta}\right)^2 + \left(\frac{dy}{d\theta}\right)^2 + \left(\frac{dz}{d\theta}\right)^2 \right\}^{\frac{1}{2}} = (a^2 \sin^2 \theta + b^2 \cos^2 \theta)^{\frac{1}{2}}.$$

Also, on $\mathscr{C}$,

$$\Omega = (a^2 \sin^2 \theta + b^2 \cos^2 \theta)^{\frac{1}{2}}.$$

Hence

$$\oint_{\mathscr{C}} \Omega \, ds = \int_0^{2\pi} \Omega(\theta) \frac{ds}{d\theta} \, d\theta$$

$$= \int_0^{2\pi} (a^2 \sin^2 \theta + b^2 \cos^2 \theta) \, d\theta$$

$$= \pi (a^2 + b^2).$$

EXAMPLE 3

Evaluate the line integral of

$$\Omega = x^2 + y^2$$

around the triangle $\mathscr{C}$ with vertices at the origin O and the points $A(1, 0, 0)$, $B(0, 1, 0)$ (Fig.5.3)

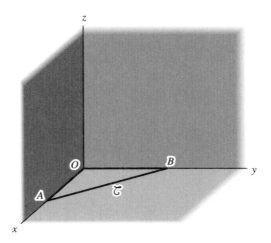

Fig. 5.3

Solution

The integral is evaluated in three parts.

(a)
$$\int_O^A \Omega\,ds = \int_0^1 x^2\,dx = \tfrac{1}{3}.$$

(b)
$$\int_O^B \Omega\,ds = \int_0^1 y^2\,dy = \tfrac{1}{3}.$$

(c) The equations of AB are

$$x + y = 1, \quad z = 0.$$

Hence we can take

$$x = 1 - t, \quad y = t, \quad z = 0,$$

where t is a parameter which varies from 0 at A to 1 at B. If s denotes arc length of AB

$$\frac{ds}{dt} = \left\{ \left(\frac{dx}{dt}\right)^2 + \left(\frac{dy}{dt}\right)^2 + \left(\frac{dz}{dt}\right)^2 \right\}^{\frac{1}{2}} = \sqrt{2}.$$

Also, on AB,

$$\Omega = (1 - t)^2 + t^2.$$

Thus

$$\int_A^B \Omega\,ds = \int_0^1 \Omega\frac{ds}{dt}\,dt$$

$$= \int_0^1 \{(1 - t)^2 + t^2\}\sqrt{2}\,dt = \tfrac{2}{3}\sqrt{2}.$$

Combining (a), (b), and (c) now gives

$$\oint_{\mathscr{C}} \Omega \, ds = \int_{O}^{A} \Omega \, ds + \int_{O}^{B} \Omega \, ds + \int_{A}^{B} \Omega \, ds$$

$$= \tfrac{2}{3}(1 + \sqrt{2}).$$

EXERCISES

5.1 Evaluate the line integral

$$\int_{0}^{2\pi a} (x + y + z) \, ds$$

along the curve with intrinsic equation

$$\mathbf{r} = (\tfrac{4}{5})^{\frac{1}{2}}(a \cos(s/2a), \, a \sin(s/2a), \, s).$$

5.2 A straight line is drawn parallel to the y-axis through the points $P(1, 0, 0), Q(1, 3, 0)$. Show that

$$\int_{P}^{Q} (x^2 + y^2 + z^2) \, ds = 12.$$

[*Hint.* In this case $y = s$.]

5.3 Evaluate

$$\int_{P}^{Q} (x + y + z) \, ds$$

along the straight line joining the points $P(1, 2, 3)$ and $Q(4, 5, 6)$.

5.4 Evaluate

$$\oint_{\mathscr{C}} \Omega \, ds,$$

where $\Omega = y^2 + z^2$ and $\mathscr{C}$ is the circle $x^2 + y^2 = a^2, \ z = 0$.

5.5 A closed curve $\mathscr{C}$ is formed as follows: first, the origin O is joined to the point $A(1, 0, 0)$ by a straight line, then the point A is joined to the point $B(1, 1, 1)$ by part of the curve with parametric equation $\mathbf{r} = (1, t, t^2)$, finally the point B is joined to the origin O by a straight line. Evaluate

$$\oint_{\mathscr{C}} xy \, ds.$$

5.2 LINE INTEGRALS OF A VECTOR FIELD

Scalar line integrals

Let the vector field $\mathbf{F}(x, y, z)$ be defined at all points on the piecewise smooth curve $\mathscr{C}$ given by equation (5.1). If $\hat{\mathbf{T}}$ denotes the unit tangent to $\mathscr{C}$ we define

$$I = \int_0^l \mathbf{F} . \hat{\mathbf{T}} \, ds \tag{5.7}$$

to be the **scalar line integral** of $\mathbf{F}$ along $\mathscr{C}$ provided, of course, that the integral exists. Since $\hat{\mathbf{T}} = d\mathbf{r}/ds$ it is usual to put

$$\hat{\mathbf{T}} \, ds = d\mathbf{r}. \tag{5.8}$$

Thus, (5.7) becomes

$$I = \int_0^l \mathbf{F} . d\mathbf{r}. \tag{5.9}$$

If $\mathscr{C}$ is a simple closed curve, the line integral around $\mathscr{C}$ is denoted by

$$K = \oint_{\mathscr{C}} \mathbf{F} . d\mathbf{r}; \tag{5.10}$$

K is called the **circulation** of $\mathbf{F}$ around $\mathscr{C}$.

   When the sense of description of a curve is reversed the direction of the unit tangent $\hat{\mathbf{T}}$ is also reversed. Thus, *the scalar line integral of a vector field along a curve changes sign when the sense of description of the curve is changed.*

EXAMPLE 4

Evaluate the scalar line integral of

$$\mathbf{F} = (z, x, y)$$

around the circle $x^2 + y^2 = 0$, $z = 0$, described in the clockwise sense relative to an observer looking along the positive z-axis.

Solution

The parametric equation of the circle $\mathscr{C}$ may be taken as

$$\mathbf{r} = (a \cos \theta, \, a \sin \theta, \, 0), \; 0 \leqslant \theta \leqslant 2\pi,$$

the range of θ being chosen so that the sense of description is as required.

On $\mathscr{C}$,

$$\mathbf{F} = (0,\, a \cos \theta,\, a \sin \theta).$$

Also,

$$\frac{d\mathbf{r}}{d\theta} = (-a \sin \theta,\, a \cos \theta,\, 0).$$

Thus,

$$\oint_{\mathscr{C}} \mathbf{F} \cdot d\mathbf{r} = \int_0^{2\pi} \mathbf{F} \cdot \frac{d\mathbf{r}}{d\theta}\, d\theta$$

$$= \int_0^{2\pi} a^2 \cos^2 \theta\, d\theta$$

$$= \pi a^2.$$

EXAMPLE 5

Show that the circulation of any constant vector field $\mathbf{A}$ around any simple closed curve $\mathscr{C}$ is zero.

Solution

Choose the x-axis parallel to $\mathbf{A}$. Then

$$\mathbf{A} \cdot d\mathbf{r} = (A, 0, 0) \cdot (dx, dy, dz)$$

$$= A\, dx.$$

Thus

$$\oint_{\mathscr{C}} \mathbf{A} \cdot d\mathbf{r} = A \oint_{\mathscr{C}} dx.$$

If P_1, P_2 are points on $\mathscr{C}$ corresponding to $x = x_1, x = x_2$ respectively, then integrating along $\mathscr{C}$,

$$\int_{P_1}^{P_2} dx = x_2 - x_1.$$

Allowing the point P_2 to move around $\mathscr{C}$ until it coincides with P_1, we have $x_2 = x_1$, and hence

$$\oint_{\mathscr{C}} dx = \int_{P_1}^{P_2} dx = 0.$$

The result required follows at once.

Vector line integrals

If the components F_1, F_2, F_3 of the vector field

$$\mathbf{F} = F_1\,\mathbf{i} + F_2\,\mathbf{j} + F_3\,\mathbf{k}$$

are integrable along a curve $\mathscr{C}$, the **vector line integral** of $\mathbf{F}$ along $\mathscr{C}$, from $s = s_1$ to $s = s_2$, is defined as

$$\int_{s_1}^{s_2} \mathbf{F}\,ds = \mathbf{i}\int_{s_1}^{s_2} F_1\,ds + \mathbf{j}\int_{s_1}^{s_2} F_2\,ds + \mathbf{k}\int_{s_1}^{s_2} F_3\,ds. \qquad (5.11)$$

The unit tangent $\hat{\mathbf{T}}$ to a curve $\mathscr{C}$ is itself a vector field defined on $\mathscr{C}$. Thus, we may use (5.11) to evaluate a **vector line integral of a scalar field** Ω, which is defined as

$$\int \Omega\,d\mathbf{r} = \int \Omega\hat{\mathbf{T}}\,ds. \qquad (5.12)$$

It is clear that $\hat{\mathbf{T}}$ may also appear as part of a vector product, leading to line integrals of the form

$$\int \mathbf{F}\times d\mathbf{r} = -\int d\mathbf{r}\times\mathbf{F} = \int \mathbf{F}\times\hat{\mathbf{T}}\,ds. \qquad (5.13)$$

The reader should experience little difficulty in working with these new forms, as they appear in the following exercises.

EXERCISES

5.6 If $\mathbf{F} = (x, 2y, 3z)$, evaluate

(i) $\displaystyle\int_O^A \mathbf{F}\cdot d\mathbf{r}$, (ii) $\displaystyle\int_O^A \mathbf{F}\times d\mathbf{r}$, (iii) $\displaystyle\int_O^A \mathbf{F}\,ds$,

along the curve $\mathbf{r} = (t, \frac{1}{\sqrt{2}}t^2, \frac{1}{3}t^3)$ from the origin O to the point $A(1, \frac{1}{2}\sqrt{2}, \frac{1}{3})$.

5.7 Evaluate

$$\int z\,d\mathbf{r}$$

along the curve $\mathbf{r} = (a\cos t,\ b\sin t,\ ct)$ from the point $t = 0$ to the point $t = 2\pi$.

5.8 If $\mathbf{F} = (y\mathbf{i} - x\mathbf{j})/(x^2 + y^2)$, and $\mathscr{C}$ is the circle $x^2 + y^2 = a^2$ in the xy-plane described in the anticlockwise sense, evaluate

$$\oint_{\mathscr{C}} \mathbf{F}\cdot d\mathbf{r}.$$

5.9 Evaluate

$$\oint_{\mathscr{C}} \mathbf{F} \cdot d\mathbf{r},$$

where $\mathbf{F} = (x - 3y,\ y - 2x,\ 0)$ and $\mathscr{C}$ is the perimeter of the ellipse $x^2/9 + y^2/4 = 1$ in the xy-plane, described in the anticlockwise sense.

5.10 Evaluate

$$\int_{\mathscr{C}} \mathbf{r} \times d\mathbf{r} \quad \text{and} \quad \int_{\mathscr{C}} \mathbf{r}\, ds$$

from the point $(a, 0, 0)$ to the point $(a, 0, 2\pi b)$ on the circular helix

$$\mathbf{r} = (a \cos t,\ a \sin t,\ bt).$$

5.3 REPEATED INTEGRALS

Before discussing surface and volume integrals, it is necessary to introduce the idea of repeated integration and (in the next section) double and triple integrals.

Let $p(x)$, $q(x)$ be functions of x, and consider the following integral of a function $f(x, y)$

$$I(x) = \int_{p(x)}^{q(x)} f(x, y)\, dy.$$

Now integrate $I(x)$ between the limits $x = a$, $x = b$. We write

$$\int_a^b \left\{ \int_{p(x)}^{q(x)} f(x, y)\, dy \right\} dx = \int_a^b \int_{p(x)}^{q(x)} f(x, y)\, dy\, dx. \qquad (5.14)$$

Such integrals are called **repeated integrals**.

Two important properties are:

(i) If $p = c$, $q = d$, where c, d are constants, then

$$\int_a^b \int_c^d f(x, y)\, dy\, dx = \int_c^d \int_a^b f(x, y)\, dx\, dy; \qquad (5.15)$$

i.e. the order of integration is immaterial when the limits of integration are constants.

(ii) If $f(x, y) = \varphi(x)\psi(y)$, then

$$\int_a^b \int_c^d \varphi(x)\psi(y)\, dy\, dx = \left(\int_a^b \varphi(x)\, dx\right)\left(\int_c^d \psi(y)\, dy\right); \qquad (5.16)$$

i.e. when the integrand is a *separable* function of x and y, and the limits of integration are constants, a repeated integral separates into the product of two single integrals. The first property is proved in the next section. The second property may be proved by observing that the first integration (with respect to y) does not affect $\varphi(x)$.

EXAMPLE 6

Evaluate

$$I = \int_0^1 \int_{\frac{1}{2}x}^{\frac{1}{2}} xy\, dy\, dx.$$

Solution

Performing the integration with respect to y, we have

$$I = \int_0^1 \left[\tfrac{1}{2} xy^2\right]_{y=\frac{1}{2}x}^{y=\frac{1}{2}} dx$$

$$= \int_0^1 (\tfrac{1}{8}x - \tfrac{1}{8}x^3)\, dx$$

$$= \tfrac{1}{32}.$$

EXAMPLE 7

Evaluate

$$I = \int_0^1 \int_0^2 (x^2 + y)\, dy\, dx,$$

and verify that the same result is obtained when the order of integration is inverted.

Solution

We have

$$I = \int_0^1 [x^2 y + \tfrac{1}{2} y^2]_{y=0}^{y=2}\, dx$$

$$= \int_0^1 (2x^2 + 2)\, dx = \tfrac{8}{3}.$$

Inverting the order of integration,

$$I = \int_0^2 \int_0^1 (x^2 + y) \, dx \, dy$$

$$= \int_0^2 [\tfrac{1}{3}x^3 + xy]_{x=0}^{x=1} \, dy$$

$$= \int_0^2 (\tfrac{1}{3} + y) \, dy = \tfrac{8}{3}, \quad \text{as before.}$$

EXERCISES

5.11 Show that

$$\int_0^{\frac{1}{2}\pi} \int_0^{\frac{1}{6}\pi} \sin x \cos y \, dx \, dy = 1 - \tfrac{1}{2}\sqrt{3}.$$

5.12 Show that

$$\int_0^1 \int_0^{1-x} (x + y)^2 \, dy \, dx = \tfrac{1}{4}.$$

5.13 Show that

$$\int_0^\pi \int_0^{\sin y} dx \, dy = 2.$$

5.4 DOUBLE AND TRIPLE INTEGRALS

The Riemann integral

We first recall the following definition of the Riemann integral of a function $f(x)$.

Let $f(x)$ be defined in the interval $a \leqslant x \leqslant b$. Subdivide the interval into m sub-intervals of lengths $\delta x_1, \delta x_2, \ldots, \delta x_m$ and let $x_1, x_2, \ldots, x_m$ be any points within the respective sub-intervals. Let the method of subdivision be such that $\delta x_1, \delta x_2, \ldots, \delta x_m$ are all less than some quantity ε_m such that

$$\varepsilon_m \to 0 \quad \text{as} \quad m \to \infty. \tag{5.17}$$

Then the Riemann integral of $f(x)$ over the interval $a \leqslant x \leqslant b$ is defined as

$$\lim_{m \to \infty} \sum_{r=1}^{m} f(x_r)\, \delta x_r = \int_a^b f(x)\, dx, \tag{5.18}$$

provided that the limit exists and is the same for all methods of subdivision satisfying (5.17).

Double integrals

Let $f(x, y)$ be defined in the rectangle $a \leqslant x \leqslant b$, $c \leqslant y \leqslant d$. Subdivide the interval $a \leqslant x \leqslant b$ into m sub-intervals of lengths $\delta x_1, \delta x_2, \ldots, \delta x_m$; and subdivide the interval $c \leqslant y \leqslant d$ into n sub-intervals of lengths $\delta y_1, \delta y_2, \ldots, \delta y_n$. Consider the sum

$$S_{mn} = \sum_{r=1}^{m} \sum_{s=1}^{n} f(x_r, y_s)\, \delta x_r\, \delta y_s, \tag{5.19}$$

where (x_r, y_s) is any point inside the sub-rectangle of area $\delta x_r \delta y_s$ (Fig. 5.4). Choose any method of subdivision such that $\delta x_1, \ldots, \delta x_m$ are all less that ε_m and $\delta y_1, \ldots, \delta y_n$ are all less than η_n, where

$$\left.\begin{aligned} \varepsilon_m &\to 0 \quad \text{as} \quad m \to \infty \\ \eta_n &\to 0 \quad \text{as} \quad n \to \infty. \end{aligned}\right\} \tag{5.20}$$

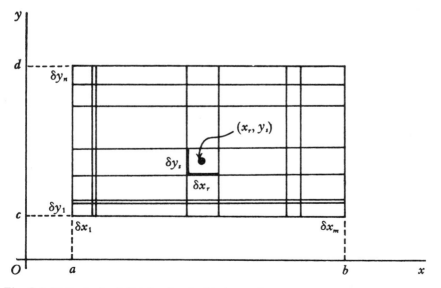

Fig. 5.4 Method of subdivision for double integrals

We then define

$$\lim_{\substack{m \to \infty \\ n \to \infty}} S_{mn} = \iint_{\mathcal{R}} f(x, y) \, dx \, dy \qquad (5.21)$$

as the double integral of $f(x, y)$, where $\mathcal{R}$ denotes the rectangle $a \leqslant x \leqslant b$, $c \leqslant y \leqslant d$, provided that the limit exists and is the same for all methods of subdivision satisfying (5.20).

The double integral defined in this way exists in nearly all cases of practical interest.

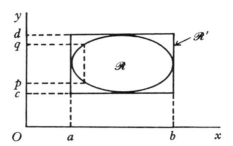

Fig. 5.5 Double integration over an non-rectangular region.

The double integral of $f(x, y)$ over a closed region $\mathcal{R}$ which is not rectangular is defined as follows. Let $\mathcal{R}'$ be a rectangle, with sides parallel to the x-, y-axes, enclosing $\mathcal{R}$ (Fig. 5.5). Define

$$g(x, y) = \begin{cases} f(x, y) & \text{if } (x, y) \text{ is in } \mathcal{R} \\ 0 & \text{if } (x, y) \text{ is not in } \mathcal{R}. \end{cases} \qquad (5.22)$$

Then we define

$$\iint_{\mathcal{R}} f(x, y) \, dx \, dy = \iint_{\mathcal{R}'} g(x, y) \, dx \, dy, \qquad (5.23)$$

where the latter integral, being over a rectangular region, is as defined by (5.21).

Evaluation of double integrals

A double integral is usually evaluated by expressing it as a repeated integral. A detailed proof that a double integral can be expressed as a repeated integral lies outside the scope of this book.* However, the following argument (which is by no means rigorous) shows that the result is plausible.

* See, for example, Courant, R. (1936–7) *Differential and Integral Calculus* Vol. II, pp. 236-9, Blackie, London.

Consider first the case when $f(x, y)$ is defined over the rectangle $a \leqslant x \leqslant b$, $c \leqslant y \leqslant d$. Then

$$\iint_{\mathcal{R}} f(x, y) \, dx \, dy = \lim_{m \to \infty} \lim_{n \to \infty} \sum_{r=1}^{m} \sum_{s=1}^{n} f(x_r, y_s) \, \delta x_r \, \delta y_s$$

$$= \lim_{n \to \infty} \sum_{s=1}^{n} \left\{ \lim_{m \to \infty} \sum_{r=1}^{m} f(x_r, y_s) \, \delta x_r \right\} \delta y_s$$

$$= \lim_{n \to \infty} \sum_{s=1}^{n} \left\{ \int_a^b f(x, y_s) \, dx \right\} \delta y_s, \quad \text{using (5.18)},$$

$$= \int_c^d \left\{ \int_a^b f(x, y) \, dx \right\} dy,$$

again using the definition (5.18) of the Riemann integral. Thus

$$\iint_{\mathcal{R}} f(x, y) \, dx \, dy = \int_c^d \int_a^b f(x, y) \, dx \, dy. \tag{5.24}$$

A similar argument gives

$$\iint_{\mathcal{R}} f(x, y) \, dx \, dy = \int_a^b \int_c^d f(x, y) \, dy \, dx. \tag{5.25}$$

We note that, combining (5.24) and (5.25), (5.15) follows. Thus the order of integration in the repeated integral is immaterial provided the corresponding double integral exists.

When $\mathcal{R}$ is not a rectangular region, enclose it by the rectangle $a \leqslant x \leqslant b$, $c \leqslant y \leqslant d$, as shown in Fig. 5.5. Assume that any line parallel to Oy cuts $\mathcal{R}$ in at most two points (the extension to more general cases is easy and is left to the reader). Then, by the definition (5.23) and the result (5.25),

$$I = \iint_{\mathcal{R}} f(x, y) \, dy \, dx = \int_a^b \int_c^d g(x, y) \, dy \, dx.$$

At a fixed value of x, let $p(x)$, $q(x)$ be the extreme values of y in $\mathcal{R}$ (Fig. 5.5). Then, since $g(x, y) = f(x, y)$ in the interval $p(x) \leqslant y \leqslant q(x)$ and $g(x, y) = 0$ in the intervals $c \leqslant y \leqslant p(x)$ and $q(x) \leqslant y \leqslant d$,

$$I = \int_a^b \int_{p(x)}^{q(x)} f(x, y) \, dx \, dy. \tag{5.26}$$

If the order of integration is inverted, the limits of the inner integral will be functions of y, and the limits of the outer integral will be c, d.

EXAMPLE 8

Express

$$I = \int_0^1 \int_{\frac{1}{2}x}^{\frac{1}{2}} xy \, dy \, dx$$

as a double integral. Evaluate I by inverting the order of integration.

Solution

We observe that x ranges from 0 to 1, and that, for any x, the extreme values of y are $\frac{1}{2}x$ and $\frac{1}{2}$. Thus the triangular region $\mathcal{R}$, shown in Fig. 5.6, is covered. It follows that, expressed as a double integral,

$$I = \iint_{\mathcal{R}} xy \, dx \, dy.$$

Now $\mathcal{R}$ is also covered if y ranges from 0 to $\frac{1}{2}$ and, for every value of y, x ranges from 0 to $2y$. Thus

$$I = \int_0^{\frac{1}{2}} \int_0^{2y} xy \, dx \, dy$$

$$= \int_0^{\frac{1}{2}} [\tfrac{1}{2} x^2 y]_{x=0}^{x=2y} \, dy$$

$$= \int_0^{\frac{1}{2}} 2y^3 \, dy$$

$$= \tfrac{1}{32}.$$

We note that this integral was evaluated in Example 6, section 5.3, by integrating first with respect to y and then with respect to x; the two results obtained agree.

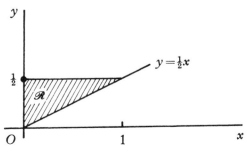

Fig. 5.6

Geometrical interpretation of double integrals

In rectangular cartesian coordinates x, y, z, the equation

$$z = f(x, y) \qquad (5.27)$$

represents a surface (we deal with surfaces in more detail in section 5.5).

Let $f(x, y) > 0$ for all points (x, y) in a region $\mathcal{R}$ of the xy-plane. From a rectangular element $\mathcal{R}$, of area $\delta x_r \delta y_s$ erect perpendiculars from the xy-plane to meet the surface, as shown in Fig. 5.7. Then, if (x_r, y_s) is any point in the rectangle, the volume of the columnar element so formed is approximately $f(x_r, y_s) \delta x_r \delta y_s$. As the double integral

$$\iint_{\mathcal{R}} f(x, y)\, dx\, dy$$

is the limit of a sum of terms of this form, it is intuitively evident that the double integral represents the volume 'underneath the surface'; that is, the volume bounded by the surface, its projection $\mathcal{R}$ on the xy-plane and the cylinder with generators through the boundary of $\mathcal{R}$ and parallel to Oz.

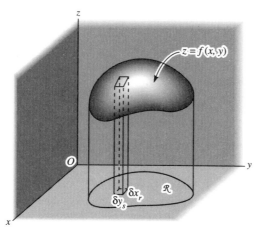

Fig. 5.7 Volume under a surface

Similar reasoning also leads to the conclusion that the double integral

$$\iint_{\mathcal{R}} dx\, dy$$

represents the area of the region $\mathcal{R}$.

Change of variables

Evaluation of a double integral is often made easier by a suitable change of variables.

If variables u, v are such that

$$x = x(u, v), \quad y = y(u, v),$$ (5.28)

it can be shown* that

$$\iint_{\mathcal{R}} f(x, y) \, dx \, dy = \iint_{\mathcal{R}'} f\{x(u, v), y(u, v)\} \, |J| \, du \, dv,$$ (5.29)

where $\mathcal{R}'$ is the region in the uv-plane consisting of points corresponding to the points of $\mathcal{R}$ in the xy-plane, and

$$J = \frac{\partial(x, y)}{\partial(u, v)} = \begin{vmatrix} \dfrac{\partial x}{\partial u} & \dfrac{\partial x}{\partial v} \\ \dfrac{\partial y}{\partial u} & \dfrac{\partial y}{\partial v} \end{vmatrix}.$$ (5.30)

The determinant J is called the **Jacobian of the transformation** (5.28). Various restrictions must be placed on the functions $x(u, v)$, $y(u, v)$, but we shall omit the details.

A very common change of variables is from rectangular cartesian coordinates x, y to plane polar coordinates r, θ, defined by

$$x = r \cos \theta, \quad y = r \sin \theta.$$ (5.31)

The Jacobian of this transformation is

$$\frac{\partial(x, y)}{\partial(r, \theta)} = \begin{vmatrix} \cos \theta & -r \sin \theta \\ \sin \theta & r \cos \theta \end{vmatrix} = r.$$

Thus,

$$\iint_{\mathcal{R}} f(x, y) \, dx \, dy = \iint_{\mathcal{R}'} f(r \cos \theta, r \sin \theta) \, r \, dr \, d\theta,$$ (5.32)

where $\mathcal{R}'$ is the region in the $r\theta$–plane corresponding to the region $\mathcal{R}$ in the xy-plane. The plausibility of this result can be seen as follows.

When a double integral is defined using rectangular cartesian coordinates, the region of integration is subdivided into elementary regions by lines $x = \text{constant}$, $y = \text{constant}$; the area of a typical element may be denoted by $\delta x \, \delta y$. In the case of polar coordinates, the corresponding subdivision is into elements bounded by circles $r = \text{constant}$ and lines $\theta = \text{constant}$ (Fig. 5.8(a)). The area of a typical element is approximately $r \, \delta r \, \delta \theta$. Recalling the definition (5.21), we should thus expect the double integral of $f(x, y)$ to transform to the limit of a sum of terms of the form $f(r \cos \theta, r \sin \theta) \, r \, \delta r \, \delta \theta$, and this leads to the result (5.32).

---

* See, for example, Courant, R. (1936–7) *Differential and Integral Calculus* Vol. II, pp. 247-53, Blackie, London.

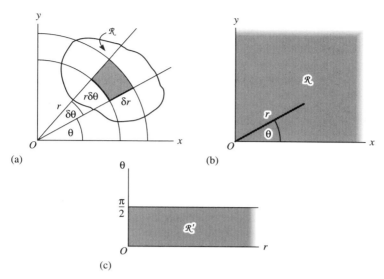

Fig. 5.8 Use of plane polar coordinates to evaluate double integrals: (a) the element $r \, \delta r \, \delta\theta$; (b) region of integration for Example 9, in the xy-plane; (c) region of integration for Example 9, in the $r\theta$-plane

EXAMPLE 9

Evaluate

$$I = \iint_{\mathcal{R}} e^{-(x^2 + y^2)} \, dx \, dy,$$

where $\mathcal{R}$ denotes the quadrant $x \geqslant 0$, $y \geqslant 0$ of the xy-plane. Deduce that

$$\int_0^\infty e^{-x^2} \, dx = \tfrac{1}{2}\sqrt{\pi}.$$

Solution

Substituting $x = r \cos \theta$, $y = r \sin \theta$, the integral becomes

$$I = \iint_{\mathcal{R}'} e^{-r^2} r \, dr \, d\theta.$$

All points of the region $\mathcal{R}$ are covered if θ ranges from 0 to $\frac{1}{2}\pi$ and if, for every value of θ in this range, r ranges from 0 to ∞ (Fig. 5.8(b)). Thus the region $\mathcal{R}'$ in the $r\theta$-plane is the semi-infinite strip $0 \leqslant r < \infty$, $0 \leqslant \theta < \frac{1}{2}\pi$ (Fig. 5.8(c)). It follows that

$$I = \int_0^{\frac{1}{2}\pi} \int_0^\infty r e^{-r^2} \, dr \, d\theta.$$

Observing that

$$\int_0^\infty r \exp(-r^2)\,dr = -[\tfrac{1}{2}\exp(-r^2)]_0^\infty = \tfrac{1}{2}$$

gives

$$I = \tfrac{1}{4}\pi .$$

In terms of rectangular coordinates,

$$I = \int_0^\infty \int_0^\infty e^{-(x^2+y^2)}\,dx\,dy.$$

Using (5.16) gives

$$I = \left(\int_0^\infty e^{-x^2}\,dx\right)\left(\int_0^\infty e^{-y^2}\,dy\right)$$

$$= \left(\int_0^\infty e^{-x^2}\,dx\right)^2 .$$

Hence

$$\int_0^\infty e^{-x^2}\,dx = \tfrac{1}{2}\sqrt{\pi},$$

the positive root being chosen since the value is clearly positive.

EXAMPLE 10

Evaluate

$$I = \iint_{\mathcal{R}} (1 - x^2/a^2 - y^2/b^2)^{\frac{1}{2}}\,dx\,dy,$$

where $\mathcal{R}$ is the region bounded by the ellipse $(x^2/a^2) + (y^2/b^2) = 1$. Hence show that the volume of the ellipsoid

$$x^2/a^2 + y^2/b^2 + z^2/c^2 = 1$$

is $\tfrac{4}{3}\pi abc.$

Solution

Consider the transformation to variables r, θ (which are not in this case polar coordinates) defined by

$$x = ar \cos \theta, \quad y = br \sin \theta.$$

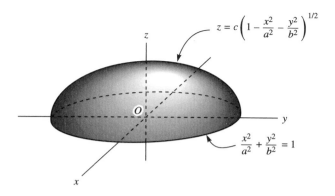

Fig. 5.9

It is easily verified that the Jacobian of this transformation is abr, and also that the region bounded by the ellipse is covered if r ranges from 0 to 1 and θ ranges from 0 to 2π. Hence we find

$$I = \int_0^1 \int_0^{2\pi} abr(1 - r^2)^{\frac{1}{2}} \, dr \, d\theta$$

which, after a short calculation, becomes

$$I = \tfrac{2}{3}\pi ab.$$

The volume of the given ellipsoid (the upper portion of which is shown in Fig. 5.9) is

$$V_0 = 2 \iint_{\mathcal{R}} z \, dx \, dy,$$

where $z = c\{1 - (x^2/a^2) - (y^2/b^2)\}^{\frac{1}{2}}$ and $\mathcal{R}$ is the region in the xy-plane bounded by the ellipse $(x^2/a^2) + (y^2/b^2) = 1$. Hence

$$V_0 = 2cI = \tfrac{4}{3}\pi abc,$$

as required.

Triple integrals

The definition of a double integral can be extended quite easily to define the triple integral of a function $f(x, y, z)$ over a closed region V. The integral is denoted thus:

$$\iiint_V f(x, y, z) \, dx \, dy \, dz.$$

Like double integrals, triple integrals are usually evaluated by repeated integration. The method is similar to that used to evaluate double integrals, and it will be sufficient to give examples.

EXAMPLE 11

If V denotes the region

$$|x| \leqslant a, \quad |y| \leqslant b, \quad |z| \leqslant c,$$

evaluate

$$I = \iiint_V (y^2 + z^2) \, dx \, dy \, dz.$$

Solution

The region V is a rectangular parallelepiped with faces in the planes $x = \pm a$, $y = \pm b$, $z = \pm c$. Thus

$$I = \int_{-c}^{c} \int_{-b}^{b} \int_{-a}^{a} (y^2 + z^2) \, dx \, dy \, dz$$

$$= \int_{-c}^{c} \int_{-b}^{b} 2a (y^2 + z^2) \, dy \, dz \quad \text{(integrating with respect to } x\text{).}$$

Hence, $\quad I = \int_{-c}^{c} 2a \left(\frac{2b^3}{3} + 2bz^2 \right) dz \quad$ (integrating with respect to y)

$$= \tfrac{8}{3} abc (b^2 + c^2).$$

EXAMPLE 12

A tetrahedron has vertices at the points $O(0,0,0)$, $A(1,0,0)$, $B(0,1,0)$, $C(0,0,1)$. If V denotes the region bounded by the tetrahedron, evaluate

$$I = \iiint_V z \, dx \, dy \, dz.$$

Solution

Let P be any point $(x, y, 0)$ inside the triangle OAB, and let the line through P parallel to Oz meet the plane ABC at Q (Fig. 5.10). The equation of the plane ABC is

$$x + y + z = 1.$$

Thus at Q, $z = 1 - x - y$.

To cover the volume of integration, we let x, y range over all points P in the

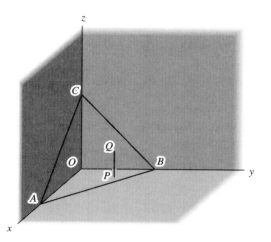

Fig. 5.10

triangle OAB, and let z range from zero (its value at P) to $1 - x - y$ (its value at Q). Thus

$$I = \iint_{\triangle OAB} \left(\int_0^{1-x-y} z \, dz \right) dx \, dy$$

$$= \iint_{\triangle OAB} \tfrac{1}{2}(1 - x - y)^2 \, dx \, dy$$

$$= \int_0^1 \int_0^{1-y} \tfrac{1}{2}(1 - x - y)^2 \, dx \, dy$$

$$= \int_0^1 \left[-\tfrac{1}{6}(1 - x - y)^3 \right]_{x=0}^{x=1-y} dy$$

$$= \int_0^1 \tfrac{1}{6}(1 - y)^3 \, dy$$

$$= \frac{1}{24}.$$

Change of variables in triple integrals

If variables u, v, w are defined by the relations

$$x = x(u, v, w), \quad y = y(u, v, w), \quad z = z(u, v, w), \tag{5.33}$$

then it can be shown that

$$\iiint_V f(x, y, z)\, dx\, dy\, dz = \iiint_{V'} f\{x(u, v, w), y(u, v, w), z(u, v, w)\}$$

$$\times |J|\, du\, dv\, dw, \tag{5.34}$$

where V' is the region in the uvw-space corresponding to the region V in the xyz-space, and

$$J = \frac{\partial(x, y, z)}{\partial(u, v, w)} = \begin{vmatrix} \dfrac{\partial x}{\partial u} & \dfrac{\partial x}{\partial v} & \dfrac{\partial x}{\partial w} \\ \dfrac{\partial y}{\partial u} & \dfrac{\partial y}{\partial v} & \dfrac{\partial y}{\partial w} \\ \dfrac{\partial z}{\partial u} & \dfrac{\partial z}{\partial v} & \dfrac{\partial z}{\partial w} \end{vmatrix}. \tag{5.35}$$

The determinant J is called the **Jacobian of the transformation**. As in the case of changes of variables in double integrals, the functions involved must satisfy certain analytical conditions, but these are omitted here.

EXERCISES

5.14 By sketching the region of integration, show that

$$\int_0^1 \int_{\sqrt{y}}^1 dx\, dy = \int_0^1 \int_0^{x^2} dy\, dx.$$

Evaluate the integrals.

5.15 By inverting the order of integration, evaluate

$$\int_0^1 \int_{y^2}^{y^{1/2}} (y/x)\, e^x\, dx\, dy.$$

5.16 A region $\mathcal{R}$ is bounded by a triangle with vertices at the origin and the points $(1, 1)$, $(-1, 1)$. Show that

$$\iint_{\mathcal{R}} e^{y^2}\, dx\, dy = (e - 1).$$

5.17 Evaluate

$$\int_0^1 \int_{\frac{1}{2}y}^y \frac{xy^2}{\sqrt{(x^3 + y^3)}}\, dx\, dy + \int_1^2 \int_{\frac{1}{2}y}^1 \frac{xy^2}{\sqrt{(x^3 + y^3)}}\, dx\, dy.$$

[*Hint.* Sketch the regions of integration.]

5.18 Evaluate

$$\iint_{\mathcal{R}} (x^2 + y^2)\, dx\, dy,$$

where $\mathcal{R}$ is the region bounded by the circles $x^2 + y^2 = a^2, x^2 + y^2 = b^2$, $(a < b)$. [*Hint.* Use plane polar coordinates r, θ.]

5.19 Find the volume bounded by the paraboloid

$$z = 4 - x^2 - y^2$$

and the xy-plane.

5.20 By transforming to plane polar coordinates r, θ, find

$$\int_0^{\frac{1}{2}\sqrt{2}} \int_x^{\sqrt{(1-x^2)}} \frac{\ln(x^2 + y^2)}{\sqrt{(x^2 + y^2)}}\, dx\, dy.$$

5.21 By means of the transformation $x = ar\cos\theta, y = br\sin\theta$, evaluate

$$\iint_{\mathcal{R}} x^2\, dx\, dy,$$

where $\mathcal{R}$ denotes the region bounded by the ellipse $x^2/a^2 + y^2/b^2 = 1$.

5.22 By means of the substitution $x = (r\cos\theta)^{\frac{1}{2}}, y = (r\sin\theta)^{\frac{1}{2}}$, prove that

$$\iint_{\mathcal{R}} x^3(1 - x^4 - y^4)\, dx\, dy = 4/45,$$

where $\mathcal{R}$ is the region defined by $x \geqslant 0, y \geqslant 0, x^4 + y^4 \leqslant 1$.

5.23 Find

$$\int_0^1 \int_0^1 \int_0^y xyz\, dx\, dy\, dz.$$

5.24 Evaluate

$$\iiint_V \exp(-x^2 - y^2 - z^2)\, dx\, dy\, dz,$$

where V denotes the whole of space.

5.25 The vertices of a triangular prism are the points

$$(0, 0, 0), (1, 0, 0), (0, 1, 0), (0, 0, 2), (0, 1, 2), (1, 0, 2).$$

Evaluate

$$\iiint_V x^2\, dx\, dy\, dz,$$

where V is the region bounded by the prism.

5.26 If the variables x, y, z are related to the variables r, θ, φ by the equations

$$x = ar \sin \theta \cos \varphi, \quad y = br \sin \theta \sin \varphi, \quad z = cr \cos \theta,$$

show that

$$\frac{\partial(x, y, z)}{\partial(r, \theta, \varphi)} = abcr^2 \sin \theta.$$

By means of the above transformation of variables, evaluate

$$\iiint_V (x^2 + y^2 + z^2) \, dx \, dy \, dz,$$

where V denotes the region bounded by the ellipsoid

$$x^2/a^2 + y^2/b^2 + z^2/c^2 = 1.$$

5.5 SURFACES

Let a variable point P have position vector

$$\overrightarrow{OP} = \mathbf{r}(u, v) = \{x(u, v), y(u, v), z(u, v)\}, \tag{5.36}$$

where: (i) u, v are continous parameters taking all values in a region $\mathcal{R}$ of the uv-plane; (ii) x, y and z are continuous, single-valued functions of u, v in $\mathcal{R}$.

If v is given fixed values, in turn, and in each case u is allowed to vary over its range of values, P will trace out a family of curves called u-**coordinate curves**. Similarly, when fixed values are assigned to u and v is varied, the loci of P are the family of v-**coordinate curves** (Fig. 5.11). The network of all

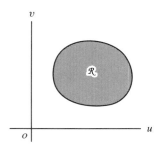

Fig. 5.11

these curves forms a surface S, say, which is thus the locus of P.

Equation (5.36) defines a **mapping** of the region $\mathcal{R}$ in the uv-plane onto the surface S in the xyz-space (Fig. 5.11). Since x, y and z are single-valued functions of u and v, each point in $\mathcal{R}$ will correspond to just one point on S. However, it would be too restrictive for our purposes to consider only mappings which are such that each point on S also corresponds to just one point of $\mathcal{R}$ (that is, one-to-one mappings). The class of mappings considered will be widened to include those which are such that most, but not necessarily all, points on S correspond to one point of $\mathcal{R}$: more precisely, we include mappings with **exceptional points** on S, each corresponding to more than one point of $\mathcal{R}$, provided that these exceptional points are either isolated from other exceptional points or (at most) constitute a finite number of curves on S. It will be found that such exceptional points cause no special difficulty.

EXAMPLE 13

Consider

$$\overrightarrow{OP} = \mathbf{r}(\varphi, z) = (a \cos \varphi, a \sin \varphi, z), \tag{5.37}$$

where $0 \leqslant \varphi \leqslant 2\pi, -\infty < z < \infty$, and a is a constant. If z takes a fixed value z_0, then since $x = a \cos \varphi$ and $y = a \sin \varphi$, P describes a circle in the plane $z = z_0$ as φ ranges from 0 to 2π. If φ is fixed and z varies, P describes a line parallel to Oz and at a distance a from it. The complete locus of P is thus an infinite circular cylinder of radius a with axis Oz (Fig. 5.12). We note that each of the points $(a, 0, z)$ on the generator which cuts the positive x-axis corresponds to the two points $(0, z)$ and $(2\pi, z)$ in the φz-plane. Thus this generator consists of exceptional points of the mapping (5.37).

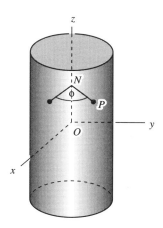

Fig. 5.12

EXAMPLE 14

If

$$\overrightarrow{OP} = \mathbf{r}(\theta, \varphi) = (a \sin \theta \cos \varphi, a \sin \theta \sin \varphi, a \cos \theta), \qquad (5.38)$$

where $0 \leqslant \theta \leqslant \pi, 0 \leqslant \varphi \leqslant 2\pi$, then

$$x^2 + y^2 + z^2 = a^2 \sin^2 \theta (\cos^2 \varphi + \sin^2 \varphi) + a^2 \cos^2 \theta = a^2.$$

In this case P lies on a sphere of radius a, centred at the origin, for all values of θ and φ, and it may be verified that with the ranges of θ and φ given, the whole sphere is covered. It is left as an exercise for the reader to verify that the semicircle $0 \leqslant \theta \leqslant \pi, \varphi = 0$ consists of exceptional points of the mapping (5.38).

Two-sided surfaces

The reader may be familiar with the fact that a surface may have only one side. For example, if a strip of paper is twisted once and the ends glued together, a one-sided surface, called a Möbius strip (Fig. 5.13) is obtained. It can be verified that this surface has only one side by tracing a continuous line along the strip and observing that all points can be reached without crossing over the edge. Such surfaces will henceforth be excluded from consideration, and we discuss only two-sided surfaces.

Fig. 5.13 A Möbius strip.

It will be sufficient for our purposes to rely upon the intuitive notion of what is meant by a two-sided surfaces, and accordingly an analytical definition is omitted.

Open and closed surfaces

A surface S is said to be **open** if every pair of points not lying on S can be joined by a continuous curve which does not cross S.

A surface S is closed if it divides space into distinct regions, $\mathcal{R}_1$ and $\mathcal{R}_2$ say, such that every continuous curve joining a point in $\mathcal{R}_1$ to a point in $\mathcal{R}_2$ crosses S at least once.

For example, a cap of a sphere is an open surface, but the complete surface of a sphere is closed.

Unit normal vector

Consider the surface S defined by (5.36). As we have seen, the locus of P when u varies and v is fixed is a u-coordinate curve on S, and when v varies and u is fixed the locus is a v-coordinate curve. At any point P on S, the vectors

$$\frac{\partial \mathbf{r}}{\partial u} = \mathbf{r}_u, \frac{\partial \mathbf{r}}{\partial v} = \mathbf{r}_v \qquad (5.39)$$

are respectively tangential to the u-, v-coordinate curves through P (Fig. 5.14).

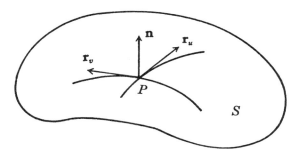

Fig. 5.14 The unit normal $\mathbf{n}$ to a surface S.

The vector

$$\mathbf{n} = \frac{\mathbf{r}_u \times \mathbf{r}_v}{|\mathbf{r}_u \times \mathbf{r}_v|} \qquad (5.40)$$

is of unit magnitude and is perpendicular to both the u- and v-coordinate curves at P. It is called a **unit normal vector** to S.

If $\mathbf{r}_u$ and $\mathbf{r}_v$ are interchanged in (5.40), the direction of $\mathbf{n}$ is reversed, and so to this extent the direction of $\mathbf{n}$ is not unique. It is convenient to avoid this ambiguity by labelling one side of S as positive, and to take the order of $\mathbf{r}_u$ and $\mathbf{r}_v$ such that $\mathbf{n}$ points away from the positive side. In the case of a closed surface bounding a region $\mathcal{R}$, it is conventional to lable the surface so that the

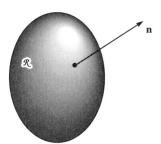

Fig. 5.15 Orientation of a surface. By convention the positive normal of a surface bounding a closed region $\mathcal{R}$ points out of $\mathcal{R}$.

unit normal points out of $\mathcal{R}$ (Fig. 5.15).

A surface is said to be **oriented** once a particular side has been labelled as positive.

To illustrate the definition (5.40), consider the circular cylinder

$$\mathbf{r} = (a \cos \varphi, a \sin \varphi, z) \quad (0 \leqslant \varphi \leqslant 2\pi, -\infty < z < \infty).$$

In this case,

$$\mathbf{r}_\varphi = (- a \sin \varphi, \ a \cos \varphi, \ 0)$$

and

$$\mathbf{r}_z = (0, 0, 1).$$

Thus

$$\mathbf{r}_\varphi \times \mathbf{r}_z = (a \cos \varphi, \ a \sin \varphi, \ 0). \tag{5.41}$$

Hence,

$$\mathbf{n} = \frac{\mathbf{r}_\varphi \times \mathbf{r}_z}{|\mathbf{r}_\varphi \times \mathbf{r}_z|} = (\cos \varphi, \ \sin \varphi, \ 0). \tag{5.42}$$

Note however, that in this simple case it is clear from a diagram (see Fig. 5.12) that the normal to the surface is parallel to the xy-plane, and hence that its z-component is zero. Further, resolving parallel to the axes Ox, Oy and remembering that $\mathbf{n}$ is of unit magnitude and parallel to NP, it is seen that the x-, y-components of $\mathbf{n}$ are $\cos \varphi, \sin \varphi$, respectively.

We have taken the outside of the cylinder to be positive.

Smooth surface

If the unit normal exists and is continuous at all points on a surface, then the surface is said to be **smooth**.

Simple surface

A **simple** surface is one which is the union of a finite number of smooth surfaces. Such surfaces are also sometimes termed **piecewise-smooth** or **regular.**

The cap of a sphere is an example of a smooth surface; and the surface of a cube is a simple, closed surface, which is the union of six smooth open surfaces.

Surface area

Let $P_0(u, v)$ be a point on the surface S with parametric equation $\mathbf{r} = \mathbf{r}(u, v)$. Let $P_1(u + du, v)$ be a neighbouring point on the u-coordinate curve through

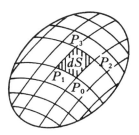

Fig. 5.16 A surface element dS.

P_0, and let $P_2(u, v + dv)$ be a neighbouring point on the v-coordinate curve through P_0. Let the u-coordinate curve through P_2 and the v-coordinate curve through P_1 meet at P_3 (Fig. 5.16). The part of S which is bounded by $P_0 P_1 P_3 P_2$ is called a **surface element.**

If we suppose that $P_0 P_1 P_3 P_2$ is approximately a parallelogram, as intuitive geometrical considerations suggest, then the area of the surface element is

$$dS \approx \left| \overrightarrow{P_0 P_1} \times \overrightarrow{P_0 P_2} \right|$$

$$\approx \left| \mathbf{r}_u \times \mathbf{r}_v \right| du \, dv, \tag{5.43}$$

using the fact that $\overrightarrow{P_0 P_1} \approx (\partial \mathbf{r}/\partial u) du$ and $\overrightarrow{P_0 P_2} \approx (\partial \mathbf{r}/\partial v) dv$. This intuitive argument leads us to *define*

$$\text{surface area of } S = \iint \left| \mathbf{r}_u \times \mathbf{r}_v \right| du \, dv, \tag{5.44}$$

where the ranges of u, v are such that the whole of S is covered. We shall refer to $\left| \mathbf{r}_u \times \mathbf{r}_v \right| du \, dv$ as an element of surface area dS, although, of course, the only precisely defined quantity is the total surface area.

To illustrate this definition, consider the circular cylinder

$$\mathbf{r} = (a \cos \varphi, a \sin \varphi, z) \quad 0 \leqslant \varphi \leqslant 2\pi, 0 \leqslant z \leqslant b,$$

which is of length b and radius a (Fig. 5.17). Its area is, of course, $2\pi ab$. Using the result (5.41),

$$\left| \mathbf{r}_\varphi \times \mathbf{r}_z \right| = a,$$

and thus, by (5.44),

$$\text{total area of the cylinder} = \int_0^b \int_0^{2\pi} a \, d\varphi \, dz$$

$$= 2\pi ab.$$

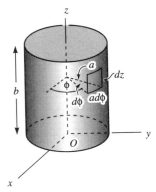

Fig. 5.17 Element of surface area on a cylinder.

Note that in this case, if φ increases by $d\varphi$, P traces out an arc of length $a\,d\varphi$; and if z increases by dz, P is displaced a distance dz (Fig. 5.17). Since these displacements are at right angles, it is evident that the area of a surface element is

$$dS = a\,d\varphi\,dz, \tag{5.45}$$

which leads to the above integral for the total surface area.

In the case of the sphere defined by (5.38), similar geometrical considerations (Fig. 5.18) lead to the conclusion that

$$dS = a^2 \sin\theta\,d\theta\,d\varphi. \tag{5.46}$$

Thus,

$$\text{total surface area of the sphere} = \int_0^{2\pi}\int_0^{\pi} a^2 \sin\theta\,d\theta\,d\varphi$$

$$= 4\pi a^2,$$

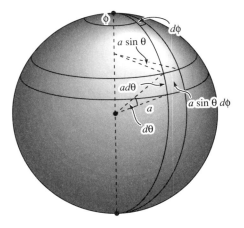

Fig. 5.18 Element of surface area on a sphere.

which again agrees with the familiar elementary result.

Coordinate surfaces

Let u, v, w be curvilinear coordinates, defined relative to a rectangular cartesian system by the relation

$$\mathbf{r} = \mathbf{r}(u, v, w). \tag{5.47}$$

If u_0, v_0, w_0 are constants, then each of the equations

$$\mathbf{r} = \mathbf{r}(u_0, v, w), \quad \mathbf{r} = \mathbf{r}(u, v_0, w), \quad \mathbf{r} = \mathbf{r}(u, v, w_0) \tag{5.48}$$

represents a surface: these surfaces are called u-, v-, w-coordinate surfaces,
respectively.

When the system of curvilinear coordinates is orthogonal, the unit normal
vectors and the elements of area of the coordinate surfaces can be conveniently
expressed in terms of quanitities defined in section 4.11. The unit normal to a
w-coordinate surface is

$$\mathbf{n} = \frac{\mathbf{r}_u \times \mathbf{r}_v}{|\mathbf{r}_u \times \mathbf{r}_v|} = \frac{\mathbf{r}_u}{|\mathbf{r}_u|} \times \frac{\mathbf{r}_v}{|\mathbf{r}_v|},$$

where we have used the fact that $\mathbf{r}_u$ and $\mathbf{r}_v$ are perpendicular. But from the
definitions (4.71), $h_1 = |\mathbf{r}_u|$ and $h_2 = |\mathbf{r}_v|$. Hence, using the definitions (4.72)
and noting that $\mathbf{e}_u, \mathbf{e}_v, \mathbf{e}_w$ form a right-handed orthonormal triad,

$$\mathbf{n} = \mathbf{e}_u \times \mathbf{e}_v = \mathbf{e}_w.$$

Similarly, the unit normals to the u- and v-coordinate surfaces are $\mathbf{e}_u$ and $\mathbf{e}_v$
respectively. (The particular orientation of the coordinate surfaces which is
implied here is conventional.)

The element of area of a w-coordinate surface is

$$dS = |\mathbf{r}_u \times \mathbf{r}_v| \, du \, dv$$

$$= h_1 h_2 \, du \, dv;$$

and similarly the element of area of the u-, v-coordinate surfaces are
$h_2 h_3 \, dv \, dw$, $h_1 h_3 \, du \, dw$, respectively.

For example, consider cylindrical polar coordinates R, φ, z, defined by

$$\mathbf{r} = (R \cos \varphi, \, R \sin \varphi, \, z).$$

The coordinate surface $R = a$ is a circular cylinder with axis Oz. By what has
been said above, and using equations (4.77), the unit normal to this surface is

$$\mathbf{n} = \mathbf{e}_R = (\cos \varphi, \, \sin \varphi, \, 0),$$

in agreement with (5.42). Further, from equations (4.76), $h_1 = 1, h_2 = R$, and so the element of area of the cylinder $R = a$ is

$$dS = h_1 h_2 \, d\varphi \, dz$$

$$= a \, d\varphi \, dz,$$

in agreement with (5.45).

Alternative form of the equation of a surface

Consider again the surface with parametric equations

$$x = x(u, v), \quad y = y(u, v), \quad z = z(u, v).$$

If we suppose that the first two equations are solved so as to give u and v as functions of x and y, and that these functions are substituted into the third equation, we obtain

$$z = f(x, y) \tag{5.49}$$

as the equation of the surface. This representation was referred to earlier in this chapter (see section 5.4). It can be shown to be a possible representation provided no portion (of non-zero area) of the surface is composed of lines parallel to the z-axis; that is, provided the normal is not parallel to the xy-plane. Clearly, similar remarks can be made about the possibilities of representing the surface in the form $x = g(y, z)$ or $y = h(z, x)$.

For example, the portion of the sphere of radius a, centre the origin, which lies in the region $z \geqslant 0$ can be represented by

$$z = (a^2 - x^2 - y^2)^{\frac{1}{2}}, \quad (x^2 + y^2 \leqslant a^2),$$

which is of the form (5.49). On the other hand, the equation of a circular cylinder of radius a and axis Oz cannot be expressed in the form (5.49). Its rectangular cartesian equation is $x^2 + y^2 = a^2$, the variable z being absent.

EXERCISES

5.27 Using the definition (5.40), verify that the unit normal to the sphere

$$\mathbf{r} = (a \sin \theta \cos \varphi, \, a \sin \theta \sin \varphi, \, a \cos \theta)$$

is

$$\mathbf{n} = (\sin \theta \cos \varphi, \, \sin \theta \sin \varphi, \, \cos \theta).$$

Also, using the definition (5.44), verify that the surface area of the lune of the sphere between the planes $\varphi = \varphi_0$, $\varphi = \varphi_0 + \alpha$ is $2\alpha a^2$.

5.28 Show that the surface

$$\mathbf{r} = (r \cos \varphi, \ r \sin \varphi, \ r\sqrt{3}),$$

where $0 \leqslant r < \infty$, $0 \leqslant \varphi \leqslant 2\pi$, is an infinite cone with vertex at the origin, with axis along Oz, and having a semi-vertical angle of $30°$. Find (i) the unit normal vector $\mathbf{n}$ at any point, and (ii) the element of surface area.

5.29 Describe (in geometrical terms) the surfaces whose parametric equations are:

(i) $\mathbf{r} = (a \cos u, \ b \sin u, \ v)$, $\ 0 \leqslant u \leqslant 2\pi$, $\ -\infty < v < \infty$;

(ii) $\mathbf{r} = (a \cosh u, \ b \sinh u, \ v)$, $\ -\infty < u < \infty$, $\ -\infty < v < \infty$.

In case (i), find, in the form of an integral with respect to u, the total surface area between the planes $z = 0$, $z = 2$.

5.30 Find the total surface area of the surface with parametric equation

$$\mathbf{r} = (u \cos v, \ u \sin v, \ u^2), \ \ 0 \leqslant u \leqslant 1, \ \ 0 \leqslant v \leqslant 2\pi.$$

5.31 Express the equation of the conical surface defined in Exercise 5.28 in the form $z = f(x, y)$.

5.32 The projection on the xy-plane of the surface $z = f(x, y)$ is the region $\mathcal{R}$. Show that the area of the surface is given by

$$\iint_{\mathcal{R}} (1 + f_x^2 + f_y^2)^{\frac{1}{2}} \, dx \, dy,$$

where suffixes denote partial derivatives with respect to x, y. Also, show that the unit normal at any point is

$$\pm \frac{(f_x, f_y, -1)}{\sqrt{(f_x^2 + f_y^2 + 1)}}.$$

[Hint. The parametric equation of the surface may be taken as $\mathbf{r} = (x, y, f(x, y))$.]

5.6 SURFACE INTEGRALS

Let S be a simple surface with parametric equation $\mathbf{r} = \mathbf{r}(u, v)$, and let $\mathcal{R}$ denote the region in the uv-plane consisting of points corresponding to the points of S. If a scalar field Ω and a vector field $\mathbf{F}$ are defined at all points on S, then, on S, $\Omega = \Omega(u, v)$ and $\mathbf{F} = \mathbf{F}(u, v)$.

We define the surface integrals of Ω and F over S as follows:

$$\iint_S \Omega \, dS = \iint_{\mathcal{R}} \Omega(u, v) \, |\mathbf{r}_u \times \mathbf{r}_v| \, du \, dv; \qquad (5.50)$$

$$\iint_S \mathbf{F} \cdot d\mathbf{S} = \iint_S \mathbf{F} \cdot \mathbf{n} \, dS = \iint_{\mathcal{R}} \mathbf{F}(u, v) \cdot (\mathbf{r}_u \times \mathbf{r}_v) \, du \, dv. \qquad (5.51)$$

Thus dS is interpreted as an element of surface area; and $d\mathbf{S}$ is an abbreviation for $\mathbf{n}dS$, where $\mathbf{n} = (\mathbf{r}_u \times \mathbf{r}_v)/|\mathbf{r}_u \times \mathbf{r}_v|$ is the unit normal vector.

If $\mathbf{F} = F_1\mathbf{i} + F_2\mathbf{j} + F_3\mathbf{k}$, in a rectangular cartesian coordinate system, we also define

$$\iint_S \mathbf{F} \, dS = \left(\iint_S F_1 \, dS \right)\mathbf{i} + \left(\iint_S F_2 \, dS \right)\mathbf{j} + \left(\iint_S F_3 \, dS \right)\mathbf{k}. \qquad (5.52)$$

The unit normal vector $\mathbf{n}$ is itself a vector field defined on S. Thus the integrals

$$\iint_S \Omega \, d\mathbf{S} = \iint_S \Omega \mathbf{n} \, dS \qquad (5.53)$$

and

$$\iint_S \mathbf{F} \times d\mathbf{S} = \iint_S \mathbf{F} \times \mathbf{n} \, dS \qquad (5.54)$$

are also now meaningful in view of the definition (5.52).

Note

Many authors use the notation

$$\int_S \Omega \, dS$$

to denote a surface integral. Likewise, in the case of volume integrals, defined in the next section, single integral signs are often used. Earlier editions of this text used single integral signs for both surface and volume integrals.

EXAMPLE 15

Evaluate

$$\iint_S \Omega \, dS$$

in the following cases:

    (i) $\Omega = x^2 + y^2$ and S is the spherical surface $x^2 + y^2 + z^2 = a^2$;

    (ii) $\Omega = x^2 + y^2$ and S is the complete surface of the cube $|x| \leq a$, $|y| \leq a$, $|z| \leq a$;

    (iii) $\Omega = (b^2 x^2/a^2 + a^2 y^2/b^2)^{\frac{1}{2}}$ and S is the curved surface of the elliptic cylinder $x^2/a^2 + y^2/b^2 = 1$, $|z| \leq c$.

Solution

(i) The parametric equations of the sphere S are

$$x = a \sin \theta \cos \varphi, \quad y = a \sin \theta \sin \varphi, \quad z = a \cos \theta,$$

where $0 \leqslant \theta \leqslant \pi$, $0 \leqslant \varphi \leqslant 2\pi$. Thus, on S

$$\Omega = a^2 \sin^2 \theta.$$

We also have $dS = a^2 \sin \theta \, d\theta \, d\varphi$. Hence

$$\iint_S \Omega \, dS = \int_0^{2\pi} \int_0^{\pi} a^4 \sin^3 \theta \, d\theta \, d\varphi$$

$$= \tfrac{8}{3} \pi a^4.$$

(ii) This integral is evaluated by considering the six separate faces of the cube.

On the two faces $z = \pm a$, parallel to the xy-plane, the element of surface area is $dS = dx \, dy$. Thus, the joint contribution to the integral from this pair of faces is

$$C_1 = 2 \int_{-a}^{a} \int_{-a}^{a} (x^2 + y^2) \, dx \, dy$$

$$= \tfrac{16}{3} a^4.$$

On the faces $x = \pm a$, $\Omega = a^2 + y^2$ and $dS = dy \, dz$. Thus the joint contribution from this pair of faces is

$$C_2 = 2 \int_{-a}^{a} \int_{-a}^{a} (a^2 + y^2) \, dy \, dz$$

$$= \tfrac{32}{3} a^4.$$

Finally, the joint contribution from the pair of faces $y = \pm a$ is

$$C_3 = 2 \int_{-a}^{a} \int_{-a}^{a} (x^2 + a^2) \, dx \, dz$$

$$= \tfrac{32}{3} a^4.$$

Hence,

$$\iint_S \Omega \, dS = C_1 + C_2 + C_3 = \tfrac{80}{3} a^4.$$

(iii) The parametric equation of the elliptic cylindrical surface can be taken as

$$\mathbf{r} = (a \cos \theta,\ b \sin \theta,\ z)$$

where $0 \leqslant \theta \leqslant 2\pi,\ -c \leqslant z \leqslant c$. Then

$$\mathbf{r}_\theta = (-a \sin \theta,\ b \cos \theta,\ 0),$$

$$\mathbf{r}_z = (0, 0, 1),$$

giving

$$\left|\mathbf{r}_\theta \times \mathbf{r}_z\right| = \left|(b \cos \theta,\ a \sin \theta,\ 0)\right|$$

$$= (b^2 \cos^2 \theta + a^2 \sin^2 \theta)^{\frac{1}{2}}.$$

Thus

$$dS = (b^2 \cos^2 \theta + a^2 \sin^2 \theta)^{\frac{1}{2}} d\theta\ dz.$$

On S,

$$\Omega = (b^2 \cos^2 \theta + a^2 \sin^2 \theta)^{\frac{1}{2}},$$

and hence

$$\iint_S \Omega\, dS = \int_{-c}^{c} \int_{0}^{2\pi} (b^2 \cos^2 \theta + a^2 \sin^2 \theta)\, d\theta\, dz$$

$$= 2\pi c(a^2 + b^2).$$

EXAMPLE 16

Let $\mathbf{r}$ denote the position vector of a point P relative to the origin O. Evaluate

$$\iint_S \mathbf{r} \cdot d\mathbf{S},$$

where S is the curved surface of the cylinder $x^2 + y^2 = a^2,\ 0 \leqslant z \leqslant 2a$.

Solution

By simple geometrical considerations (Fig. 5.19), when P lies on S,

$$\mathbf{r} \cdot \mathbf{n} = OP \cos \alpha \quad (\alpha = \text{angle between } \mathbf{r} \text{ and the unit vector } \mathbf{n})$$

$$= a.$$

Thus

$$\iint_S \mathbf{r} \cdot d\mathbf{S} = \iint_S a\, dS,$$

$$= a \times (\text{area of } S)$$

$$= 4\pi a^3.$$

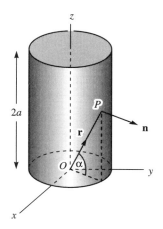

Fig. 5.19

EXAMPLE 17

If S denotes the surface of the sphere

$$\mathbf{r} = (a \sin \theta \cos \varphi, \ a \sin \theta \sin \varphi, \ a \cos \theta),$$

where $0 \leqslant \theta \leqslant \pi$, $0 \leqslant \varphi \leqslant 2\pi$, show that

$$\iint_S (x^2 + y^2) \, \mathbf{dS} = 0.$$

Solution

The unit normal to the sphere will be directed away from its centre, which is the origin O. Thus

$$\mathbf{n} = \hat{\mathbf{r}} = \mathbf{i} \sin \theta \cos \varphi + \mathbf{j} \sin \theta \sin \varphi + \mathbf{k} \cos \theta.$$

(This result can of course be deduced from the definition (5.40) of the unit normal.) Also $dS = a^2 \sin \theta \, d\theta \, d\varphi$, and on S, $x^2 + y^2 = a^2 \sin^2 \theta$. Thus

$$\iint_S (x^2 + y^2) \, \mathbf{dS} = \int_0^{2\pi} \int_0^{\pi} a^4 \sin^3 \theta (\mathbf{i} \sin \theta \cos \varphi + \mathbf{j} \sin \theta \sin \varphi + \mathbf{k} \cos \theta) \, d\theta \, d\varphi$$

$$= \mathbf{i} \int_0^{2\pi} \int_0^{\pi} a^4 \sin^4 \theta \cos \varphi \, d\theta \, d\varphi$$

$$+ \mathbf{j} \int_0^{2\pi} \int_0^{\pi} a^4 \sin^4 \theta \sin \varphi \, d\theta \, d\varphi$$

$$+ \mathbf{k} \int_0^{2\pi} \int_0^{\pi} a^4 \sin^3 \theta \cos \theta \, d\theta \, d\varphi.$$

Integrating with respect to φ first, and observing that

$$\int_0^{2\pi} \cos \varphi \, d\varphi = \int_0^{2\pi} \sin \varphi \, d\varphi = 0$$

gives

$$\iint_S (x^2 + y^2) \, \mathbf{dS} = 2\pi a^4 \left[\tfrac{1}{4} \sin^4 \theta \right]_0^\pi \mathbf{k}$$

$$= \mathbf{0}.$$

Note

The above result can be deduced by a simple symmetry argument; for Fig. 5.20 shows that the contributions to the integral from surface elements $\mathbf{dS}_1$ and $\mathbf{dS}_2$ at opposite ends of a diameter cancel. The value of $x^2 + y^2$ is the same at each end of the diameter.

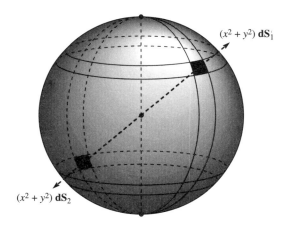

Fig. 5.20

EXAMPLE 18

Evaluate

$$\mathbf{I} = \iint_S \mathbf{r} \, dS,$$

where S denotes the hemisphere $x^2 + y^2 + z^2 = a^2$, $z \geqslant 0$, and $\mathbf{r}$ is the radius vector from the origin.

Solution

The parametric equation of S is

$$\mathbf{r} = (a \sin \theta \cos \varphi, \ a \sin \theta \sin \varphi, \ a \cos \theta),$$

where $0 \leqslant \theta \leqslant \frac{1}{2}\pi,\ 0 \leqslant \varphi \leqslant 2\pi$. Since $dS = a^2 \sin \theta\, d\theta\, d\varphi$, the three components of $\mathbf{I}$ are thus

$$I_1 = a^3 \int_0^{2\pi} \int_0^{\frac{1}{2}\pi} \sin^2 \theta \cos \varphi\, d\theta\, d\varphi,$$

$$I_2 = a^3 \int_0^{2\pi} \int_0^{\frac{1}{2}\pi} \sin^2 \theta \sin \varphi\, d\theta\, d\varphi,$$

$$I_3 = a^3 \int_0^{2\pi} \int_0^{\frac{1}{2}\pi} \sin^2 \theta \cos \theta\, d\theta\, d\varphi.$$

Hence $I_1 = 0,\ I_2 = 0$ and $I_3 = \pi a^3$, giving

$$\mathbf{I} = (0, 0, \pi a^3),$$

referred to the axes $Oxyz$.

Independence of choice of parameters

Let

$$\mathbf{r} = \mathbf{r}(u, v) \quad \text{and} \quad \mathbf{r} = \mathbf{r}(u', v') \tag{5.55}$$

be different parametric representations of the same surface S, so that $u,\ v$ are
related to $u',\ v'$ thus:

$$u = u(u', v') \quad \text{and} \quad v = v(u', v'). \tag{5.56}$$

Then

$$\mathbf{r}_{u'} \times \mathbf{r}_{v'} = (x_{u'}, y_{u'}, z_{u'}) \times (x_{v'}, y_{v'}, z_{v'})$$

$$= (y_{u'} z_{v'} - y_{v'} z_{u'},\ z_{u'} x_{v'} - z_{v'} x_{u'},\ x_{u'} y_{v'} - x_{v'} y_{u'})$$

$$= \left(\frac{\partial(y, z)}{\partial(u', v')}, \frac{\partial(z, x)}{\partial(u', v')}, \frac{\partial(x, y)}{\partial(u', v')} \right),$$

using the Jacobian notation defined by equation (5.30). By the chain rule for
Jacobians, proved in Appendix C, it follows that

$$\mathbf{r}_{u'} \times \mathbf{r}_{v'} = \left(\frac{\partial(y, z)}{\partial(u, v)}, \frac{\partial(z, x)}{\partial(u, v)}, \frac{\partial(x, y)}{\partial(u, v)} \right) \frac{\partial(u, v)}{\partial(u', v')},$$

$$= (\mathbf{r}_u \times \mathbf{r}_v) J,$$

where J denotes the Jacobian of the transformation (5.56). Using this result
and the formula (5.29) for changing the variables in a double integral, we thus

find

$$\iint_{\mathcal{R}} \varphi(u, v)\, |\mathbf{r}_u \times \mathbf{r}_v|\, du\, dv = \iint_{\mathcal{R}'} \varphi(u', v')\, |\mathbf{r}_{u'} \times \mathbf{r}_{v'}|\, du'\, dv',$$

where the regions of integration $\mathcal{R}$, $\mathcal{R}'$ each correspond to S. But, by definition, these are the surface integrals of φ over S in terms of the two parametric representations (5.55). It follows that *the integral of φ over S is independent of the particular choice of parameters on S.* The significance of this result is perhaps made clearer by considering a simple physical situation.

Let P denote the **downward component** of force per unit area acting on a surface S. On an element of surface area dS, the downward force will be $P\, dS$ and hence the total downward force on S will be

$$\iint_S P\, dS.$$

To evaluate this integral it would be necessary to choose a particular parametric representation of S, but on physical grounds the final result should be independent of the choice made. The result proved above confirms this.

A surface integral involving the vector element of area $\mathbf{dS} = \mathbf{n}\, dS$ is again independent of the choice of parameters, but if the orientation is changed the direction of $\mathbf{n}$ is reversed, and the integral changes sign.

EXERCISES

5.33 Evaluate

$$\iint_S xyz\, dS$$

over that part of a sphere of unit radius and centre at the origin which lies in the positive octant $x \geqslant 0$, $y \geqslant 0$, $z \geqslant 0$.

5.34 The ends of a circular cylinder of radius a and axis Oz are in the planes $z = 0$, $z = 2a$. If S denotes the complete surface of the cylinder (including the plane ends), evaluate

$$\iint_S z\, dS.$$

5.35 If S denotes the surface of the hemisphere $x^2 + y^2 + z^2 = 1$, $z \geqslant 0$, find

$$\text{(i)} \quad \iint_S \mathbf{F} \cdot \mathbf{dS}, \qquad \text{(ii)} \quad \iint_S (\mathbf{r} \times \mathbf{k}) \times \mathbf{dS},$$

where

$$\mathbf{F} = (x^2, y^2, z^2), \quad \mathbf{r} = (x, y, z), \quad \mathbf{k} = (0, 0, 1),$$

and the unit normal to S points away from the origin.

5.36 Evaluate

$$\text{(i)} \iint_S \mathbf{F} \, dS, \quad \text{(ii)} \iint_S \mathbf{F} \cdot \mathbf{dS}, \quad \text{(iii)} \iint_S \mathbf{F} \times \mathbf{dS},$$

where

$$\mathbf{F} = (y + z, \ z + x, \ x + y),$$

and S is the square

$$0 \leqslant x \leqslant 1, \quad 0 \leqslant y \leqslant 1, \quad z = 0,$$

positively oriented along the positive z-axis.

5.37 If S denotes that part of the conical surface

$$\mathbf{r} = (r \cos \varphi, \ r \sin \varphi, \ r\sqrt{3})$$

lying between the planes $z = 0, \ z = 1$, evaluate

$$\iint_S (1 - z) \, \mathbf{dS}.$$

[*Hint.* Cf. Exercise 5.28, section 5.5.]

5.38 If S denotes that part of the plane $2x + y + 2z = 6$ which lies in the positive octant, and $\mathbf{F} = (4x, y, z)$, evaluate

$$\iint_S \mathbf{F} \cdot \mathbf{dS}.$$

Take the unit normal to S pointing away from the origin. [*Hint.* The parametric equation of the plane can be taken as

$$\mathbf{r} = (x, y, 3 - x - \tfrac{1}{2} y).]$$

5.7 VOLUME INTEGRALS

Let u, v, w be a system of curvilinear coordinates such that, relative to a rectangular cartesian coordinate system, the position vector of a point is

$$\mathbf{r} = \mathbf{r}(u, v, w). \tag{5.57}$$

Let V be a region in the rectangular cartesian space, and let $\mathcal{R}$ be the region in the uvw-space consisting of points corresponding to the points of V. Then the volume V_0 of V is defined as

$$V_0 = \iiint_{\mathcal{R}} |(\mathbf{r}_u \times \mathbf{r}_v) \cdot \mathbf{r}_w| \, du \, dv \, dw. \qquad (5.58)$$

Further, if Ω is a scalar field defined on V, the volume integral of Ω over V is defined as

$$\iiint_V \Omega \, dV = \iiint_{\mathcal{R}} \Omega(u, v, w) \, |(\mathbf{r}_u \times \mathbf{r}_v) \cdot \mathbf{r}_w| \, du \, dv \, dw. \qquad (5.59)$$

It can be verified, by using the transformation formula (5.34) for triple integrals, that the integrals (5.58), (5.59) are invariant under a change of curvilinear coordinates (cf. the corresponding result for surface integrals, proved in the previous section).

The motivation for the above definition of volume is as follows.

If u is given a small (positive) increment du, the point $P_0(u, v, w)$ is displaced along a u-coordinate curve to $P_1(u + du, v, w)$, and $\overrightarrow{P_0 P_1} \approx \mathbf{r}_u \, du$.

Similarly, corresponding to increments dv in v and dw in w, P is displaced to P_2 and P_3 respectively, where $\overrightarrow{P_0 P_2} \approx \mathbf{r}_v \, dv$ and $\overrightarrow{P_0 P_3} \approx \mathbf{r}_w \, dw$ (Fig. 5.21). The element bounded by the various coordinate surfaces through P_0, P_1, P_2, P_3 is approximately a parallelepiped of volume

$$dV = \left| (\overrightarrow{P_0 P_1} \times \overrightarrow{P_0 P_2}) \cdot \overrightarrow{P_0 P_3} \right|$$

$$= |(\mathbf{r}_u \times \mathbf{r}_v) \cdot \mathbf{r}_w| \, du \, dv \, dw. \qquad (5.60)$$

As the triple integral (5.58) is (by definition) the limit as $du \to 0$, $dv \to 0$, $dw \to 0$ of a sum of terms of the form (5.60), the definition of volume agrees with the intuitive notion.

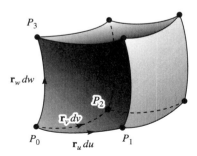

Fig. 5.21 An element of volume; $\overrightarrow{P_0 P_1} \approx \mathbf{r}_u \, du$, etc.

When u, v, w are orthogonal curvilinear coordinates,

$$(\mathbf{r}_u \times \mathbf{r}_v) \cdot \mathbf{r}_w = h_1 h_2 h_3 (\mathbf{e}_u \times \mathbf{e}_v) \cdot \mathbf{e}_w = h_1 h_2 h_3,$$

in the usual notation (section 4.11), and so

$$\iiint_V \Omega \, dV = \iiint_{\mathcal{R}} \Omega(u, v, w) \, h_1 h_2 h_3 \, du \, dv \, dw. \tag{5.61}$$

There are three important particular cases.

(i) Rectangular cartesian coordinates (x, y, z)

In this case $h_1 = h_2 = h_3 = 1$, and hence

$$\iiint_V \Omega \, dV = \iiint_V \Omega(x, y, z) \, dx \, dy \, dz. \tag{5.62}$$

Also, the volume of V is simply

$$V_0 = \iiint_V dx \, dy \, dz. \tag{5.63}$$

Note that these results are often taken as the definitions of volume and volume integral.

(ii) Cylindrical polar coordinates (R, φ, z)

From equations (4.76), $h_1 = 1$, $h_2 = R$, $h_3 = 1$, and so in this case

$$\iiint_V \Omega \, dV = \iiint_{\mathcal{R}} \Omega(R, \varphi, z) \, R \, dR \, d\varphi \, dz. \tag{5.64}$$

Also

$$V_0 = \iiint_{\mathcal{R}} R \, dR \, d\varphi \, dz \tag{5.65}$$

gives the volume of V.

(iii) Spherical polar coordinates (r, θ, φ)

In this case equations (4.74) give $h_1 = 1$, $h_2 = r$, $h_3 = r \sin \theta$, and hence

$$\iiint_V \Omega \, dV = \iiint_{\mathcal{R}} \Omega(r, \theta, \varphi) \, r^2 \sin \theta \, dr \, d\theta \, d\varphi. \tag{5.66}$$

Also

$$V_0 = \iiint_{\mathcal{R}} r^2 \sin\theta \, dr \, d\theta \, d\varphi. \tag{5.67}$$

EXAMPLE 19

Evaluate

$$I = \iiint_V \exp\left[-(x^2 + y^2 + z^2)\right] dV,$$

where V denotes the whole of space.

Solution

To cover the whole of space, let each of x, y, z range from $-\infty$ to ∞. Then

$$I = \int_{-\infty}^{\infty} \int_{-\infty}^{\infty} \int_{-\infty}^{\infty} \exp\left[-(x^2 + y^2 + z^2)\right] dx \, dy \, dz$$

$$= \left(\int_{-\infty}^{\infty} e^{-x^2} dx\right)\left(\int_{-\infty}^{\infty} e^{-y^2} dy\right)\left(\int_{-\infty}^{\infty} e^{-z^2} dz\right).$$

But, using a result proved in Example 9, section 5.4,

$$\int_{-\infty}^{\infty} e^{-x^2} dx = 2\int_0^{\infty} e^{-x^2} dx = \sqrt{\pi}.$$

Thus,

$$I = \pi^{3/2}.$$

EXAMPLE 20

Let V denote the octant of a sphere defined by $x^2 + y^2 + z^2 \leqslant a^2$, $x \geqslant 0$, $y \geqslant 0$, $z \geqslant 0$. Evaluate

$$I = \iiint_V r^2 \, dV.$$

Solution

The region of integration V is shown in Fig. 5.22, and is covered by letting r range from 0 to a, θ range from 0 to $\frac{1}{2}\pi$, and φ range from 0 to $\frac{1}{2}\pi$. Thus, as

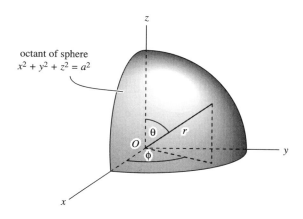

octant of sphere
$x^2 + y^2 + z^2 = a^2$

Fig. 5.22

$dV = r^2 \sin\theta \, dr \, d\theta \, d\varphi,$

$$I = \int_0^{\frac{1}{2}\pi} \int_0^{\frac{1}{2}\pi} \int_0^a r^4 \sin\theta \, dr \, d\theta \, d\varphi$$

$$= \left(\int_0^a r^4 \, dr \right) \left(\int_0^{\frac{1}{2}\pi} \sin\theta \, d\theta \right) \left(\int_0^{\frac{1}{2}\pi} d\varphi \right)$$

$$= \frac{1}{10} \pi a^5.$$

EXERCISES

5.39 If V denotes the region $R \leqslant a, \, 0 \leqslant z \leqslant b$ in cylindrical polar coordinates $R, \, \varphi, \, z$, evaluate

$$\iiint_V (R^2 + bz)^{\frac{1}{2}} \, dV.$$

5.40 If V denotes the region inside the semicircular cylinder $0 \leqslant x \leqslant \sqrt{(a^2 - y^2)}, \, 0 \leqslant z \leqslant 2a$, show that

$$\iiint_V x \, dV = \frac{4}{3} a^4.$$

5.41 Using spherical polar coordinates $r, \, \theta, \, \varphi$, evaluate

$$\iiint_V r^2 \sin^2\theta \, dV,$$

where V is the region in the positive octant bounded by the sphere $r = a$.

5.42 Parabolic coordinates u, v, w are defined so that the position vector is

$$\mathbf{r} = (\tfrac{1}{2}(u^2 - v^2), \ uv, \ -w).$$

Show that an element of volume in this coordinate system is $dV = (u^2 + v^2) \, du \, dv \, dw$.

Show that the region V bounded by the parabolic sheets $y^2 = 1 + 2x$, $y^2 = 1 - 2x$ and the planes $z = 0$, $z = 1$ is given in parabolic coordinates by

$$-1 \leqslant u \leqslant 1, \quad 0 \leqslant v \leqslant 1, \quad -1 \leqslant w \leqslant 0.$$

Evaluate

$$\iiint_V z \, dV.$$

Integral theorems

6

6.1 INTRODUCTION

In Chapter 4, scalar and vector fields were defined and the properties of gradient, divergence, and curl were discussed; and in Chapter 5 we were concerned with various integrals of scalar and vector fields, and the techniques whereby such integrals are evaluated. The ground has thus been prepared for the two central theorems in vector analysis: (i) **the divergence theorem** (also called **Gauss's theorem**), which relates the integral of a vector field **F** over a closed surface S to the volume integral of div **F** over the region bounded by S; and (ii) **Stokes's theorem** which relates the integral of a vector field **F** around a closed curve $\mathscr{C}$ to the integral of curl **F** over any open surface S bounded by $\mathscr{C}$. In this chapter we shall prove these theorems and some related results.

6.2 THE DIVERGENCE THEOREM (GAUSS'S THEOREM)

Definitions

1. A simple closed surface S is said to be **convex** if any straight line meeting it does so in at most two points.

2. A simple closed surface S will be called a **semi-convex surface** if axes $Oxyz$ can be so chosen that any line drawn parallel to one of the axes and meeting S either (i) does so in just one or just two points, or (ii) has a portion of finite length in common with S.

The divergence theorem

A closed region V is bounded by a simple closed surface S. If the vector field **F** is defined and continuously differentiable throughout V, then

$$\iint_S \mathbf{F} \cdot \mathbf{dS} = \iiint_V \operatorname{div} \mathbf{F} \, dV. \qquad (6.1)$$

Proof

The result will first be proved, in part (a), for a region V bounded by a convex surface S. The theorem will then be extended to more general regions in parts (b), (c) and (d).

(a) Take rectangular axes $Oxyz$ and denote by $\mathcal{R}$ the projection of S on to the xy-plane. Divide S into upper and lower parts, S_U and S_L, as viewed from $\mathcal{R}$ (Fig. 6.1). Let a line drawn parallel to Oz from a point in $\mathcal{R}$ cut S_L at P and S_U at Q, and denote the z-coordinates of P, Q by $z_P, z_Q (z_P < z_Q)$. All points (x, y, z) in V are covered if x and y range over all values in $\mathcal{R}$ and the corresponding values of z range from z_P to z_Q.

Denoting the components of $\mathbf{F}$ relative to the axes $Oxyz$ by F_1, F_2, F_3, we have

$$\iiint_V \frac{\partial F_3}{\partial z}\, dV = \iiint_V \frac{\partial F_3}{\partial z}\, dx\, dy\, dz.$$

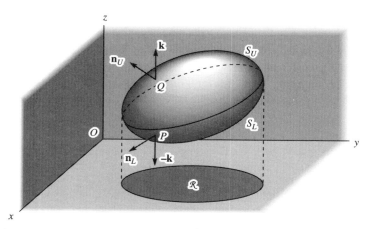

Fig. 6.1

Performing the integration with respect to z, this becomes

$$\iiint_V \frac{\partial F_3}{\partial z}\, dV = \iint_{\mathcal{R}} [F_3(x, y, z)]_{z=z_P}^{z=z_Q}\, dx\, dy$$

$$= \iint_{\mathcal{R}} \{F_3(x, y, z_Q) - F_3(x, y, z_P)\}\, dx\, dy. \qquad (6.2)$$

The integral on the right-hand side of (6.2) will now be expressed as an integral over the surface S.

Suppose that the equation of S_U is $z = f(x, y)$ (cf. section 5.5). Then a particular parametric representation of S_U is

$$\mathbf{r} = [x, y, f(x, y)],$$

where x, y range over values in $\mathcal{R}$. The unit normal to S_U is therefore

$$\mathbf{n}_U = \pm\, (\mathbf{r}_x \times \mathbf{r}_y)/|\mathbf{r}_x \times \mathbf{r}_y|$$

$$= \pm\, (1, 0, f_x) \times (0, 1, f_y)/|\mathbf{r}_x \times \mathbf{r}_y|$$

$$= \pm\, (-f_x, -f_y, 1)/(f_x^2 + f_y^2 + 1)^{\frac{1}{2}},$$

where suffixes refer to partial derivatives with respect to x, y. As we noted in section 5.5, it is conventional to choose the sign of the unit normal to a closed surface such as S so that it points away from the region enclosed. The z-component of the unit normal to the part S_U must therefore be positive, and so

$$\mathbf{n}_U = (-f_x, -f_y, 1)/(f_x^2 + f_y^2 + 1)^{\frac{1}{2}}.$$

Denoting the element of area of S_U by dS_U, it follows that

$$\mathbf{dS}_U = \mathbf{n}_U\, dS_U = \mathbf{r}_x \times \mathbf{r}_y\, dx\, dy$$

$$= (-f_x, -f_y, 1)\, dx\, dy.$$

Hence

$$\mathbf{k} \cdot \mathbf{dS}_U = dx\, dy,$$

where, as usual, $\mathbf{k} = (0, 0, 1)$ is the unit vector parallel to the z-axis. Noting that the z-component of the unit normal to the lower part S_L will be negative, it also follows by similar reasoning that

$$\mathbf{k} \cdot \mathbf{dS}_L = - dx\, dy.$$

Substituting these expressions into (6.2) and then noting that $S_U \cup S_L = S$ (see footnote), we thus have

$$\iiint_V \frac{\partial F_3}{\partial z}\, dV = \iint_{S_U} F_3 \mathbf{k} \cdot \mathbf{dS}_U + \iint_{S_L} F_3 \mathbf{k} \cdot \mathbf{dS}_L$$

$$= \iint_S F_3 \mathbf{k} \cdot \mathbf{dS}. \tag{6.3}$$

Similarly,

$$\iiint_V \frac{\partial F_2}{\partial y}\, dV = \iint_S F_2 \mathbf{j} \cdot \mathbf{dS} \tag{6.4}$$

and

$$\iiint_V \frac{\partial F_1}{\partial x}\, dV = \iint_S F_1 \mathbf{i} \cdot \mathbf{dS}. \tag{6.5}$$

The expression $S_U \cup S_L = S$ means that S is 'the union of' S_U and S_L; or, in other words, that S consists of the two parts S_U and S_L.

By adding (6.3), (6.4) and (6.5) we obtain

$$\iiint_V \left(\frac{\partial F_1}{\partial x} + \frac{\partial F_2}{\partial y} + \frac{\partial F_3}{\partial z} \right) dV = \iint_S (F_1 \mathbf{i} + F_2 \mathbf{j} + F_3 \mathbf{k}) \cdot \mathbf{dS};$$

that is

$$\iiint_V \operatorname{div} \mathbf{F} \, dV = \iint_S \mathbf{F} \cdot \mathbf{dS},$$

which is the result required.

(b) Consider next a region V bounded by a semi-convex surface S; then parts of S may consist of lines parallel to one of the axes. It is sufficient to consider how the proof in part (a) is modified when a portion S' of S consists of lines parallel to the z-axis; we denote the remainder of S by S'' (Fig. 6.2).

Following the lines of the argument in part (a), we obtain instead of (6.3),

$$\iint_{S''} F_3 \mathbf{k} \cdot \mathbf{dS''} = \iiint_V \frac{\partial F_3}{\partial z} \, dV. \tag{6.6}$$

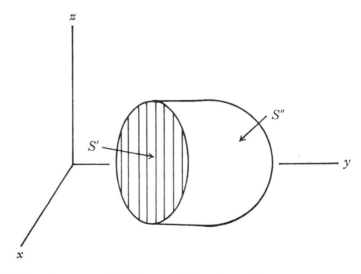

Fig. 6.2 S' here is composed of lines parallel to the z-axis.

But on S', the unit vector $\mathbf{k}$ is everywhere perpendicular to the vector element of area $\mathbf{dS'}$, and hence

$$\iint_{S'} F_3 \mathbf{k} \cdot \mathbf{dS'} = 0. \tag{6.7}$$

Since $S' \cup S'' = S$, we obtain (6.3) by adding (6.6) and (6.7). The remainder of the proof requires no modification, and the theorem follows as before.

(c) If the region V does not fall into one of the classes considered in parts (a) and (b), but can be divided into regions V_1, V_2 which do, we have the situation shown in Fig. 6.3. The theorem still holds, and may be established as follows.

Let S_1, S_2 denote the bounding surfaces of V_1, V_2 respectively. By what we have proved so far, the divergence theorem holds for V_1 and V_2 considered separately. Thus,

$$\iint_{S_1} \mathbf{F} \cdot d\mathbf{S}_1 = \iiint_{V_1} \operatorname{div} \mathbf{F}\, dV_1$$

and

$$\iint_{S_2} \mathbf{F} \cdot d\mathbf{S}_2 = \iiint_{V_2} \operatorname{div} \mathbf{F}\, dV_2.$$

By addition, as $V_1 \cup V_2 = V$,

$$\iint_{S_1} \mathbf{F} \cdot d\mathbf{S}_1 + \iint_{S_2} \mathbf{F} \cdot d\mathbf{S}_2 = \iiint_{V} \operatorname{div} \mathbf{F}\, dV. \tag{6.8}$$

Denote by S_1', S_2' the parts of S_1, S_2 which also form part of S (the bounding surface of V); and denote the remaining parts of S_1, S_2 by S_1'', S_2'' (Fig. 6.3). These latter parts actually coincide, and at each point their unit normals (which are directed away from V_1, V_2, respectively) point in opposite directions. Thus

$$\mathbf{F} \cdot d\mathbf{S}_1'' = -\mathbf{F} \cdot d\mathbf{S}_2'' \tag{6.9}$$

at all points on S_1'', S_2''.

Since $S_1 = S_1' \cup S_1''$ and $S_2 = S_2' \cup S_2''$, the left-hand side of (6.8) can be expressed as

$$\iint_{S_1'} \mathbf{F} \cdot d\mathbf{S}_1' + \iint_{S_1''} \mathbf{F} \cdot d\mathbf{S}_1'' + \iint_{S_2'} \mathbf{F} \cdot d\mathbf{S}_2' + \iint_{S_2''} \mathbf{F} \cdot d\mathbf{S}_2''.$$

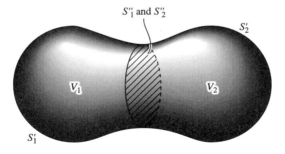

Fig. 6.3 Here S_1'' and S_2'' are opposite sides of the shaded surface separating V_1 from V_2.

Using (6.9), the second and fourth integrals are seen to cancel each other, and as $S_1' \cup S_2' = S$, the sum of the remaining two integrals is

$$\iint_S \mathbf{F} \cdot \mathbf{dS}.$$

Thus (6.8) becomes

$$\iint_S \mathbf{F} \cdot \mathbf{dS} = \iiint_V \operatorname{div} \mathbf{F} \, dV,$$

which proves the theorem for this case.

(d) The most general case that we contemplate here is that of a region V which can be subdivided into a finite number of regions $V_i (i = 1, 2, \ldots, n)$ bounded by semi-convex surfaces S_i. The proof given above for $n = 2$ can be extended to cover the general case; we leave the details to the reader as an exercise.

Extension to functions with discontinuous derivatives

In the statement of Gauss's theorem, the condition imposed upon the vector field $\mathbf{F}$ was that it should be defined and continuously differentiable throughout the closed region V. However, the theorem holds under less restrictive conditions.

A particularly important case is that of a vector field $\mathbf{F}$ which is continuous in V, but whose partial derivatives have finite discontinuities on a finite number of simple surfaces in V. The proof already given actually covers this case. The only point which need be drawn to the reader's attention is that the integration which leads to equation (6.2) is still valid, because in the interval $z_P \leqslant z \leqslant z_Q$ the integrand $\partial F_3/\partial z$ will have at most a finite number of finite discontinuities; the Riemann integral of such a function certainly exists.*

Although div $\mathbf{F}$ will not be defined on a surface where the partial derivatives of $\mathbf{F}$ are not continuous, it is nevertheless convenient to leave the statement (6.1) unchanged. More accurately, on a surface of discontinuity we should replace div $\mathbf{F}$ by $\partial F_1/\partial x + \partial F_2/\partial y + \partial F_3/\partial z$.

Examples of regions covered by the proof

Two particular regions to which the divergence theorem applies are worthy of mention.

First, the theorem applies to the region bounded by a torus (or 'anchor ring'). For, take axes $Oxyz$ such that O is at the centre of the torus and Oz is perpendicular to its central plane (Fig. 6.4). The quadrant of the torus which

* See, for example, Phillips, E. G. (1939) *A Course of Analysis*, p. 174, Cambridge University Press.

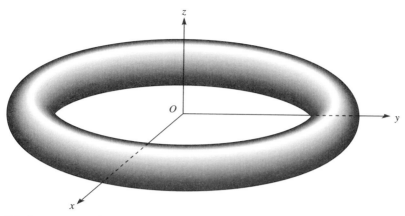

Fig. 6.4 A torus or anchor ring.

lies in the region $x \geqslant 0$, $y \geqslant 0$ is bounded by a semi-convex surface; and the remaining portion can be cut into three similar quadrants, each piece being bounded by a semi-convex surface. By part (d) of the proof, it follows that the divergence theorem is valid for the torus.

Second, the divergence theorem applies to the region V bounded by two concentric spheres. However, the integral

$$\iint_S \mathbf{F} \cdot d\mathbf{S}$$

is taken over both the inner and the outer bounding surfaces, and on each of these surfaces the unit normal is directed away from the enclosed region V (Fig. 6.5). These facts can be established by taking axes $Oxyz$ with origin at the common centre of the spheres, and applying the divergence theorem to the portion of V lying in each octant. We leave the verification to the reader.

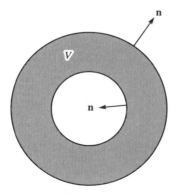

Fig. 6.5 The region V between two spheres.

EXAMPLE 1

The vector fields F, G have continuous first derivatives throughout a closed region V bounded by a simple closed surface S. At each point on S, $\mathbf{F} \times \mathbf{G}$ is tangential to the surface. Prove that

$$\iiint_V \mathbf{F} \cdot \operatorname{curl} \mathbf{G} \, dV = \iiint_V \mathbf{G} \cdot \operatorname{curl} \mathbf{F} \, dV.$$

Solution

Consider the identity (4.52), namely

$$\operatorname{div}(\mathbf{F} \times \mathbf{G}) = \mathbf{G} \cdot \operatorname{curl} \mathbf{F} - \mathbf{F} \cdot \operatorname{curl} \mathbf{G}.$$

As the conditions for the divergence theorem are fulfilled,

$$\iint_S (\mathbf{F} \times \mathbf{G}) \cdot \mathbf{dS} = \iiint_V \operatorname{div}(\mathbf{F} \times \mathbf{G}) \, dV$$

$$= \iiint_V (\mathbf{G} \cdot \operatorname{curl} \mathbf{F} - \mathbf{F} \cdot \operatorname{curl} \mathbf{G}) \, dV.$$

But $\mathbf{dS} = \mathbf{n} \, dS$, and as $\mathbf{F} \times \mathbf{G}$ is everywhere tangential to S (and so normal to $\mathbf{n}$) we have $(\mathbf{F} \times \mathbf{G}) \cdot \mathbf{n} = 0$. Thus the surface integral on the left-hand side vanishes, and the required result follows.

EXAMPLE 2

Important corollary to the divergence theorem
If the scalar field Ω together with its gradient are defined throughout a closed region V bounded by a simple closed surface S, then

$$\iint_S \Omega \mathbf{dS} = \iiint_V \operatorname{grad} \Omega \, dV. \tag{6.10}$$

Proof

Let $\mathbf{a}$ be any non-zero constant vector field. From the identity (4.49), as $\mathbf{a}$ is a constant vector,

$$\operatorname{div}(\Omega \mathbf{a}) = \mathbf{a} \cdot \operatorname{grad} \Omega.$$

Thus applying the divergence theorem to the vector field $\Omega \mathbf{a}$,

$$\iint_S \Omega \mathbf{a} \cdot \mathbf{dS} = \iiint_V \operatorname{div}(\Omega \mathbf{a}) \, dV$$

$$= \iiint_V \mathbf{a} \cdot \operatorname{grad} \Omega \, dV.$$

Since $\mathbf{a}$ is constant, we may rewrite this as

$$\mathbf{a} \cdot \left(\iint_S \Omega\, \mathbf{dS} - \iiint_V \operatorname{grad} \Omega\, dV \right) = 0.$$

Putting

$$\iint_S \Omega\, \mathbf{dS} - \iiint_V \operatorname{grad} \Omega\, dV = \mathbf{F},$$

it follows that the component of $\mathbf{F}$ parallel to $\mathbf{a}$ is zero. But the direction of $\mathbf{a}$ can be chosen arbitrarily, and so $\mathbf{F} = \mathbf{0}$, which gives the required result.

EXERCISES

6.1 Using the divergence theorem, evaluate

$$\iint_S \mathbf{F} \cdot \mathbf{dS},$$

where $\mathbf{F} = (x^3, y^3, z^3)$ and S is the sphere $x^2 + y^2 + z^2 = a^2$. Verify the result by evaluating the surface integral directly.

6.2 If the scalar field Ω has continuous second order derivatives in the closed region V bounded by a simple closed surface S, prove that

$$\iint_S \Omega \nabla \Omega \cdot \mathbf{dS} = \iiint_V \{ \Omega \nabla^2 \Omega + (\nabla \Omega)^2 \}\, dV.$$

If $\Omega = x + y + z$, deduce that

$$\iint_S \Omega \nabla \Omega \cdot \mathbf{dS} = 3V_0,$$

where V_0 is the volume of the region V.

6.3 Show that

$$\iint_S \mathbf{F} \times \mathbf{dS} = - \iiint_V \operatorname{curl} \mathbf{F}\, dV,$$

stating the conditions to be satisfied by $\mathbf{F}$ and the relationship of S and V. [*Hint*. Apply the divergence theorem to the vector field $\mathbf{F} \times \mathbf{a}$, where $\mathbf{a}$ is an arbitrary constant vector field.]

6.4 If the region V is bounded by a simple closed surface S, and $\mathbf{F}$, $\mathbf{G}$ are continuously differentiable in V, prove that

$$\iint_S \mathbf{F}(\mathbf{G} \cdot \mathbf{dS}) = \iiint_V (\mathbf{G} \cdot \nabla)\mathbf{F}\, dV + \iiint_V \mathbf{F} \operatorname{div} \mathbf{G}\, dV.$$

[*Hint*. Consider a single component of this vector equation.]

6.5 A region V is bounded by simple closed surfaces S_1 and S_2. If the origin lies outside V, prove that

$$\iint_{S_1} \frac{\mathbf{r} \cdot \mathbf{dS}_1}{r^3} + \iint_{S_2} \frac{\mathbf{r} \cdot \mathbf{dS}_2}{r^3} = 0.$$

6.6 A given continuously differentiable scalar function λ is positive in a region V bounded by a simple closed surface S. Continuously differentiable vector fields $\mathbf{u}$, $\mathbf{v}$, $\mathbf{A}$, defined in V, are such that:

(i) $\mathbf{v} = \mathbf{u} + \mathrm{curl}\,\mathbf{A}$;
(ii) $\mathrm{curl}\,\lambda \mathbf{u} = \mathbf{0}$;
(iii) the tangential components of $\mathbf{A}$ vanish on S.

Prove that

$$\iiint_V \lambda v^2 \, dV \geqslant \iiint_V \lambda u^2 \, dV.$$

[*Hint*. Consider the integral over S of $2\lambda \mathbf{u} \times \mathbf{A}$.]

6.7 The scalar field Ω satisfies Laplace's equation in a region V bounded by a simple closed surface S. Show that

$$\iint_S \mathbf{n} \cdot \mathrm{grad}\,\Omega \, dS = 0.$$

If S_r is a sphere of radius r and fixed centre O lying in V, show that

$$\frac{d}{dr}\left(\frac{1}{r^2} \iint_{S_r} \Omega \, dS\right) = \frac{1}{r^2} \iint_{S_r} \frac{\partial \Omega}{\partial r} \, dS.$$

Taking spherical polar coordinates r, θ, φ with origin at O, and writing $\Omega(r, \theta, \varphi) = \Omega_r$, deduce that

$$\int_0^{2\pi} \int_0^{\pi} (\Omega_r - \Omega_0) \sin\theta \, d\theta \, d\varphi = 0.$$

Hence prove that the maximum and minimum values of Ω occur on the boundary S.

6.3 GREEN'S THEOREMS

Let scalar fields Φ and Ψ, together with $\nabla^2 \Phi$ and $\nabla^2 \Psi$, be defined throughout a closed region V, bounded by a simple closed surface S, and suppose that any discontinuities in the second derivatives of Φ and Ψ are finite and are confined to a finite number of simple surfaces in V. **Green's first theorem** is that

$$\iint_S \Phi \frac{\partial \Psi}{\partial n} \, dS = \iiint_V (\Phi \nabla^2 \Psi + \operatorname{grad} \Phi \,.\, \operatorname{grad} \Psi) \, dV. \qquad (6.11)$$

Here, $\partial/\partial n$ denotes the directional derivative (see section 4.5) along the outward normal to S. Interchanging Φ and Ψ in (6.11) we obtain

$$\iint_S \Psi \frac{\partial \Phi}{\partial n} \, dS = \iiint_V (\Psi \nabla^2 \Phi + \operatorname{grad} \Psi \,.\, \operatorname{grad} \Phi) \, dV;$$

whence, by substraction from (6.11),

$$\iint_S \left(\Phi \frac{\partial \Psi}{\partial n} - \Psi \frac{\partial \Phi}{\partial n} \right) dS = \iiint_V (\Phi \nabla^2 \Psi - \Psi \nabla^2 \Phi) \, dV, \qquad (6.12)$$

which is **Green's second theorem**. The first theorem is proved as follows.

Proof

Take $\mathbf{H} = \Phi \operatorname{grad} \Psi$. Then using identity (4.49),

$$\operatorname{div} \mathbf{H} = \Phi \operatorname{div}(\operatorname{grad} \Psi) + \operatorname{grad} \Psi \,.\, \operatorname{grad} \Phi$$

$$= \Phi \nabla^2 \Psi + \operatorname{grad} \Phi \,.\, \operatorname{grad} \Psi.$$

Applying the divergence theorem to the vector field $\mathbf{H}$, we thus have

$$\iint_S \Phi \operatorname{grad} \Psi \,.\, \mathbf{dS} = \iiint_V (\Phi \nabla^2 \Psi + \operatorname{grad} \Phi \,.\, \operatorname{grad} \Psi) \, dV.$$

But $\Phi \operatorname{grad} \Psi \,.\, \mathbf{dS} = \Phi \mathbf{n} \,.\, \operatorname{grad} \Psi \, dS = \Phi(\partial \Psi/\partial n) \, dS$. Hence (6.11) follows.

Laplace's equation: a uniqueness theorem

Let V be a closed region, bounded by a simple surface S, and suppose that a scalar field Ω is such that:
   (i) Ω satisfies Laplace's equation in V;   (6.13)
   (ii) the values of Ω on S are given.   (6.14)
Then Ω is unique.

Proof

Suppose there are two scalar fields Ω, say Ω_1 and Ω_2, satisfying conditions (i) and (ii). Then

$$\nabla^2 \Omega_1 = 0, \quad \nabla^2 \Omega_2 = 0 \quad \text{throughout } V;$$

and

$$\Omega_1 = \Omega_2 \quad \text{at all points on } S.$$

Consider

$$\Phi = \Omega_1 - \Omega_2. \tag{6.15}$$

Then, since ∇^2 is a linear operator,

$$\nabla^2 \Phi = \nabla^2 \Omega_1 - \nabla^2 \Omega_2$$

$$= 0 \quad \text{throughout} \quad V;$$

and

$$\Phi = 0 \text{ on } S.$$

Applying Green's theorem (6.11) with $\Psi = \Phi$, we thus have

$$\iiint_V (\text{grad } \Phi)^2 \, dV = 0. \tag{6.16}$$

Now $(\text{grad } \Phi)^2 \geq 0$ at all points. Suppose that $(\text{grad } \Phi)^2 > 0$ at some point P in the interior of V. Then as $\text{grad } \Phi$ is continuous throughout the interior of V (the existence of $\text{grad } \Phi$ is sufficient to ensure this) there must be an open region, V_1 say, containing P and in which $(\text{grad } \Phi)^2 > 0$. We then have

$$\iiint_{V_1} (\text{grad } \Phi)^2 \, dV_1 > 0.$$

As the integral of $(\text{grad } \Phi)^2$ over the remainder of V cannot be negative, it follows that

$$\iiint_V (\text{grad } \Phi)^2 \, dV > 0,$$

which contradicts (6.16). We conclude that at all points in V, $\text{grad } \Phi = \mathbf{0}$. In components this becomes

$$\frac{\partial \Phi}{\partial x} = 0, \quad \frac{\partial \Phi}{\partial y} = 0, \quad \frac{\partial \Phi}{\partial z} = 0,$$

showing that Φ is independent of x, y and z. Thus, in V, $\Phi = \text{constant}$. Since $\Phi = 0$ on the boundary of V, the constant is zero, and so $\Phi \equiv 0$ in V. Returning to (6.15), we see that this implies that $\Omega_1 \equiv \Omega_2$, which establishes the required result that there cannot be two different scalar fields each satisfying conditions (6.13) and (6.14).

Extension to infinite regions

Suppose that V is the infinite region outside a simple closed surface S. The scalar field Ω is again unique if, in addition to conditions (6.13) and (6.14),

$$\Omega = O\left(\frac{1}{r}\right) \text{ at large distances } r \text{ (see footnote).} \tag{6.17}$$

Expression (6.17) means that Ω tends to zero at least as rapidly as $1/r$ when $r \to \infty$. More precisely there exist a value R of r and a positive constant K, independent of r, such that $|\Omega| \leq K/r$ whenever $r \geq R$.

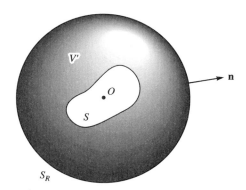

Fig. 6.6 The sphere S_R of radius R and centre the origin completely surrounds S. As $R \to \infty$ the sphere S_R becomes infinitely large.

Proof

We proceed as before, and consider scalar fields Ω_1 and Ω_2, each satisfying conditions (6.13), (6.14), and (6.17). Since $\Omega_1 = O(1/r)$ and $\Omega_2 = O(1/r)$ as $r \to \infty$, we also have

$$\Phi = \Omega_1 - \Omega_2 = O\left(\frac{1}{r}\right) \text{ as } r \to \infty. \tag{6.18}$$

Applying Green's theorem (6.11), with $\Phi = \Psi$, to the region V' bounded internally by S and externally by a sphere S_R of radius R and centre at the origin O (Fig. 6.6),

$$\iiint_{V'} \{\Phi \nabla^2 \Phi + (\text{grad}\,\Phi)^2\}\, dV' = \iint_S \Phi \frac{\partial \Phi}{\partial n}\, dS + \iint_{S_R} \Phi \frac{\partial \Phi}{\partial n}\, dS_R.$$

As before, $\nabla^2 \Phi = 0$ in V' and $\Phi = 0$ on S. Also, on S_R, $\partial/\partial n \equiv \partial/\partial r$, because the outward unit normal to S_R is in the same direction as the unit vector $\hat{\mathbf{r}}$ from the origin. Thus

$$\iiint_{V'} (\text{grad}\,\Phi)^2\, dV' = \iint_{S_R} \left[\Phi \frac{\partial \Phi}{\partial r}\right]_{r=R} dS_R.$$

If R is sufficiently large, then by (6.18) there exists a positive constant K such that for all $r \geqslant R$,

$$\left|\Phi \frac{\partial \Phi}{\partial r}\right| \leqslant \frac{K}{r} \times \frac{K}{r^2} = \frac{K^2}{r^3}.$$

Thus

$$\left|\iiint_{V'} (\text{grad}\,\Phi)^2\, dV'\right| \leqslant \iint_{S_R} (K^2/R^3)\, dS_R$$

$$= 4\pi K^2/R,$$

since the area of S_R is $4\pi R^2$. When $R \to \infty$, V' becomes the given infinite region V, and so in the limit

$$\iiint_V (\operatorname{grad} \varPhi)^2 \, dV = 0,$$

which is the same as equation (6.16). From this point the proof is completed as before.

EXERCISES

6.8 Assuming that V is a closed region bounded by a simple closed surface S, obtain from Green's second theorem the following results:

(i) For any scalar field $\varPhi$ having continuous second derivatives throughout V,

$$\iint_S \frac{\partial \varPhi}{\partial n} \, dS = \iiint_V \nabla^2 \varPhi \, dV.$$

(ii) For any scalar fields $\varPhi$ and $\varPsi$ which satisfy Laplace's equation in V,

$$\iint_S \varPhi \frac{\partial \varPsi}{\partial n} \, dS = \iint_S \varPsi \frac{\partial \varPhi}{\partial n} \, dS.$$

6.9 Scalar fields $\varPhi, \varPhi'$ satisfy Laplace's equation in a region V bounded by n simple closed surfaces $S_1, S_2, \ldots, S_n$. On the surface $S_i (i = 1, 2, \ldots, n)$,

$$\iint_{S_i} \frac{\partial \varPhi}{\partial n} \, dS_i = \alpha'_i, \quad \iint_{S_i} \frac{\partial \varPhi'}{\partial n} \, dS_i = \alpha'_i,$$

and

$$\varPhi = C_i, \quad \varPhi' = C'_i,$$

where C_i, C'_i are constants. Using Green's second theorem, show that

$$\sum_{i=1}^{n} C_i \alpha'_i = \sum_{i=1}^{n} C'_i \alpha_i.$$

6.10 The scalar field $\varOmega$ satisfies Laplace's equation in the closed region V bounded by a simple closed surface S. If $\varOmega = C$ (a constant) on S, prove by means of Green's first theorem (with $\Phi = \Psi = \Omega$) that $\Omega = C$ throughout V. [*Note.* This result can also be deduced at once from the uniqueness theorem, proved in the text of this section.]

6.11 A closed region V is bounded by a simple closed surface S. A continuously differentiable scalar field Ω is such that:

    (i) $\nabla^2 \Omega = 0$ in V;

    (ii) $\partial\Omega/\partial n = f$ on S, where f is a given function and $\partial/\partial n$ denotes the derivative along the normal to S;

    (iii) Ω takes a given value at one point in V.

Prove that Ω is unique.

If V is an infinite region bounded internally by a simple closed surface S, and if condition (iii) is replaced by:

    (iii)$'$ $\Omega = O(1/r)$ at large distances from S,

prove that Ω is again unique. [The proofs of these results follow closely the lines of the proof of the uniqueness theorem in the text of this section.]

6.4 STOKES'S THEOREM

Consider a simple closed curve $\mathscr{C}$ spanned by a surface S, as in Fig. 6.7. Then, briefly (a complete statement is given later), Stokes's theorem states that, for a vector field **F**,

$$\oint_{\mathscr{C}} \mathbf{F} \cdot d\mathbf{r} = \iint_{S} \text{curl } \mathbf{F} \cdot d\mathbf{S}. \tag{6.19}$$

For the theorem to hold, the orientation of the surface S and the curve $\mathscr{C}$ must be correctly related, and **F** must satisfy certain analytical conditions. The theorem will be proved first for curves spanned by flat surfaces and later will be extended to twisted curves spanned by surfaces of arbitrary shape.

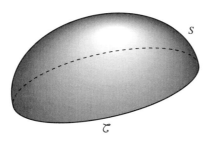

Fig. 6.7 A simple closed curve $\mathscr{C}$ spanned by a surface S.

Orientation of a surface spanning a closed curve

Let Q, R be neighbouring points on the boundary $\mathscr{C}$ of a simple open surface S, such that $\mathscr{C}$ is described in the sense of the smaller arc from Q to R (Fig. 6.8). Let P be a point on S near to Q and R. The surface S and its boundary $\mathscr{C}$ are said to be correspondingly oriented if $\overrightarrow{PQ}$, $\overrightarrow{PR}$ and the unit normal $\mathbf{n}$ to S (in that order) form a right-handed triad.

An equivalent statement is that S and $\mathscr{C}$ are correspondingly oriented if an observer on the positive side of S travelling around $\mathscr{C}$ in its positive sense always has the surface S on his left.

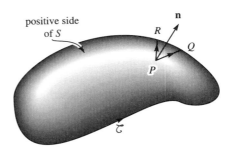

Fig. 6.8 The orientation of a surface S spanning a curve $\mathscr{C}$.

Stokes's theorem in the plane

Let $F_1(x, y)$ and $F_2(x, y)$ be defined and have continuous first derivatives throughout a closed region $\mathscr{R}$ in the xy-plane. Let the boundary of $\mathscr{R}$ be a simple closed curve described in the anticlockwise sense (Fig. 6.9). Then

$$\oint_{\mathscr{C}} \left(F_1 \frac{dx}{ds} + F_2 \frac{dy}{ds} \right) ds = \int\int_{\mathscr{R}} \left(\frac{\partial F_2}{\partial x} - \frac{\partial F_1}{\partial y} \right) dx\, dy, \qquad (6.20)$$

where s denotes arc length along $\mathscr{C}$. This is Stokes's theorem in the plane. The reader should verify that (6.19) reduces to (6.20) when the surface S lies entirely in the xy-plane.*

---

* Stokes's theorem in the plane is often expressed more briefly as

$$\oint_{\mathscr{C}} (F_1\, dx + F_2\, dy) = \int\int_{\mathscr{R}} \left(\frac{\partial F_2}{\partial x} - \frac{\partial F_1}{\partial y} \right) dx\, dy,$$

which implies that the curve $\mathscr{C}$ can be expressed in both the forms $y = f(x)$ and $x = g(y)$. However as the curve is closed, f and g would not be single-valued functions. Therefore we shall retain the parameter s.

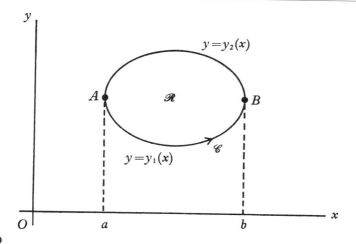

Fig. 6.9

Proof

We consider various cases in turn.

(a) *The case when $\mathscr{C}$ is such that any line parallel to the x- or y-axis has at most two points in common with $\mathscr{C}$*

Suppose that, in $\mathscr{R}$, x ranges from a to b, and let the ordinates through the points $(a, 0)$, $(b, 0)$ touch $\mathscr{C}$ at A, B, respectively. Let $y = y_1(x)$ and $y = y_2(x)$ be the equations of the lower and upper parts of $\mathscr{C}$ (between A and B) respectively (Fig. 6.9). Then

$$\iint_{\mathscr{R}} -\frac{\partial F_1}{\partial y} \, dx \, dy = \int_a^b \int_{y_1(x)}^{y_2(x)} -\frac{\partial F_1}{\partial y} \, dy \, dx$$

$$= \int_a^b \{F_1[x, y_1(x)] - F_1[x, y_2(x)]\} \, dx$$

$$= \int_A^B F_1 \frac{dx}{ds} \, ds + \int_B^A F_1 \frac{dx}{ds} \, ds, \qquad (6.21)$$

where the first line integral is taken along the lower part of $\mathscr{C}$ and the second along the upper part. Thus

$$\iint_{\mathscr{R}} -\frac{\partial F_1}{\partial y} \, dx \, dy = \oint_{\mathscr{C}} F_1 \frac{dx}{ds} \, ds. \qquad (6.22)$$

Similarly, we can show that

$$\iint_{\mathscr{R}} \frac{\partial F_2}{\partial x} \, dx \, dy = \oint_{\mathscr{C}} F_2 \frac{dy}{ds} \, ds. \qquad (6.23)$$

Addition of the two results (6.22) and (6.23) thus gives (6.20), as required.

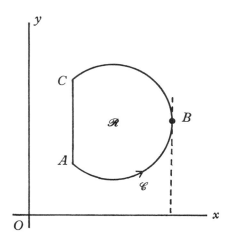

Fig. 6.10 The portion AC of $\mathcal{C}$ is parallel to the y-axis.

(b) *The case when part of $\mathcal{C}$ is parallel to an axis*
Suppose the curve $\mathcal{C}$ is as shown in Fig. 6.10, and let A, B, C be the points indicated on the diagram. We now obtain instead of (6.21)

$$\iint_{\mathcal{R}} -\frac{\partial F_1}{\partial y}\, dx\, dy = \int_A^B F_1 \frac{dx}{ds}\, ds + \int_B^C F_1 \frac{dx}{ds}\, ds$$

$$= \oint_{\mathcal{C}} F_1 \frac{dx}{ds}\, ds - \int_C^A F_1 \frac{dx}{ds}\, ds,$$

where all the line integrals are taken along $\mathcal{C}$. But on CA, $dx/ds = 0$, and so the final integral vanishes. We thus have (6.22) once again, and the theorem follows as before.

(c) *The case when some lines parallel to the axes cut the boundary of $\mathcal{R}$ more than twice*
For simplicity we shall consider only the situation shown in Fig. 6.11, but the method can readily be extended to deal with still more complicated boundaries.

Subdivide $\mathcal{R}$ into regions $\mathcal{R}_1$, $\mathcal{R}_2$, $\mathcal{R}_3$ by drawing the line ABC, as in the diagram. Each of $\mathcal{R}_1$, $\mathcal{R}_2$, $\mathcal{R}_3$ is a region of the type considered in part (b). Denoting the boundary of $\mathcal{R}_1$ by $\mathcal{C}_1$, application of part (b) gives

$$\iint_{\mathcal{R}_1} -\frac{\partial F_1}{\partial y}\, dx\, dy = \oint_{\mathcal{C}_1} F_1 \frac{dx}{ds}\, ds$$

$$= \int_A^C F_1 \frac{dx}{ds}\, ds, \quad \text{taken along } \mathcal{C},$$

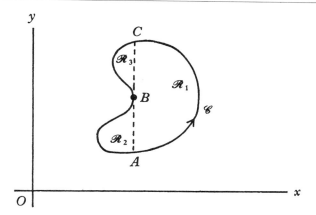

Fig. 6.11 Showing how to deal with more complicated boundaries.

since $dx/ds = 0$ on the line CA. Similarly,

$$\iint_{\mathcal{R}_2} -\frac{\partial F_1}{\partial y} \, dx \, dy = \int_C^B F_1 \frac{dx}{ds} \, ds, \quad \text{taken along } \mathcal{C};$$

and

$$\iint_{\mathcal{R}_3} -\frac{\partial F_1}{\partial y} \, dx \, dy = \int_B^A F_1 \frac{dx}{ds} \, ds, \quad \text{taken along } \mathcal{C}.$$

Since $\mathcal{R}_1$, $\mathcal{R}_2$, $\mathcal{R}_3$ taken together form $\mathcal{R}$, addition of these three results gives (6.22). The proof is completed as before.

Lemma

If the vector fields $\mathbf{F} = \mathbf{F}(u, v)$ and $\mathbf{r} = \mathbf{r}(u, v)$ have continuous first partial derivatives, then

$$(\text{curl } \mathbf{F}) \cdot (\mathbf{r}_u \times \mathbf{r}_v) = \mathbf{r}_v \cdot \mathbf{F}_u - \mathbf{r}_u \cdot \mathbf{F}_v, \tag{6.24}$$

where suffixes refer to partial derivatives with respect to u, v.

Proof

Using the notation of Section 4.9,

$$(\text{curl } \mathbf{F}) \cdot (\mathbf{r}_u \times \mathbf{r}_v) = \left(\mathbf{e}_i \times \frac{\partial \mathbf{F}}{\partial x_i} \right) \cdot \left(\frac{\partial \mathbf{r}}{\partial u} \times \frac{\partial \mathbf{r}}{\partial v} \right)$$

$$= \mathbf{e}_i \cdot \left\{ \frac{\partial \mathbf{F}}{\partial x_i} \times \left(\frac{\partial \mathbf{r}}{\partial u} \times \frac{\partial \mathbf{r}}{\partial v} \right) \right\}$$

$$= \mathbf{e}_i \cdot \left\{ \left(\frac{\partial \mathbf{r}}{\partial v} \cdot \frac{\partial \mathbf{F}}{\partial x_i} \right) \frac{\partial \mathbf{r}}{\partial u} - \left(\frac{\partial \mathbf{r}}{\partial u} \cdot \frac{\partial \mathbf{F}}{\partial x_i} \right) \frac{\partial \mathbf{r}}{\partial v} \right\}.$$

Now

$$\mathbf{r} = x_1\mathbf{e}_1 + x_2\mathbf{e}_2 + x_3\mathbf{e}_3 = x_j\mathbf{e}_j,$$

and so

$$\begin{aligned}
\mathbf{e}_i \cdot (\partial\mathbf{r}/\partial u) &= \mathbf{e}_i \cdot \mathbf{e}_j \, \partial x_j/\partial u \\
&= \delta_{ij} \, \partial x_j/\partial u \\
&= \partial x_i/\partial u.
\end{aligned}$$

Similarly,

$$\mathbf{e}_i \cdot (\partial\mathbf{r}/\partial v) = \partial x_i/\partial v.$$

Thus

$$(\text{curl } \mathbf{F}) \cdot (\mathbf{r}_u \times \mathbf{r}_v) = \left(\mathbf{r}_v \cdot \frac{\partial\mathbf{F}}{\partial x_i}\right)\frac{\partial x_i}{\partial u} - \left(\mathbf{r}_u \cdot \frac{\partial\mathbf{F}}{\partial x_i}\right)\frac{\partial x_i}{\partial v}.$$

Expanding the first term on the right of the above expression, and using the chain rule (4.3),

$$\left(\mathbf{r}_v \cdot \frac{\partial\mathbf{F}}{\partial x_i}\right)\frac{\partial x_i}{\partial u} = \mathbf{r}_v \cdot \left(\frac{\partial\mathbf{F}}{\partial x_1}\frac{\partial x_1}{\partial u} + \frac{\partial\mathbf{F}}{\partial x_2}\frac{\partial x_2}{\partial u} + \frac{\partial\mathbf{F}}{\partial x_3}\frac{\partial x_3}{\partial u}\right)$$

$$= \mathbf{r}_v \cdot \mathbf{F}_u.$$

Using this and a similar expression with u and v interchanged, it follows that

$$(\text{curl } \mathbf{F}) \cdot (\mathbf{r}_u \times \mathbf{r}_v) = \mathbf{r}_v \cdot \mathbf{F}_u - \mathbf{r}_u \cdot \mathbf{F}_v,$$

as required.

With the help of the above lemma and the knowledge of Stokes's theorem in the plane, we can now proceed to the general case, as follows.

Stokes's theorem

Let the vector field $\mathbf{F}$, together with curl $\mathbf{F}$, be defined everywhere on a simple open surface S with correspondingly oriented boundary $\mathscr{C}$. Then

$$\oint_{\mathscr{C}} \mathbf{F} \cdot d\mathbf{r} = \iint_S \text{curl } \mathbf{F} \cdot d\mathbf{S}. \tag{6.25}$$

Proof

Let the parametric equation of S be

$$\mathbf{r} = \mathbf{r}(u, v), \tag{6.26}$$

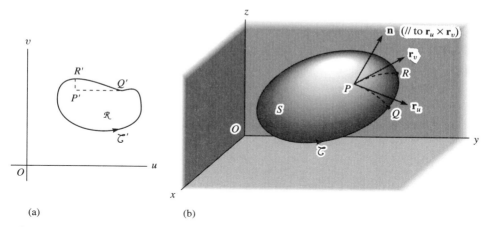

Fig. 6.12 (a) The mapping onto the uv-plane of the surface S shown in (b). If the surface normal $\mathbf{n}$ to S is chosen parallel to $\mathbf{r}_u \times \mathbf{r}_v$, $\mathscr{C}'$ will be traversed in the conventional anticlockwise sense.

where u, v range over a closed region $\mathscr{R}$ in the uv-plane. Apart from the usual analytical restrictions on $\mathbf{r}(u, v)$ and its first partial derivatives, we shall find it necessary in the course of the proof to assume that the second partial derivatives of $\mathbf{r}$ with respect to u, v exist and are continuous. This is a condition on the nature of the surface S slightly stronger than usual, but in practical applications this will cause no difficulty. We denote by $\mathscr{C}'$ the boundary of $\mathscr{R}$ (Fig. 6.12).

The boundary $\mathscr{C}'$ of $\mathscr{R}$ is mapped into the boundary $\mathscr{C}$ of S by (6.26). For, as the mapping is single-valued and continuous, if $\mathscr{C}'$ is shrunk continuously to a point P' in $\mathscr{R}$ the corresponding curve $\mathscr{C}$ will sweep out in a continuous manner the whole of S, and will finally reduce to a point P on S.

If $\mathscr{C}'$ is traversed in the anti-clockwise sense (Fig. 6.12), the surface S and the boundary $\mathscr{C}$ will be correspondingly oriented if we take the unit normal to S as

$$\mathbf{n} = \mathbf{r}_u \times \mathbf{r}_v / |\mathbf{r}_u \times \mathbf{r}_v|.$$

To see this, take any point P' in $\mathscr{R}$ close to $\mathscr{C}'$, such that the lines through P' in the directions u increasing and v increasing meet $\mathscr{C}'$ at neighbouring points Q' and R' respectively (Fig. 6.12). Let P, Q, R be the corresponding points on S. When $\mathscr{C}'$ is described in the anticlockwise sense from Q' to R', $\mathscr{C}$ will be described in the sense of the smaller arc from Q to R. Also, P and Q lie on the same u-coordinate curve, with u increasing towards Q; and P and R lie on the same v-coordinate curve, with v increasing towards R. As $\mathbf{r}_u$, $\mathbf{r}_v$ are tangential to the u-, v-coordinate curves and in the directions u, v increasing, respectively, and as $\mathbf{r}_u$, $\mathbf{r}_v$ and $\mathbf{n}$ (in that order) form a right-handed triad, it follows that $\overrightarrow{PQ}$, $\overrightarrow{PR}$ and $\mathbf{n}$ also form a right-handed triad. This is the required condition for $\mathscr{C}$ and S to be correspondingly oriented.

Consider the right-hand side of (6.25). Substituting $\mathbf{dS} = (\mathbf{r}_u \times \mathbf{r}_v)\,du\,dv$, using the lemma (6.24), and assuming that the second derivatives of $\mathbf{r}$ are continuous (so that $\mathbf{r}_{uv} = \mathbf{r}_{vu}$), we have

$$\iint_S \operatorname{curl} \mathbf{F} \cdot \mathbf{dS} = \iint_{\mathscr{R}} (\mathbf{r}_v \cdot \mathbf{F}_u - \mathbf{r}_u \cdot \mathbf{F}_v)\,du\,dv$$

$$= \iint_{\mathscr{R}} \left\{ \frac{\partial}{\partial u}(\mathbf{F} \cdot \mathbf{r}_v) - \frac{\partial}{\partial v}(\mathbf{F} \cdot \mathbf{r}_u) \right\} du\,dv.$$

Applying Stokes's theorem in the plane, as given by (6.20), it follows that

$$\iint_S \operatorname{curl} \mathbf{F} \cdot \mathbf{dS} = \oint_{\mathscr{C}'} \left(\mathbf{F} \cdot \mathbf{r}_u \frac{du}{ds'} + \mathbf{F} \cdot \mathbf{r}_v \frac{dv}{ds'} \right) ds'$$

$$= \oint_{\mathscr{C}} \mathbf{F} \cdot \left(\frac{\partial \mathbf{r}}{\partial u} \frac{du}{ds} + \frac{\partial \mathbf{r}}{\partial v} \frac{dv}{ds} \right) ds,$$

where s' and s denote arc lengths along $\mathscr{C}'$ and $\mathscr{C}$ respectively. But the chain rule (4.3) gives

$$\frac{\partial \mathbf{r}}{\partial u} \frac{du}{ds} + \frac{\partial \mathbf{r}}{\partial v} \frac{dv}{ds} = \frac{d\mathbf{r}}{ds}.$$

Hence, as required,

$$\iint_S \operatorname{curl} \mathbf{F} \cdot \mathbf{dS} = \oint_{\mathscr{C}} \mathbf{F} \cdot d\mathbf{r}.$$

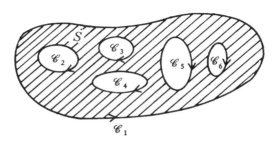

Fig. 6.13

Corollary: surfaces with several boundaries

If the simple open surface S is bounded by n correspondingly oriented simple closed curves, $\mathscr{C}_1, \mathscr{C}_2, \ldots, \mathscr{C}_n$ (Fig. 6.13 illustrates the case $n = 6$) we have instead of (6.25):

$$\oint_{\mathscr{C}_1} \mathbf{F} . d\mathbf{r} + \oint_{\mathscr{C}_2} \mathbf{F} . d\mathbf{r} + \ldots + \oint_{\mathscr{C}_n} \mathbf{F} . d\mathbf{r} = \iint_S \operatorname{curl} \mathbf{F} . d\mathbf{S}. \qquad (6.27)$$

Proof

Consider the case when there are just two boundaries $\mathscr{C}_1$, $\mathscr{C}_2$ (Fig. 6.14). Let AB be a simple curve on S (called a **cross cut**) joining a point A on $\mathscr{C}_1$ to a point B on $\mathscr{C}_2$. If S is cut along AB, we may regard the surface as having a single boundary consisting of $\mathscr{C}_1$, AB, $\mathscr{C}_2$ and BA as shown. Applying Stokes's theorem in the form already proved,

$$\oint_{\mathscr{C}_1} \mathbf{F} . d\mathbf{r} + \int_{AB} \mathbf{F} . d\mathbf{r} + \oint_{\mathscr{C}_2} \mathbf{F} . d\mathbf{r} + \int_{BA} \mathbf{F} . d\mathbf{r} = \iint_S \operatorname{curl} \mathbf{F} . d\mathbf{S}.$$

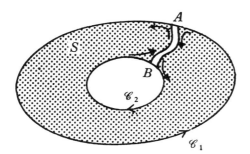

Fig. 6.14

But

$$\int_{AB} \mathbf{F} . d\mathbf{r} = -\int_{BA} \mathbf{F} . d\mathbf{r}.$$

Thus

$$\oint_{\mathscr{C}_1} \mathbf{F} . d\mathbf{r} + \oint_{\mathscr{C}_2} \mathbf{F} . d\mathbf{r} = \iint_S \operatorname{curl} \mathbf{F} . d\mathbf{S},$$

which proves (6.27) for the case $n = 2$.

The general case can be treated by introducing $n - 1$ suitable cross cuts.

EXAMPLE 3

A simple open surface S is bounded by a correspondingly oriented simple closed curve $\mathscr{C}$. If $\mathbf{a}$ is a constant vector field, and $\mathbf{r}$ denotes the position vector relative to the origin, prove that

$$\oint_{\mathscr{C}} (\mathbf{a} \times \mathbf{r}) \, . \, d\mathbf{r} = 2 \iint_{S} \mathbf{a} \, . \, d\mathbf{S}.$$

Deduce that

$$\iint_{S} d\mathbf{S} = \tfrac{1}{2} \oint_{\mathscr{C}} \mathbf{r} \times d\mathbf{r}.$$

Solution

By Stokes's theorem,

$$\oint_{\mathscr{C}} (\mathbf{a} \times \mathbf{r}) \, . \, d\mathbf{r} = \iint_{S} \mathrm{curl}\, (\mathbf{a} \times \mathbf{r}) \, . \, d\mathbf{S} = I \quad \text{(say)}.$$

Using identity (4.53) and the fact that $\mathbf{a}$ is constant,

$$I = \iint_{S} \{ \mathbf{a} \, \mathrm{div}\, \mathbf{r} - (\mathbf{a} \, . \, \nabla) \mathbf{r} \} \, . \, d\mathbf{S}.$$

If $\mathbf{a} = (a_2, a_2, a_3)$ relative to the axes $Oxyz$, we have

$$(\mathbf{a} \, . \, \nabla) \mathbf{r} = \left(a_1 \frac{\partial}{\partial x} + a_2 \frac{\partial}{\partial y} + a_3 \frac{\partial}{\partial z} \right) (x, y, z)$$

$$= (a_1, a_2, a_3)$$

$$= \mathbf{a}.$$

Also, $\mathrm{div}\, \mathbf{r} = \mathrm{div}(x, y, z) = 3$. Thus

$$\oint_{\mathscr{C}} (\mathbf{a} \times \mathbf{r}) \, . \, d\mathbf{r} = 2 \iint_{S} \mathbf{a} \, . \, d\mathbf{S}.$$

Since $\mathbf{a}$ takes the same value at all points on $\mathscr{C}$ and S, the above result may be expressed as

$$\mathbf{a} \cdot \left(\oint_{\mathscr{C}} \mathbf{r} \times d\mathbf{r} - 2 \int\int_{S} d\mathbf{S} \right) = 0.$$

By an argument similar to that used in Example 2, in section 6.2, it follows that the bracketed expression vanishes, which gives the result required.

EXERCISES

6.12 Verify Stokes's theorem for the vector field $\mathbf{F} = (x^2y, z, 0)$ and the hemisphere

$$x^2 + y^2 + z^2 = a^2, \quad z \geqslant 0.$$

6.13 If S is a simple open surface, bounded by a correspondingly oriented curve $\mathscr{C}$, and Φ, Ψ are continuously differentiable scalar fields, prove from Stokes's theorem that

$$\oint_{\mathscr{C}} \Phi \operatorname{grad} \Psi \cdot d\mathbf{r} = \int\int_{S} \operatorname{grad} \Phi \times \operatorname{grad} \Psi \cdot d\mathbf{S}.$$

6.14 A simple open surface S is bounded by a correspondingly oriented curve $\mathscr{C}$. If Ω is a continuously differentiable scalar field, prove the following corollary of Stokes's theorem:

$$\oint_{\mathscr{C}} \Omega \, d\mathbf{r} = - \int\int_{S} \operatorname{grad} \Omega \times d\mathbf{S}.$$

[*Hint.* Consider $\Omega\mathbf{a}$, where $\mathbf{a}$ is a constant vector field, and proceed as in Example 2, section 6.2.]

6.15 Prove the following corollary of Stokes's theorem, and state the conditions to be satisfied by $\mathbf{F}$, S and $\mathscr{C}$:

$$\oint_{\mathscr{C}} \mathbf{F} \times d\mathbf{r} = - \int\int_{S} (\mathbf{n} \times \nabla) \times \mathbf{F} \, dS.$$

[*Hint.* Put $\mathbf{F} = F_1 \mathbf{e}_1 + F_2 \mathbf{e}_2 + F_3 \mathbf{e}_3$, where $\mathbf{e}_1, \mathbf{e}_2, \mathbf{e}_3$ are the basic unit vectors of a rectangular cartesian coordinate system, and use the result of Exercise 6.14.]

6.16 Two non-intersecting simple closed curves $\mathscr{C}_1$, $\mathscr{C}_2$ lie in the xy-plane, with $\mathscr{C}_1$ enclosing $\mathscr{C}_2$, and are described in the anticlockwise sense. Show that, for the vector field $\mathbf{F} = (zy^2, 2xyz, x)$,

$$\oint_{\mathscr{C}_1} \mathbf{F} \cdot d\mathbf{r} = \oint_{\mathscr{C}_2} \mathbf{F} \cdot d\mathbf{r}.$$

6.17 Consider the remarks made in Section 6.2 about the extension of Gauss's theorem to functions whose partial derivatives are not continuous. Verify that Stokes's theorem (6.25) extends in a similar way, provided that any discontinuities in the derivatives of $\mathbf{F}$ are finite and are confined to a finite number of simple curves on the surface S.

6.5 LIMIT DEFINITIONS OF DIV F AND CURL F

Div F
Let P be a point in a region V bounded by a simple closed surface S, throughout which the vector field $\mathbf{F}$ and its divergence are defined. If V_0 denotes the volume of V, the mean value theorem for integrals* shows that there exists a point P' in V such that

$$\frac{1}{V_0} \iiint_V \operatorname{div} \mathbf{F} \, dV = \operatorname{div} \mathbf{F}, \quad \text{evaluated at } P'.$$

Thus, by the divergence theorem,

$$\frac{1}{V_0} \iint_S \mathbf{F} \cdot d\mathbf{S} = \operatorname{div} \mathbf{F}, \quad \text{evaluated at } P'.$$

If the surface S is now allowed to collapse towards P', the linear dimensions of V become arbitrarily small, and P' must approach P (Fig. 6.15(a)). It follows that at P

$$\operatorname{div} \mathbf{F} = \lim_{V_0 \to 0} \frac{1}{V_0} \iint_S \mathbf{F} \cdot d\mathbf{S}. \qquad (6.28)$$

This relationship is very often used to define div $\mathbf{F}$.

Curl F
Let $\mathbf{F}$ and curl $\mathbf{F}$ be defined throughout a region V containing a given point P. Draw a plane surface S through P and contained in V, with its normal in the direction of a given unit vector $\mathbf{n}$ and bounded by a correspondingly oriented simple closed curve $\mathscr{C}$ (Fig. 6.15(b)). Let A be the area of the surface S. Then, using Stokes's theorem followed by the mean value theorem, there exists a point P' on S such that

* See, for example, Courant, R. (1936–7) *Differential and Integral Calculus*, Vol. II, p. 232, Blackie, London.

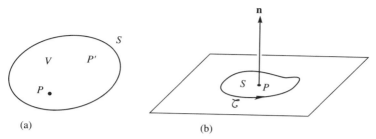

Fig. 6.15

$$\oint_{\mathscr{C}} \mathbf{F} \cdot d\mathbf{r} = \iint_{S} \mathbf{n} \cdot \text{curl } \mathbf{F} \, dS$$

$$= A \, \mathbf{n} \cdot \text{curl } \mathbf{F}, \quad \text{evaluated at } P'.$$

Allowing $\mathscr{C}$ to shrink towards P, it follows that at P

$$\mathbf{n} \cdot \text{curl } \mathbf{F} = \lim_{A \to 0} \frac{1}{A} \oint_{\mathscr{C}} \mathbf{F} \cdot d\mathbf{r}. \qquad (6.29)$$

This gives the component of curl $\mathbf{F}$ at P in the direction of $\mathbf{n}$. As the direction of the unit vector $\mathbf{n}$ may be chosen arbitrarily, we can obtain the components of curl $\mathbf{F}$ along three non-coplanar axes, and (6.29) thus effectively specifies the value of curl $\mathbf{F}$ at P completely.

EXERCISES

6.18 Let S denote the surface of a sphere of variable radius, centred at the origin, and let V_0 be the volume enclosed by the sphere. Verify that

$$(\text{div } \mathbf{r})_O = \lim_{V_0 \to 0} \frac{1}{V_0} \iint_{S} \mathbf{r} \cdot d\mathbf{S},$$

where the suffix O indicates the value at the origin.

6.19 Let $\mathscr{C}$ be the rectangular contour $x = \pm h$, $y = \pm h$, lying in the xy-plane and described in the anticlockwise sense. Verify that for the vector field $\mathbf{F} = (z^2, x, y^2)$,

$$\mathbf{k} \cdot (\text{curl } \mathbf{F})_O = \lim_{h \to 0} \frac{1}{4h^2} \oint_{\mathscr{C}} \mathbf{F} \cdot d\mathbf{r},$$

where the suffix O indicates the value at the origin.

6.6 GEOMETRICAL AND PHYSICAL SIGNIFICANCE OF DIVERGENCE AND CURL

Let S be a closed surface bounding a region V of volume V_0, and consider

$$I = \frac{1}{V_0} \iint_S \mathbf{F} \cdot d\mathbf{S}.$$

If, at each point on S, $\mathbf{F}$ is directed away from the enclosed region V, then $\mathbf{F} \cdot \mathbf{n} > 0$ at each point and $I > 0$. Similarly, if $\mathbf{F}$ points into V at each point on S, then $I < 0$. In the first case $\mathbf{F}$ is diverging from V and in the second case $\mathbf{F}$ is converging towards V. (Fig. 6.16).

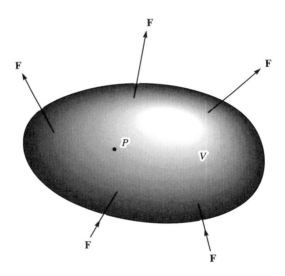

Fig. 6.16

More generally, $\mathbf{F} \cdot \mathbf{n}$ may take some positive and some negative values on S. The sign of I indicates whether the average field on S is divergent or convergent. Allowing S to shrink towards a fixed point P and using (6.28), it follows that the sign of div $\mathbf{F}$ shows whether the mean field in the neighbourhood of P is divergent or convergent; the magnitude of div $\mathbf{F}$ gives an indication of the strength of the divergence or convergence. In the early days of vector analysis, the expression

$$-\frac{\partial F_1}{\partial x} - \frac{\partial F_2}{\partial y} - \frac{\partial F_3}{\partial z}$$

was often referred to as the **convergence** of the vector field $\mathbf{F} = (F_1, F_2, F_3)$.

The meaning of divergence can be clearly visualized in the theory of fluid mechanics. In a continuous motion of an **incompressible** fluid, there can be no net flow of fluid into or out of any fixed closed surface (assuming that no cavities occur and that no fluid is being created or destroyed). Denoting the fluid velocity by **v**,

$$\iint_S \mathbf{v} \cdot \mathbf{dS}$$

gives the net flux across a surface S. We should thus expect div **v** to vanish at all points, and this is indeed one of the field equations governing the flow of an incompressible fluid.

To understand the significance of curl **F**, take an arbitrarily small circular disc centred at a point P and with its axis in the direction of curl **F** at P. Let $\mathscr{C}$ denote the boundary of the disc, and consider

$$I = \oint_{\mathscr{C}} \mathbf{F} \cdot d\mathbf{r} = \oint_{\mathscr{C}} \mathbf{F} \cdot \hat{\mathbf{T}} \, ds,$$

where $\hat{\mathbf{T}}$ is the unit tangent to $\mathscr{C}$. Since **n** (the unit normal to the disc) and curl **F** are in the same direction, $\mathbf{n} \cdot \text{curl} \, \mathbf{F} > 0$, and so $I > 0$. Averaged over the boundary of the disc, the value of $\mathbf{F} \cdot \hat{\mathbf{T}}$ is thus positive, and this indicates that the field in the neighbourhood of P has (on the average) a rotational component directed in the positive sense of $\mathscr{C}$ (Fig. 6.17). The field thus rotates about an axis parallel to curl **F**. When curl $\mathbf{F} = \mathbf{0}$, there is no net rotation.

In fluid mechanics, the vanishing of curl **v** (**v** = velocity) implies that the fluid elements are not in a state of rotation. Such flows are termed **irrotational**.

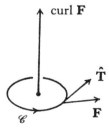

Fig. 6.17

EXERCISES

6.20 Illustrate the remarks made in this section about the significance of divergence by reference to the vector field $\mathbf{F} = (x, y, z)$, taking a spherical surface centred at the origin. Discuss in a similar way the divergence of the vector field $\mathbf{F} = (x^2, y^2, z^2)$ at the origin O.

6.21 Illustrate the remarks made in this section about curl **F** by reference to the vector fields

$$\text{(i) } \mathbf{F} = (x, y, z), \qquad \text{(ii) } \mathbf{F} = (y, 0, 0),$$

in each case choosing a rectangular circuit in the xy-plane.

6.22 A rigid body spins about a fixed axis through its centre O with angular velocity $\boldsymbol{\omega}$. If **v** is the velocity of any point in the body, prove that

$$\boldsymbol{\omega} = \tfrac{1}{2} \operatorname{curl} \mathbf{v}.$$

[*Note.* Because of this connection with rotation, curl **v** is sometimes denoted by rot **v**.]

Applications in potential theory | 7

The applications of vector analysis lie mainly in electromagnetism, fluid mechanics, gravitation, and elasticity, and in each case scalar and vector potentials play an important part in the theory. In this chapter a brief account is given of potential functions and some closely related topics.

7.1 CONNECTIVITY

Let $\mathscr{C}$ and $\mathscr{C}'$ be simple closed curves drawn in a region $\mathscr{R}$ of three-dimensional space. The curves are said to be **reconcilable** if it is possible to continuously distort one curve (without 'opening' it) so as to bring it into coincidence with the other, without the curve ever leaving $\mathscr{R}$.

A region is said to be **singly connected** (or **simply connected**) if all simple closed curves drawn in the region are reconcilable with each other.

If the simple closed curves which can be drawn in a region fall into two distinct classes, such that all the curves in one class are reconcilable with each other but not with curves in the other class, then the region is said to be **doubly connected** or to have **connectivity two**.

In general, a region $\mathscr{R}$ has **connectivity n** if there are just n classes of simple closed curves in $\mathscr{R}$, the curves in any class being reconcilable with each other but not with curves in other classes.

Examples
The interior of a sphere and the interior of an infinitely long circular cylinder are singly connected regions. However, the region outside a circular cylinder is doubly connected; and so is the region inside a torus (or anchor ring). In two dimensions, the region between two concentric circles is doubly connected. The reader should satisfy himself that these statements are in accordance with the definitions.

Reduction of a multiply connected region

A region which is multiply connected is often reduced to a singly connected region by removing all points on suitable surfaces.

For example, consider the region $\mathcal{R}$ between two coaxial cylinders C_1, C_2 (Fig. 7.1). Curves in $\mathcal{R}$ which surround the inner cylinder, C_1, are reconcilable with each other, and closed curves in $\mathcal{R}$ which do not surround C_1 are reconcilable with each other. As the curves in one class are not reconcilable with curves in the other, the region $\mathcal{R}$ is doubly connected. However, suppose that all the points on a plane AB joining a generator of C_1 to a generator of C_2 are removed. The remaining region, $\mathcal{R}'$ say, is then singly connected, because the only closed curves that lie entirely in $\mathcal{R}'$ do not surround C_1 (a closed curve surrounding C_1 passes outside $\mathcal{R}'$ where it crosses AB).

When a multiply connected region arises in practice, it is nearly always sufficient to confine attention to a corresponding singly connected region by introducing suitable 'cuts'. We shall therefore discuss only singly connected regions.

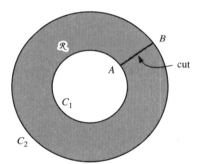

Fig. 7.1 The region $\mathcal{R}$ between coaxial cylinders can be made singly connected by removing from $\mathcal{R}$ all points of a plane joining the cylinders.

EXERCISE

7.1 State the degree of connectivity (one, two, three, etc.) of the following:

(i)   the region outside a torus;

(ii) the region outside two spheres, neither of which encloses or intersects the other;

(iii) the two-dimensional region outside two coplanar circles, neither of which encloses or intersects the other.

7.2 THE SCALAR POTENTIAL

It was shown in section 4.9, equation (4.47) that if a vector field $\mathbf{F}$ is such that $\mathbf{F} = \text{grad } \Omega$, then curl $\mathbf{F} = \mathbf{0}$. We now prove the converse of this result, namely, that if a vector field $\mathbf{F}$ defined in a singly connected region $\mathcal{R}$ is such that

$$\text{curl } \mathbf{F} \equiv \mathbf{0} \text{ in } \mathcal{R}, \tag{7.1}$$

then there exists a scalar field Ω, defined in $\mathcal{R}$, such that

$$\mathbf{F} \equiv \text{grad } \Omega. \tag{7.2}$$

The function Ω is called a **scalar potential** of the vector field $\mathbf{F}$.

Proof

Choose any point O in $\mathcal{R}$ as origin, and denote by P the point with coordinates (x, y, z). Consider the line integral

$$\Omega = \int_O^P \mathbf{F} \cdot d\mathbf{r}, \tag{7.3}$$

taken along any simple curve in $\mathcal{R}$ joining O and P. We shall show that Ω is a scalar field with the required property that $\mathbf{F} = \text{grad } \Omega$.

Let $\mathcal{C}$ be any simple closed curve in $\mathcal{R}$, passing through both O and P. Since curl $\mathbf{F} \equiv \mathbf{0}$ in $\mathcal{R}$, it follows from Stokes' theorem (Section 6.4) that

$$\oint_{\mathcal{C}} \mathbf{F} \cdot d\mathbf{r} = 0.$$

If $\mathcal{C}$ consists of two portions $\mathcal{C}_1$, $\mathcal{C}_2$ as shown in Fig. 7.2(a), this becomes

$$\int_{\mathcal{C}_1}{}_O^P \mathbf{F} \cdot d\mathbf{r} + \int_{\mathcal{C}_2}{}_P^O \mathbf{F} \cdot d\mathbf{r} = 0.$$

However, if $\mathcal{C}_2$ is described in the sense from O to P, as in Fig. 7.2(b), we obtain

$$\int_{\mathcal{C}_1}{}_O^P \mathbf{F} \cdot d\mathbf{r} - \int_{\mathcal{C}_2}{}_O^P \mathbf{F} \cdot d\mathbf{r} = 0.$$

(a)

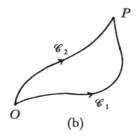

(b)

Fig. 7.2

It follows that Ω, as defined by (7.3), is independent of the curve joining O to P, and is therefore a scalar function of x, y, z (the coordinates of P).

To show that $\mathbf{F} = \text{grad } \Omega$, let $P(x_0, y_0, z_0)$ and $Q(x, y_0, z_0)$ be two points such that PQ lies wholly in $\mathcal{R}$. Then

$$\Omega(x, y_0, z_0) - \Omega(x_0, y_0, z_0) = \int_O^Q \mathbf{F} \cdot d\mathbf{r} - \int_O^P \mathbf{F} \cdot d\mathbf{r}$$

$$= -\int_Q^O \mathbf{F} \cdot d\mathbf{r} - \int_O^P \mathbf{F} \cdot d\mathbf{r}.$$

But

$$\int_O^P \mathbf{F} \cdot d\mathbf{r} + \int_P^Q \mathbf{F} \cdot d\mathbf{r} + \int_Q^O \mathbf{F} \cdot d\mathbf{r} = 0,$$

by Stokes's theorem. Thus

$$\Omega(x, y_0, z_0) - \Omega(x_0, y_0, z_0) = \int_P^Q \mathbf{F} \cdot d\mathbf{r}.$$

The curve along which the integral is taken can be the straight line PQ, which is parallel to Ox, and thus we have

$$\Omega(x, y_0, z_0) - \Omega(x_0, y_0, z_0) = \int_{x_0}^x F_1 dx,$$

where F_1 denotes the x-component of $\mathbf{F}$.* Differentiating with respect to x, as x_0, y_0, z_0 are constant,

$$F_1 = \partial\Omega/\partial x;$$

here we have used the fact that F_1 is continuous, which is implied by the existence of curl $\mathbf{F}$. We may similarly prove that the y- and z-components of $\mathbf{F}$ are respectively $F_2 = \partial\Omega/\partial y$ and $F_3 = \partial\Omega/\partial z$. Hence

$$\mathbf{F} = \text{grad } \Omega,$$

as required.

Uniqueness

Suppose that Ω and Ω' are two scalar potentials such that

$$\mathbf{F} = \text{grad } \Omega, \quad \mathbf{F} = \text{grad } \Omega'$$

and write

* The use of x as a variable in the integrand and in the limits is convenient and should cause no confusion here. We could, of course, replace x in the integrand by some other dummy variable, say t.

$$U = \Omega - \Omega'.$$

By subtraction, grad $U = \mathbf{0}$, and hence, taking components,

$$\partial U/\partial x = 0, \quad \partial U/\partial y = 0, \quad \partial U/\partial z = 0.$$

These equations show that U is independent of x, y and z, giving $U = \text{constant}$. Thus

$$\Omega' = \Omega + \text{constant}.$$

This shows that *the scalar potential is unique apart from an arbitrary additive constant*. Some degree of arbitrariness in Ω should be expected, because the integral (7.3) depends upon the origin O, whose position was chosen arbitrarily.

Irrotational vector fields

A vector field $\mathbf{F}$ such that curl $\mathbf{F} \equiv \mathbf{0}$ in a region $\mathcal{R}$ is said to be **irrotational** in $\mathcal{R}$.

Conservative vector fields

A vector field $\mathbf{F}$ such that, for any two points P and Q,

$$\int_P^Q \mathbf{F} \cdot d\mathbf{r}$$

is independent of the path of integration between P and Q is said to be **conservative**.

EXERCISES

7.2 Show that, in a singly connected region, an irrotational vector field is conservative; and, conversely, a conservative vector field is irrotational.

7.3 In Newtonian mechanics, when a force $\mathbf{F}$ is a function of position only, the **potential energy** of the force at any point P is defined as

$$-\int_O^P \mathbf{F} \cdot d\mathbf{r},$$

where O is any origin. Deduce that the potential energy of the gravitational force on a particle of mass m is mgz, where g denotes acceleration due to gravity and z denotes height.

7.4 Find scalar potentials of the vector fields

$$\mathbf{F} = r^{-3}\,\mathbf{r} \text{ and } \mathbf{F} = \mathbf{r}.$$

7.5 Verify that a scalar potential of the vector field

$$\mathbf{F} = 3(\boldsymbol{\mu} \cdot \mathbf{r}) r^{-5} \, \mathbf{r} - \boldsymbol{\mu} r^{-3}$$

is

$$\Omega = - (\boldsymbol{\mu} \cdot \mathbf{r})/r^3,$$

where $\boldsymbol{\mu}$ is a constant vector.

7.3 THE VECTOR POTENTIAL

In section 4.9, equation (4.46), we showed that if, in a region $\mathcal{R}$, a vector field $\mathbf{F}$ having continuous second derivatives is such that $\mathbf{F} \equiv \operatorname{curl} \mathbf{A}$, then $\operatorname{div} \mathbf{F} \equiv 0$. We now prove the converse of this result, namely, that if

$$\operatorname{div} \mathbf{F} \equiv 0 \quad \text{in } \mathcal{R}, \tag{7.4}$$

then there exists a vector field $\mathbf{A}$ such that

$$\mathbf{F} \equiv \operatorname{curl} \mathbf{A} \quad \text{in } \mathcal{R}. \tag{7.5}$$

The function $\mathbf{A}$ is called a **vector potential** of the vector field $\mathbf{F}$.

Proof

Denote by

$$\xi(y, z)$$

the value of F_1, the x-component of $\mathbf{F}$, at the point $(0, y, z)$ of the plane $x = 0$. If F_2 and F_3 are the y- and z-components of $\mathbf{F}$, consider

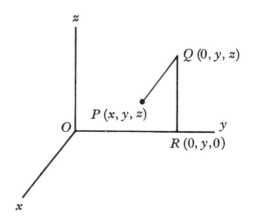

Fig. 7.3

$$\mathbf{A} = \left(\int_0^x F_3 \, dx - \int_0^z \xi \, dz \right) \mathbf{j} - \left(\int_0^x F_2 \, dx \right) \mathbf{k}, \tag{7.6}$$

where

$$\int_0^x F_2 \, dx \quad \text{and} \quad \int_0^x F_3 \, dx$$

are the integrals of $F_2(x, y, z)$ and $F_3(x, y, z)$ with respect to x (y and z being fixed) from the point $Q(0, y, z)$ to the point $P(x, y, z)$; and

$$\int_0^z \xi \, dz$$

is the integral of $\xi(y, z) = F_1(0, y, z)$ with respect to z from the point $R(0, y, 0)$ to the point $Q(0, y, z)$ (Fig. 7.3).* From (7.6), we have

$$\operatorname{curl} \mathbf{A} = \left(-\frac{\partial}{\partial y} \int_0^x F_2 \, dx - \frac{\partial}{\partial z} \int_0^x F_3 \, dx + \xi \right) \mathbf{i} + F_2 \mathbf{j} + F_3 \mathbf{k}.$$

Differentiating through the integrals (which is permissible because the existence of div $\mathbf{F}$ implies that the functions F_1, F_2, F_3 are continuously differentiable), we obtain

$$\operatorname{curl} \mathbf{A} = -\left\{ \int_0^x \left(\frac{\partial F_2}{\partial y} + \frac{\partial F_3}{\partial z} \right) dx - \xi \right\} \mathbf{i} + F_2 \mathbf{j} + F_3 \mathbf{k}. \tag{7.7}$$

But div $\mathbf{F} \equiv 0$, so that

$$\frac{\partial F_2}{\partial y} + \frac{\partial F_3}{\partial z} = -\frac{\partial F_1}{\partial x}.$$

Substituting this into (7.7) and integrating gives

$$\operatorname{curl} \mathbf{A} = ([F_1]_0^x + \xi) \, \mathbf{i} + F_2 \mathbf{j} + F_3 \mathbf{k}$$

$$= F_1 \mathbf{i} + F_2 \mathbf{j} + F_3 \mathbf{k} = \mathbf{F}$$

as required. Thus (7.6) is a particular vector potential.

Uniqueness

Suppose that two vector fields $\mathbf{A}$ and $\mathbf{A}'$ are such that, in a singly connected region $\mathcal{R}$,

$$\mathbf{F} = \operatorname{curl} \mathbf{A} \quad \text{and} \quad \mathbf{F} = \operatorname{curl} \mathbf{A}'.$$

---

* Here there is an implicit assumption about the region $\mathcal{R}$, namely that axes $Oxyz$ can be so chosen that the straight lines RQ and QP lie entirely in $\mathcal{R}$. A proof for more general regions can be given, but the assumption made here covers most singly connected regions which are met in practice.

Write

$$\mathbf{B} = \mathbf{A} - \mathbf{A}'.$$

Then, by subtraction,

$$\operatorname{curl} \mathbf{B} \equiv \mathbf{0} \text{ in } \mathcal{R}.$$

It follows from the preceding section that there exists a scalar field Ω such that

$$\mathbf{B} = \operatorname{grad} \Omega.$$

Thus

$$\mathbf{A} = \mathbf{A}' + \operatorname{grad} \Omega, \qquad (7.8)$$

showing that *apart from the addition of the gradient of an arbitrary scalar field, the vector potential* $\mathbf{A}$ *is unique.*

Comment

From (7.8),

$$\operatorname{div} \mathbf{A} = \operatorname{div} \mathbf{A}' + \nabla^2 \Omega,$$

and so we have some latitude in the choice of div$\mathbf{A}$. In electromagnetism, this fact is exploited to simplify the form of the basic equations. For example, in electrostatics, it is usual to specify that div $\mathbf{A} \equiv 0$.

Solenoidal vector fields

A vector field $\mathbf{F}$ is said to be **solenoidal** in a region $\mathcal{R}$ if div $\mathbf{F} \equiv 0$ in $\mathcal{R}$.

EXERCISES

7.6 By using formula (7.6), find a vector field $\mathbf{A}$ such that $\mathbf{F} = \operatorname{curl} \mathbf{A}$ when $\mathbf{F} = (y - x, z, x)$.

7.7 Using spherical polar coordinates r, θ, φ, verify that the vector field $\mathbf{F} = \mathbf{e}_r/r^2$ is solenoidal. Find the function $\psi(r, \theta)$ such that

$$\mathbf{A} = \frac{\psi(r, \theta)}{r \sin \theta} \mathbf{e}_\varphi$$

is the vector potential of $\mathbf{F}$, satisfying the condition div $\mathbf{A} \equiv 0$.

7.4 POISSON'S EQUATION

In many applications, the scalar potential Ω satisfies the partial differential equation

$$\nabla^2 \Omega = f(x, y, z), \qquad (7.9)$$

where $f(x, y, z)$ is a given bounded scalar function of position. This is known as **Poisson's equation**, except in the particular case $f(x, y, z) \equiv 0$, when it reduces to Laplace's equation.

The solution may be needed in a region V bounded by a simple closed surface S for various **boundary conditions**. (For simplicity, we suppose that there is only one bounding surface, but the argument could easily be extended to the general case.) Typical boundary conditions are:

(i) *the Dirichlet condition*

$$\Omega = g(x, y, z) \quad \text{on } S, \qquad (7.10)$$

where $g(x, y, z)$ is a given function of position;

(ii) *the Neumann condition*

$$\frac{\partial \Omega}{\partial n} = h(x, y, z) \quad \text{on } S, \qquad (7.11)$$

where $h(x, y, z)$ is a given function of position, and $\partial \Omega / \partial n = \hat{\mathbf{n}} \cdot \nabla \Omega$, where $\hat{\mathbf{n}}$ denotes the unit normal on S.

Uniqueness

Generally, the solution of Poisson's equation is uniquely determined by either condition (7.10) or condition (7.11). This may be proved by supposing that Ω' is a second solution of (7.9), satisfying the same boundary conditions as Ω. Then writing

$$U = \Omega - \Omega'$$

we find that

$$\nabla^2 \Omega = f \quad \text{and} \quad \nabla^2 \Omega' = f,$$

giving by subtraction

$$\nabla^2 U = 0.$$

The boundary condition (7.10) gives

$$\Omega = g \quad \text{and} \quad \Omega' = g \text{ on } S;$$

therefore

$$U = 0 \quad \text{on } S.$$

It follows from the uniqueness theorem of section 6.3 that, throughout V,

$$U \equiv 0.$$

Thus, the solution of Poisson's equation is uniquely determined by the boundary condition (7.10).

Similarly, it may be shown (the proof is set as an exercise) that the solution of Poisson's equation is unique, apart from an arbitrary additive constant, under a Neumann boundary condition.

It follows that the solution of the equation with *two* boundary conditions is possible only in very special circumstances. Problems with two boundary conditions are said to be **overdetermined**. In the cases when a solution exists satisfying both boundary conditions, the boundary conditions are said to be **compatible** or **dependent**.

Solution of Poisson's equation

Earlier (Section 6.3) we proved Green's theorem that for the scalar fields Φ and Ψ,

$$\iint_S \left(\Phi \frac{\partial \Psi}{\partial n} - \Psi \frac{\partial \Phi}{\partial n} \right) dS = \iiint_V (\Phi \nabla^2 \Psi - \Psi \nabla^2 \Phi) \, dV. \tag{7.12}$$

Choose the origin O inside V, and take

$$\Phi = \Omega, \quad \Psi = \frac{1}{r}. \tag{7.13}$$

Then, using the formula for ∇^2 in spherical polar coordinates,

$$\nabla^2 \Psi \equiv \frac{1}{r^2} \frac{d}{dr} \left\{ r^2 \frac{d}{dr} \left(\frac{1}{r} \right) \right\} \equiv 0 \tag{7.14}$$

except at the origin ($r = 0$). As $\nabla^2 \Psi$ is not defined at $r = 0$, we exclude this point and apply Green's theorem to the region $V - V_\varepsilon$ bounded by $S \cup S_\varepsilon$ as in Fig. 7.4, where S_ε denotes the surface of a small sphere of radius ε surrounding O, and V_ε denotes the region bounded by S_ε. Substituting (7.13) and (7.14) into (7.12) gives

$$\iint_{S \cup S_\varepsilon} \left\{ \frac{1}{r} \frac{\partial \Omega}{\partial n} - \Omega \frac{\partial}{\partial n} \left(\frac{1}{r} \right) \right\} dS = \iiint_{V - V_\varepsilon} \frac{1}{r} \nabla^2 \Omega \, dV;$$

that is

$$\iint_S \left\{ \frac{1}{r} \frac{\partial \Omega}{\partial n} - \Omega \frac{\partial}{\partial n} \left(\frac{1}{r} \right) \right\} dS + I_\varepsilon = \iiint_V \frac{1}{r} \nabla^2 \Omega \, dV - J_\varepsilon, \tag{7.15}$$

where

$$I_\varepsilon = \iint_{S_\varepsilon} \left\{ \frac{1}{r} \frac{\partial \Omega}{\partial n} - \Omega \frac{\partial}{\partial n} \left(\frac{1}{r} \right) \right\} dS \tag{7.16}$$

and

$$J_\varepsilon = \iiint_{V_\varepsilon} \frac{1}{r} \nabla^2 \Omega \, dV. \tag{7.17}$$

We shall find the limiting form of (7.15) when $\varepsilon \to 0$.

Fig. 7.4 S_ε is a small sphere, centre O; and $V - V_\varepsilon$ is the region between S_ε and the surface S.

By convention, the surface normal points out of $V - V_\varepsilon$ and hence, on the boundary S_ε, $\partial/\partial n \equiv - \partial/\partial r$. Thus using spherical polar coordinates r, θ, φ we obtain

$$I_\varepsilon = \int_0^{2\pi} \int_0^\pi \left(-\frac{1}{\varepsilon} \frac{\partial \Omega}{\partial r} - \frac{\Omega}{\varepsilon^2} \right) \varepsilon^2 \sin\theta \, d\theta \, d\varphi$$

$$= -\varepsilon \int_0^{2\pi} \int_0^\pi \frac{\partial \Omega}{\partial r} \sin\theta \, d\theta \, d\varphi - \int_0^{2\pi} \int_0^\pi \Omega(\varepsilon, \theta, \varphi) \sin\theta \, d\theta \, d\varphi.$$

Hence, denoting by Ω_o the value of Ω at the origin O and noting that Ω is assumed to be continuous at this point,

$$\lim_{\varepsilon \to 0} I_\varepsilon = -\Omega_o \int_0^{2\pi} \int_0^\pi \sin\theta \, d\theta \, d\varphi = -4\pi \, \Omega_o.$$

Also,

$$J_\varepsilon = \int_0^{2\pi} \int_0^\pi \int_0^\varepsilon r\nabla^2\Omega \, \sin\theta \, dr \, d\theta \, d\varphi,$$

and if M denotes the upper bound of $\left| \nabla^2\Omega \right|$ (f is bounded in equation (7.9)), we have

$$|J_\varepsilon| \leq M \left| \int_0^{2\pi} \int_0^\pi \int_0^\varepsilon r\sin\theta \, dr \, d\theta \, d\varphi \right|$$

$$= M \, 2\pi \, \varepsilon^2 \, .$$

Thus,

$$\lim_{\varepsilon \to 0} J_\varepsilon = 0.$$

Allowing ε to tend to zero in (7.15), it follows that

$$\iint_S \left\{ \frac{1}{r} \frac{\partial \Omega}{\partial n} - \Omega \frac{\partial}{\partial n} \left(\frac{1}{r} \right) \right\} dS - 4\pi \, \Omega_o = \iiint_V \frac{1}{r} \nabla^2 \Omega \, dV. \tag{7.18}$$

Substituting $\nabla^2 \Omega = f$ from equation (7.9), the value of Ω at the origin O is thus

$$\Omega_o = -\frac{1}{4\pi} \iiint_V \frac{f}{r} \, dV + \frac{1}{4\pi} \iint_S \left\{ \frac{1}{r} \frac{\partial \Omega}{\partial n} - \Omega \frac{\partial}{\partial n} \left(\frac{1}{r} \right) \right\} dS. \tag{7.19}$$

By choosing the origin O to be any field point, (7.19) gives the value of Ω at that point. However, this result does not give an immediate solution of Poisson's equation, because the right-hand side of (7.19) requires prior knowledge of Ω and $\partial\Omega/\partial n$ on S, and we have already seen that only one of these quantities is usually given there. The result is nevertheless useful because analytical and numerical methods are available for solving some equations of this type (called **integral equations**).

In the particular case when V is a region of infinite extent and $\Omega = O(1/r)$ at large distances (in physical applications this condition is nearly always satisfied), expression (7.19) leads at once to the solution for Ω. For, take S to be a sphere of radius R centred at the origin. Then

$$\iint_S \left\{ \frac{1}{r} \frac{\partial \Omega}{\partial n} - \Omega \frac{\partial}{\partial n} \left(\frac{1}{r} \right) \right\} dS = \int_0^{2\pi} \int_0^\pi \left(\frac{1}{r} \frac{\partial \Omega}{\partial r} + \frac{\Omega}{r^2} \right)_{r=R} R^2 \sin\theta \, d\theta \, d\varphi$$

$$= \int_0^{2\pi} \int_0^\pi O\left(\frac{1}{R} \right) \sin\theta \, d\theta \, d\varphi,$$

at large distances R. Thus, allowing R to tend to infinity, the surface integral vanishes, and (7.19) becomes

$$\Omega_o = -\frac{1}{4\pi} \iiint \frac{f}{r} \, dV, \tag{7.20}$$

where the integration is now taken over all space and

$$r = (x^2 + y^2 + z^2)^{\frac{1}{2}}. \tag{7.21}$$

Changing the origin to the point (x, y, z), we obtain the alternative form

$$\Omega(x, y, z) = -\frac{1}{4\pi} \iiint \frac{f(x', y', z')}{\{(x-x')^2 + (y-y')^2 + (z-z')^2\}^{\frac{1}{2}}} \, dx' \, dy' \, dz'. \tag{7.22}$$

This is the required solution of Poisson's equation.

EXERCISES

7.8 Prove that the solution of Poisson's equation under a Neumann boundary condition is unique, apart from an arbitrary additive constant.

7.9 The solution of Poisson's equation $\nabla^2\Omega = f(x, y, z)$, where $f(x, y, z)$ is a given scalar field, is required in a region V bounded by a simple closed surface S. The boundary condition is

$$\Omega = g(x, y, z) \quad \text{on } S,$$

where $g(x, y, z)$ is a given scalar function of position. By taking $f(x, y, z) \equiv 0$ outside S, reduce the problem to that of solving Laplace's equation $\nabla^2 U = 0$ under the boundary condition

$$U = g(x, y, z) + \frac{1}{4\pi} \iiint \frac{f(x', y', z')}{\sqrt{\{(x-x')^2 + (y-y')^2 + (z-z')^2\}}} \, dx' \, dy' \, dz' \quad \text{on } S,$$

where the integral is taken over all space.

7.10 In spherical polar coordinates r, θ, φ when Ω is a function of r only, the equation $\nabla^2\Omega = 1$ becomes

$$\frac{1}{r^2} \frac{d}{dr}\left(r^2 \frac{d\Omega}{dr}\right) = 1.$$

Deduce that the continuous solution of the equation $\nabla^2\Omega = 1$ under the boundary condition $\Omega = 1$ on the sphere $r = 1$ is

$$\Omega = (5 + r^2)/6.$$

7.11 If $\nabla^2\Omega = 1$ for $r \leqslant 1$ and $\nabla^2\Omega = 0$ for $r > 1$, show that equation (7.22) yields the solution

$$\Omega = \frac{1}{6} r^2 - \frac{1}{2}, \quad r \leqslant 1.$$

Using the method of Exercise 7.9, obtain from this result the solution to Exercise 7.10.

7.5 POISSON'S EQUATION IN VECTOR FORM

In electromagnetism a vector potential $\mathbf{A}$ occurs and satisfies Poisson's equation in vector form, namely

$$\nabla^2\mathbf{A} = \mathbf{F}, \tag{7.23}$$

where $\mathbf{F}$ is a known vector field. If the rectangular cartesian components of $\mathbf{A}$

are A_1, A_2, A_3 and those of $\mathbf{F}$ are F_1, F_2, F_3, the components of (7.23) are

$$\nabla^2 A_i = F_i \quad (i = 1, 2, 3), \tag{7.24}$$

which is the scalar form of Poisson's equation.

If the region under consideration is unbounded, and $A_i = O(1/r)$ at large distance r from the origin, the solution of (7.24) is

$$A_i(x, y, z) = -\frac{1}{4\pi} \iiint \frac{F_i(x', y', z')}{\{(x - x')^2 + (y - y')^2 + (z - z')^2\}^{\frac{1}{2}}} \, dx' \, dy' \, dz'. \tag{7.25}$$

Combining the components to recover $\mathbf{A}$ gives

$$\mathbf{A}(x, y, z) = -\frac{1}{4\pi} \iiint \frac{\mathbf{F}(x', y', z')}{\{(x - x')^2 + (y - y')^2 + (z - z')^2\}^{\frac{1}{2}}} \, dx' \, dy' \, dz'. \tag{7.26}$$

7.6 HELMHOLTZ'S THEOREM

This theorem states that for any continuously differentiable vector field $\mathbf{H}$ there exists a scalar field Ω and a vector field $\mathbf{A}$ such that

$$\mathbf{H} = \text{grad}\, \Omega + \text{curl}\, \mathbf{A}. \tag{7.27}$$

Here we shall suppose that the region under consideration is unbounded, and that $|\mathbf{H}| = O(1/r^2)$ at large distances.

Proof

Consider the differential equation

$$\nabla^2 \Omega = \text{div}\, \mathbf{H}$$
$$= f(x, y, z), \quad \text{say}. \tag{7.28}$$

Imposing the condition $\Omega = O(1/r)$ at large distances, an explicit formula for Ω is given by (7.22), namely,

$$\Omega(x, y, z) = -\frac{1}{4\pi} \iiint \frac{f(x', y', z')}{\{(x - x')^2 + (y - y')^2 + (z - z')^2\}^{\frac{1}{2}}} \, dx' \, dy' \, dz'. \tag{7.29}$$

Equation (7.28) can also be expressed as

$$\text{div}\, (\mathbf{H} - \text{grad}\, \Omega) = 0.$$

It follows from section 7.3, that there exists a vector field $\mathbf{A}$ such that

$$\mathbf{H} - \text{grad}\, \Omega = \text{curl}\, \mathbf{A}.$$

This proves the theorem.

An explicit formula for $\mathbf{A}$ can be obtained as follows. Taking the curl of (7.27) gives

$$\text{curl curl } \mathbf{A} = \text{curl } \mathbf{H}. \tag{7.30}$$

As we remarked at the end of Section 7.3, the value of div $\mathbf{A}$ can be chosen arbitrarily. We now use this freedom to impose the condition div $\mathbf{A} = 0$. Then, as $\nabla^2 \mathbf{A} \equiv \text{grad}(\text{div } \mathbf{A}) - \text{curl curl } \mathbf{A}$, (7.30) becomes

$$\nabla^2 \mathbf{A} = -\text{curl } \mathbf{H}$$

$$= \mathbf{F}(x, y, z) \quad \text{say}, \tag{7.31}$$

Using the solution (7.26), we thus have

$$\mathbf{A}(x, y, z) = -\frac{1}{4\pi} \iiint \frac{\mathbf{F}(x', y', z')}{\{(x - x')^2 + (y - y')^2 + (z - z')^2\}^{\frac{1}{2}}} \, dx' \, dy' \, dz'. \tag{7.32}$$

which is the required explicit formula for $\mathbf{A}$. That this result is consistent with the condition div $\mathbf{A} = 0$ can be seen as follows.

Take the divergence of (7.31). Then

$$\nabla^2(\text{div } \mathbf{A}) = 0. \tag{7.33}$$

At large distances, $\mathbf{A}$ (as given by (7.32)) is seen to be $O(1/r)$, and so div $\mathbf{A} = O(1/r^2)$. As there are no internal boundaries, a particular solution is evidently div $\mathbf{A} \equiv 0$. By the uniqueness theorem proved in section 6.3, it follows that this is the only possible solution, and hence $\mathbf{A}$ as given by (7.32) must be such that div $\mathbf{A} = 0$.

Comments

(1) The vector fields grad Ω and curl $\mathbf{A}$ in in (7.27) are, respectively, irrotational and solenoidal; that is curl (grad Ω) $\equiv \mathbf{0}$ and div (curl $\mathbf{A}$) $\equiv 0$. Helmoltz's theorem thus shows that the vector field $\mathbf{H}$ can be resolved into the sum of an irrotational part and a solenoidal part.

(2) Expressions (7.29) and (7.32) show how to determine Ω and $\mathbf{A}$ from div $\mathbf{H}$ and curl $\mathbf{H}$. Regions where div $\mathbf{H} \neq 0$ are called **sources** of $\mathbf{H}$; and regions where curl $\mathbf{H} \neq 0$ are called **vortices** of $\mathbf{H}$. This terminology originates in hydrodynamics.

7.7 SOLID ANGLES

Finally, we give a brief account (mainly in geometrical terms) of solid angles, since these occur frequently in potential theory.

Let dS be an element of surface area at a point P of a surface S, and let O be any other point. Denote by θ the angle between $\overrightarrow{OP}$ and the unit normal $\mathbf{n}$ to S at P (Fig. 7.5). If $\overrightarrow{OP} = \mathbf{r}$, we define

$$d\omega = \frac{\cos\theta \, dS}{r^2} \tag{7.34}$$

to be the **solid angle** subtended by dS at O. The total solid angle subtended by S at O is

$$\omega = \iint_S \frac{\cos\theta\, dS}{r^2} = \iint_S \frac{\mathbf{r}\cdot\mathbf{dS}}{r^3}. \tag{7.35}$$

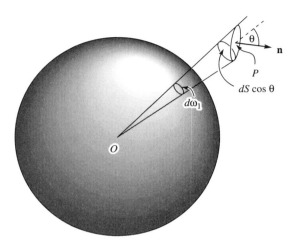

Fig. 7.5

Consider a cone with base dS and vertex O. Draw a sphere with centre O and of unit radius, and let $d\omega_1$ be the area of surface cut off by the cone (Fig. 7.5). The projection of the area dS onto the plane perpendicular to OP is $dS\,|\cos\theta|$. The ratio of this area to $d\omega_1$ is $OP^2 : 1$ and so

$$d\omega_1 = \frac{dS\,|\cos\theta|}{r^2} = |d\omega|. \tag{7.36}$$

Thus, if θ is an acute angle (so that $|\cos\theta| = \cos\theta$), the solid angle subtended by dS at O is just the area cut off by the cone on the sphere of unit radius and center at O; if θ is an obtuse angle, the solid angle is $(-1)\times$ area cut off. This intuitive geometrical argument often enables a solid angle to be determined more easily than by performing the integrations in (7.35).

Special cases

(1) Consider a closed surface S which is such that all straight lines radiating from points inside it cut the surface once only (Fig. 7.6). If O is any point inside S and the unit normals $\mathbf{n}$ at points P on S are directed away from the interior (in the conventional way), the angles θ between $\overrightarrow{OP}$ and $\mathbf{n}$ are all acute. The solid angle subtended by S at O is therefore equal to the surface area of a sphere of unit radius, which is 4π.

Suppose next that O lies outside S, and let a line radiating from O cut S at P and Q (Fig. 7.7). The angles between OPQ and the unit normals at P and Q (φ and θ in the diagram) are respectively obtuse and acute. Draw a slender cone with vertex at O, cutting off elements of area at P and Q. The solid angles

Fig. 7.6

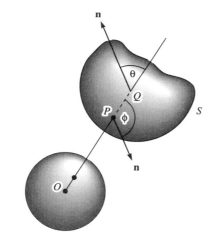

Fig. 7.7

subtended at O by these elements are equal in magnitude but have opposite signs; their net contribution to the total solid angle subtended by S at O is therefore zero. This holds for all lines OPQ, so the solid angle subtended by S at O is zero.

If O lies on S, a similar argument may be used to show that the solid angle is 2π.

Summarizing, *the solid angle subtended by S at O is 4π, 2π or zero according as O lies inside, on or outside S, respectively.*

The same results hold for more general surfaces S, such as that shown in Fig. 7.8. When O lies inside S in the position indicated, and a line radiating from O cuts S at P, Q and R, the contributions to the solid angle from surface elements at P, Q, R are equal in magnitude, but two are positive and one is negative; hence the net contribution is the same as if $OPQR$ cut S only once. Similarly, any line $\mathscr{L}$ radiating from a point outside S cuts the surface an even

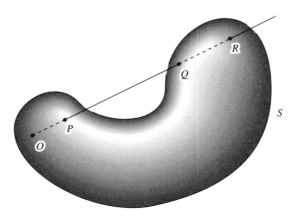

Fig. 7.8

number of times and so the total contribution to the solid angle from the surface elements on $\mathcal{L}$ is zero. The general result for any closed surface can therefore be established.

(2) Let O be a point on the axis of a circular disc, and distant a from it, and suppose the disc is oriented so that the unit normal $\mathbf{n}$ points away from O, as in Fig. 7.9. Let α be the semi-vertical angle of the cone which has the disc as base and vertex at O. Taking cylindrical polar coordinates with origin at O and axis coinciding with the axis of the disc, the solid angle subtended at O is

$$\omega = \iint_S \frac{\cos\theta\,dS}{r^2} = \int_{-\pi}^{\pi} \int_0^{a\tan\alpha} \frac{\cos\theta\,R\,dR\,d\varphi}{a^2 \sec^2\theta}$$

$$= 2\pi \int_0^{a\tan\alpha} \cos^3\theta\,R\,dR/a^2 .$$

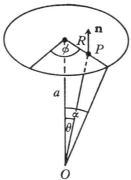

Fig. 7.9

Putting $R = a\tan\theta$, this becomes

$$\omega = 2\pi \int_0^\alpha \sin\theta \, d\theta$$

$$= 2\pi(1 - \cos\alpha).$$

By taking $\alpha = \frac{1}{2}\pi$, it also follows from this that an infinite plane subtends a solid angle of magnitude 2π at any point which does not lie on the plane.

EXERCISES

7.12 By using the divergence theorem, prove that the solid angle subtended by a simple closed surface at an exterior point is zero.

7.13 Points P, Q are situated on opposite sides of a simple open surface S and are arbitrarily close to the surface. The solid angles subtended by S at P and Q are denoted by ω_P^S and ω_Q^S, respectively. If P is on the positive side of S, prove that

$$\omega_P^S - \omega_Q^S \to -4\pi \quad \text{as} \quad PQ \to 0.$$

[*Hint.* Join to S a simple open surface S', such that $S \cup S'$ is a simple closed surface surrounding Q; and observe that

$$\omega_P^{S'} \to \omega_Q^{S'} \text{ as } PQ \to 0.]$$

7.14 A square has sides of length $2a$, and a point P is situated at a distance a from the square on a normal through the intersection of its diagonals. If the unit normal points away from P, find the solid angle subtended by the square at this point.

8 | Cartesian tensors

8.1 INTRODUCTION

We have seen that the essential feature of a vector is that, in any rectangular cartesian coordinate system, it may be represented by three components, each associated with a particular axis. These components depend only on the orientation of the axes and transform according to the rules (2.1) when the axes are rotated. Tensor analysis may be regarded as a generalization of vector analysis to certain mathematical and physical entities known as tensors which require more than three components for their complete specification. There are again physically meaningful rules for transforming the components of tensors when the axes are changed. To give some motivation for the study of tensors, we shall first provide a specific example.

When an elastic body is deformed, certain internal forces will, in general, be set up. At any particular point P within such a body, consider a small plane area whose normal is in the direction of the x_1-axis of a rectangular cartesian system $Ox_1 x_2 x_3$. The force per unit area on this element due to the material in contact with it is a vector having three rectangular cartesian components, which may be denoted by σ_{11}, σ_{12} and σ_{13}. Similarly, the forces per unit area on small plane areas at P whose normals are in the directions Ox_2 and Ox_3, respectively, will have components which may be denoted by $\sigma_{21}, \sigma_{22}, \sigma_{23}$ and $\sigma_{31}, \sigma_{32}, \sigma_{33}$, respectively. It may be shown that the nine components σ_{ij} $(i, j = 1, 2, 3)$ are sufficient to specify fully the 'state of stress' at the point P. Further, by applying Newton's laws of motion to an elementary part of the body, it is possible to deduce just how these components transform under a rotation of the axes.

There are many other mathematical and physical entities which may be represented in a similar way to stress, and this provides the motivation for setting up a comprehensive and general theory.

The purpose of this chapter is to give an introduction to **cartesian tensors**, that is tensors which are expressed in terms of components referred to rectangular cartesian coordinate systems. However, it is worth noting that, as with vectors, orthogonal curvilinear coordinates can often be handled with this framework – see the remark following Example 1, section 8.3. General tensor analysis, which is an indispensable tool in the general theory of relativity and the study of non-Euclidean geometries, uses general coordinates; the concept of space itself is generalized. It is therefore a much more sophisticated theory and falls well outside the intended scope of this volume.

8.2 CARTESIAN TENSORS: BASIC ALGEBRA

Throughout the remainder of this chapter, $Ox_1 x_2 x_3$ and $Ox'_1 x'_2 x'_3$ denote rectangular cartesian coordinate systems whose relative orientations are given, as in Chapter 1, by the transformation matrix

	x_1	x_2	x_3
x'_1	l_{11}	l_{12}	l_{13}
x'_2	l_{21}	l_{22}	l_{23}
x'_3	l_{31}	l_{32}	l_{33}

$$(8.1)$$

The summation convention will also be used throughout.

Definition

Suppose that A is a mathematical or physical entity which, when it is associated with a rectangular cartesian coordinate system $Ox_1 x_2 x_3$, may be represented by a set of 3^n scalars $a_{ij\ldots}$, where there are $n (\geqslant 0)$ suffixes attached to a. Each suffix takes one of the values 1, 2 or 3 and the suffixes $i, j, \ldots$ are ordered. Let $a'_{rs\ldots}$ (n suffixes) be the corresponding set of scalars representing A when it is referred to the axes $Ox'_1 x'_2 x'_3$, and suppose that

$$a'_{rs\ldots} = l_{ri} l_{sj} \ldots a_{ij\ldots}. \qquad (8.2)$$

If the sets of scalars representing A are also invariant under a translation of the axes (and hence depend only on the orientation of the axes), A is said to be a **tensor of order (or rank)** n; the scalars $a_{ij\ldots}$ are called components of A.

A tensor of order zero is evidently a single scalar a having the same value in all coordinate systems; it is therefore a **scalar invariant**.

In the particular case when $n = 1$, A has three components, a_1, a_2 and a_3, and the transformation (8.2) reduces to

$$a'_r = l_{ri} a_i. \qquad (8.3)$$

This is the same as the *vector* transformation law, and so *a tensor of order one is a vector*. This confirms our earlier statement that cartesian tensor analysis may be regarded as a generalization of vector analysis.

When $n = 2$, the tensor A is of second order and has nine components which may be conveniently assembled to form the matrix

$$\begin{pmatrix} a_{11} & a_{12} & a_{13} \\ a_{21} & a_{22} & a_{23} \\ a_{31} & a_{32} & a_{33} \end{pmatrix} \qquad (8.4)$$

It should be noted that the elements of this matrix may change from one set of axes to another. The transformation rule (8.2) becomes

$$a'_{rs} = l_{ri}l_{sj}a_{ij}. \qquad (8.5)$$

Observe that the components are ordered in the sense that a_{ij} is associated with Ox_i and Ox_j in that order; a_{11} is therefore associated with Ox_1 twice, a_{12} with Ox_1 and Ox_2 in that order, a_{21} with Ox_2 and Ox_1 in that order, and so on. In the particular case of the stress tensor with components σ_{ij}, mentioned in section 8.1, the first suffix indicates the direction of the normal to an element under consideration and the second suffix the direction of the component of the stress acting on that element.

Apart from scalar invariants and vectors, the most commonly occurring tensors are those of second order; the main emphasis in the remainder of this chapter will be on tensors of this type.

Elementary algebraic operations

The definitions of multiplication of a tensor by a scalar, and addition and subtraction of tensors *of the same order* are analogous to the corresponding definitions of these operations in sections 2.3 and 2.4 for vectors. For example, if A is a second order tensor with components a_{ij} $(i, j = 1, 2, 3)$ and λ is a scalar, then λA is a tensor with components λa_{ij}. Further, if B is another second order tensor with components b_{ij}, then $A + B$ is a tensor with components $a_{ij} + b_{ij}$. Addition (and subtraction) of tensors of different orders is not defined.

It may be readily proved from the definition of a tensor that the operation of addition is commutative and associative (cf. section 2.4).

Product of tensors

Let A and B be tensors of orders m and n, respectively, whose components referred to the axes $Ox_1 x_2 x_3$ are $a_{ij\ldots}$ and $b_{pq\ldots}$, there being m suffixes attached to a and n suffixes attached to b. Then the 3^{m+n} scalars

$$c_{ij\ldots pq\ldots} = a_{ij\ldots}b_{pq\ldots} \qquad (8.6)$$

form a tensor C, say, of order $m + n$. This is defined to be the **product** of A and B and is written as $C = AB$.

Proof

Under the rotation of axes from the set $Ox_1 x_2 x_3$ to the set $Ox_1' x_2' x_3'$, defined by the transformation matrix (8.1), the transformed components of A and B are

$$a_{rs\ldots}' = l_{ri}\, l_{sj} \ldots a_{ij\ldots}$$

and

$$b_{uv\ldots}' = l_{up}\, l_{vq} \ldots b_{pq\ldots}$$

respectively. Hence, relative to the new axes, the components of C are

$$c_{rs\ldots uv\ldots}' = a_{rs\ldots}'\, b_{uv\ldots}'$$
$$= l_{ri}\, l_{sj} \ldots l_{up}\, l_{vq} \ldots a_{ij\ldots}\, b_{pq\ldots}$$
$$= l_{ri}\, l_{sj} \ldots l_{up}\, l_{vq} \ldots c_{ij\ldots pq\ldots}.$$

This is the tensor law of transformation, and the tensorial character of C is therefore established.

As a particular example of this important operation, consider the vectors $\mathbf{a} = (a_1, a_2, a_3)$ and $\mathbf{b} = (b_1, b_2, b_3)$. As said before, these are first order tensors, and by the result just proved their product is a second order tensor with the following nine components:

$$c_{ij} = a_i b_j \quad (i, j = 1, 2, 3).$$

The operation of tensor multiplication is associative, i.e.

$$(AB)C = A(BC). \tag{8.7}$$

The distributive law

$$A(B + C) = AB + AC \tag{8.8}$$

is also obeyed. The proofs of these results are left for the reader as exercises.

Contraction

The operation of making two suffixes in the components of a tensor equal and then summing over the repeated suffix is called **contraction**. If a tensor is of order $n\,(\geqslant 2)$, contraction over any pair of suffixes gives a tensor of order $n - 2$. The proof of this is left as an exercise.

As an example, consider the second order tensor with components $a_i b_j$, which is the (tensor) product of the vectors $\mathbf{a} = (a_1, a_2, a_3)$ and $\mathbf{b} = (b_1, b_2, b_3)$. Contracting over the suffixes i and j gives

$$a_i b_i = a_1 b_1 + a_2 b_2 + a_3 b_3,$$

which is a tensor of order zero (a scalar invariant). It is, of course, the scalar product of **a** and **b**.

Symmetry and anti-symmetry

If a tensor A with components a_{ij} is such that

$$a_{ij} = a_{ji}$$

for all i and j, then A is said to be **symmetrical**. If

$$a_{ij} = -a_{ji}$$

for all i and j, then A is said to be **anti-symmetrical**. Observe that when A is anti-symmetrical, the components a_{11}, a_{22} and a_{33} must be zero.

This definition may be extended to tensors of higher order, the symmetry or anti-symmetry being defined with respect to a particular pair of suffixes.

Every second order tensor can be represented as the sum of a symmetrical and an anti-symmetrical tensor. For, if a tensor has components a_{ij}, then

$$a_{ij} = \tfrac{1}{2}(a_{ij} + a_{ji}) + \tfrac{1}{2}(a_{ij} - a_{ji}),$$

and it is readily seen that the two expressions in parentheses on the right-hand side are, respectively, the components of symmetrical and anti-symmetrical tensors.

Symmetry and anti-symmetry are *intrinsic* properties of a tensor, being independent of the coordinate system in which it is represented. To prove this, let A have components a_{ij} relative to axes $Ox_1 x_2 x_3$. Then, relative to the axes $Ox'_1 x'_2 x'_3$,

$$a'_{rs} = l_{ri} l_{sj} a_{ij}$$
$$= l_{rj} l_{si} a_{ji},$$

where in the last line, the dummy suffixes i and j over which the expression is summed have been interchanged. If A is symmetrical with respect to the axes $Ox_1 x_2 x_3$, $a_{ji} = a_{ij}$. It follows that

$$a'_{rs} = l_{si} l_{rj} a_{ij} = a'_{sr},$$

which shows that A is also symmetrical with respect to the axes $Ox'_1 x'_2 x'_3$. Similarly, the property of anti-symmetry is preserved when the axes are changed.

The quotient rule

Let A be a mathematical or physical entity which, when it is associated with any set of rectangular cartesian coordinate axes $Ox_1 x_2 x_3$, may be represented by an ordered set of nine scalars a_{ij} (i, $j = 1, 2, 3$). Suppose that for all vectors $\mathbf{v} = (v_1, v_2, v_3)$, the scalars

$$u_i = a_{ij} v_j \quad (i = 1, 2, 3) \tag{8.9}$$

are the components of a vector $\mathbf{u}$. Then A is a second order tensor.

Proof

Under a translation of the axes let the components a_{ij} transform to a'_{ij}. Since the components of $\mathbf{u}$ and $\mathbf{v}$ are invariant under such a translation

$$u_i = a'_{ij} v_j.$$

Subtracting this from (8.9) gives

$$(a_{ij} - a'_{ij})v_j = 0,$$

and since this holds for all vectors $\mathbf{v}$, it follows that $a'_{ij} = a_{ij}$. The components of A are therefore invariant under a translation of the axes.

Consider next a rotation to new axes $Ox'_1 x'_2 x'_3$ and let the components of $\mathbf{u}$, $\mathbf{v}$ and A transform to u'_i, v'_i and a'_{ij}, respectively. Then

$$a'_{r\alpha} v'_\alpha = u'_r \quad (r = 1, 2, 3)$$

$$= l_{ri} u_i$$

$$= l_{ri} a_{ij} v_j.$$

But

$$v'_\alpha = l_{\alpha j} v_j,$$

and hence $(a'_{r\alpha} l_{\alpha j} - l_{ri} a_{ij})v_j = 0.$

As this holds for all vectors $\mathbf{v}$, the coefficients of v_1, v_2 and v_3 must vanish and hence

$$a'_{r\alpha} l_{\alpha j} = l_{ri} a_{ij}.$$

Multiplying throughout by l_{sj} and summing over the suffix j gives

$$a'_{r\alpha} l_{\alpha j} l_{sj} = l_{ri} l_{sj} a_{ij}.$$

But

$$l_{\alpha j} l_{sj} = \delta_{\alpha s}.$$

and hence

$$a'_{rs} = l_{ri} l_{sj} a_{ij}.$$

The components of A therefore satisfy the tensor laws under a translation and rotation of the axes, which proves that A is a second order tensor.

The result just proved extends to tensors of higher order and is usually called the **quotient rule**. In the general case, (8.9) is replaced by

$$u_{ij\ldots} = a_{ij\ldots rs\ldots} v_{rs\ldots}, \tag{8.10}$$

where $u_{ij\ldots}$ and $v_{rs\ldots}$ are components of tensors U and V of orders m and n, respectively, and the inference which may be drawn is that, if (8.10) holds for all tensors V of order n, then $a_{ij\ldots rs\ldots}$ are the components of a tensor of order $m + n$. The proof is similar to that above.

EXERCISES

8.1 Relative to rectangular cartesian coordinate axes $Ox_1 x_2 x_3$, the second order tensor A has components a_{ij}. Another rectangular cartesian coordinate system $Ox_1' x_2' x_3'$ is chosen such that Ox_1' coincides with Ox_2 and Ox_3' coincides with Ox_3. Find the components of A relative to the new axes, and verify that the tensor remains symmetrical if it is so initially.

8.2 Show from first principles that the scalar obtained by contracting any second order tensor over its two suffixes is invariant under a rotation of the axes [*Hint.* Use the orthonormality conditions $l_{ri} l_{rj} = \delta_{ij}$].

8.3 The tensors A and B with components a_{ij} and b_{ij} are given to be symmetrical and anti-symmetrical respectively. Show that $a_{ij} b_{ij} = 0$.

8.4 The set of scalars u_{ijkm} is such that for all second order tensors with components E_{km} the scalars

$$\sigma_{ij} = u_{ijkm} E_{km} \qquad (i, j = 1, 2, 3)$$

are the components of a second order tensor. Prove that the scalars u_{ijkm} are the components of a fourth order tensor.

[*Hint.* This is a particular case of the quotient rule (8.10) and may be proved by a method similar to that used to prove the special case (8.9). To establish that the components u_{ijkm} transform correctly under a rotation of the axes, begin by considering $u_{rs\alpha\beta}' E_{\alpha\beta}' = \sigma_{rs}'$.]

8.3 ISOTROPIC TENSORS

A cartesian tensor is said to be **isotropic** if its components are identical in all rectangular cartesian coordinate systems. Such tensors play a fundamental role in theoretical studies of materials whose physical properties are independent of direction. This section is concerned with the isotropic tensors of second, third and fourth orders, which are of particular importance.

The delta tensor

In the second order tensor law

$$a'_{rs} = l_{ri} l_{sj} a_{ij},$$

set $a_{ij} = \delta_{ij}$ where δ_{ij} is the Kronecker delta defined in section 1.6. Then using the orthonormality conditions (1.28),

$$a'_{rs} = l_{ri} l_{si} = \delta_{rs}.$$

This shows that when the numbers $\delta_{11}, \delta_{12}, \ldots$ are associated with a set of rectangular cartesian coordinate axes, they transform into themselves under the tensor rotation law. If the association with the axes is made independent of the choice of origin, so that invariance under a translation of the axes is assured, the numbers δ_{ij} evidently form a second order isotropic tensor. It is called the **delta tensor** and is the most important of all the isotropic tensors.

Even and odd permutations

Let i, j, k be a permutation of the numbers 1, 2, 3. The permutation is said to be **even** if i, j, k are three consecutive numbers of the set 1, 2, 3, 1, 2, and is said to be **odd** otherwise. For example, 2, 3, 1 is an even permutation of the numbers 1, 2, 3, and 2, 1, 3 is an odd permutation. It may be readily verified that all even permutations of the numbers 1, 2, 3 can be brought about by an even number of interchanges of pairs of these numbers and all odd permutations by an odd number of interchanges.

The alternating tensor

Suppose that each of the suffixes i, j and k in the symbol ε_{ijk} can take any one of the values 1, 2 or 3. Let

$$\varepsilon_{ijk} = \begin{cases} 0 \text{ if two or more of the suffixes } i, j, k \text{ are equal} \\ 1 \text{ if } i, j, k \text{ is an even permutation of the numbers } 1, 2, 3 \\ -1 \text{ if } i, j, k \text{ is an odd permutation of the numbers } 1, 2, 3 \end{cases} \quad (8.11)$$

The symbol ε_{ijk} is called the **alternator**.

Set $a_{ijk} = \varepsilon_{ijk}$ in the third order tensor rotation law

$$a'_{rst} = l_{ri} l_{sj} l_{tk} a_{ijk}.$$

Then

$$a'_{rst} = l_{ri} l_{sj} l_{tk} \varepsilon_{ijk} = l_{r1} l_{s2} l_{t3} + l_{r2} l_{s3} l_{t1} + l_{r3} l_{s1} l_{t2}$$

$$- l_{r1} l_{s3} l_{t2} - l_{r2} l_{s1} l_{t3} - l_{r3} l_{s2} l_{t1}$$

$$= \begin{vmatrix} l_{r1} & l_{s1} & l_{t1} \\ l_{r2} & l_{s2} & l_{t2} \\ l_{r3} & l_{s3} & l_{t3} \end{vmatrix} ; \tag{8.12}$$

the latter expression follows from the definition of a third order determinant, give in Appendix A.

If any pair of the suffixes r, s, t are equal, the determinant (8.12) has two equal columns and hence vanishes. If $r = 1$, $s = 2$ and $t = 3$, the determinant reduces to the transpose of that given in equation (1.17) and it then has the value 1. Interchanging any pair of columns in a determinant changes its sign (Appendix A), and hence if $r = 2$, $s = 1$ and $t = 3$, the determinant has the value -1. Now, all even permutations of the numbers 1, 2, 3 can be brought about by an even number of interchanges of pairs of these numbers, and all permutations which are odd can be brought about by an even number of interchanges of pairs of the numbers 2, 1, 3. The sign (and value) of a determinant is left unchanged by an even number of interchanges of pairs of its columns, and it thus follows that, for all possible values of r, s and t the determinant (8.12) has the same value as ε_{rst}. Hence

$$a'_{rst} = \varepsilon_{rst}$$

which shows that, under a rotation of the axes, the tensor law is satisfied and each one of the set of numbers ε_{ijk} transforms into itself. If the association with the axes is defined to be independent of the choice of origin, then ε_{ijk} are the components of a third order isotropic tensor called the **alternating tensor**.

Relation between the alternating tensor and delta tensor

A very useful and important relation between the alternating and delta tensors is that

$$\varepsilon_{ijk}\,\varepsilon_{rsk} = \delta_{ir}\delta_{js} - \delta_{is}\delta_{jr} \tag{8.13}$$

This may be established as follows.

If $i = j$ or $r = s$, the right-hand side of (8.13) is zero and the left-hand side also vanishes by the definition of the alternator.

Consider the case when $i \neq j$ and $r \neq s$. Without loss of generality we may choose $i = 1$ and $j = 2$. Using the definition of the alternator, the left-hand side of (8.13) then becomes

$$\varepsilon_{121}\,\varepsilon_{rs1} + \varepsilon_{122}\,\varepsilon_{rs2} + \varepsilon_{123}\,\varepsilon_{rs3} = \varepsilon_{rs3}.$$

The right-hand side of (8.13) becomes

$$\delta_{1r}\delta_{2s} - \delta_{1s}\delta_{2r} = \Delta, \text{ say.}$$

As $r \neq s$, there are just the following possibilities to consider :

$r = 3$ in which case $\Delta = 0$ for all s ;

$s = 3$ in which case $\Delta = 0$ for all r ;

$r = 1$, $s = 2$, giving $\Delta = 1$;

$r = 2$, $s = 1$ giving $\Delta = -1$.

Hence $\Delta = \varepsilon_{rs3}$, and the identity is proved.

Another way of proving the identity is indicated in Exercise 8.13, section 8.3.

Relation of the alternating tensor to the vector product

The i-th component of the vector product of $\mathbf{a} = (a_1, a_2, a_3)$ with $\mathbf{b} = (b_1, b_2, b_3)$ is related to ε_{ijk} by the formula

$$(\mathbf{a} \times \mathbf{b})_i = \varepsilon_{ijk} a_j b_k . \tag{8.14}$$

To prove this, consider the x_1-component. When $i = 1$, ε_{ijk} is non-zero only when $j = 2$ and $k = 3$ or when $j = 3$ and $k = 2$. Hence, according to (8.14),

$$(\mathbf{a} \times \mathbf{b})_1 = \varepsilon_{123} a_2 b_3 + \varepsilon_{132} a_3 b_2 = a_2 b_3 - a_3 b_2,$$

which verifies that the formula gives the correct x_1-component. Likewise the x_2- and x_3-components are correctly represented.

Observe that $\varepsilon_{ijk} a_j b_k$ is the product $\varepsilon_{ijk} a_r b_s$ contracted over the suffixes j, r and k, s. Since the alternating tensor is of third order and $\mathbf{a}$ and $\mathbf{b}$ are first order tensors, their product is a fifth order tensor. Contraction of this over two pairs of suffixes reduces the order by 4 (see section 8.2) yielding a first order tensor which is the vector $\mathbf{a} \times \mathbf{b}$. This provides an alternative to the proof in section 2.7 that $\mathbf{a} \times \mathbf{b}$ is a vector.

Another formula, closely associated with (8.14), is that for any vector field $\mathbf{F}$,

$$(\mathrm{curl}\ \mathbf{F})_i = \varepsilon_{ijk} \partial F_k / \partial x_j. \tag{8.15}$$

The reader should have no difficulty in proving this.

EXAMPLE 1

In an incompressible fluid, the stress on an element whose normal is in the direction of a unit vector $\mathbf{n}$ has components $t_i(\mathbf{n}) = \sigma_{ij} n_j$ $(i = 1, 2, 3)$, where σ_{ij} is the stress tensor. Given that

$$\sigma_{ij} = -p\delta_{ij} + \mu\left(\frac{\partial v_i}{\partial x_j} + \frac{\partial v_j}{\partial x_i}\right), \tag{8.16}$$

verify that

$$\mathbf{t(n)} = -p\mathbf{n} + 2\mu(\mathbf{n} \cdot \nabla)\mathbf{v} + \mu\mathbf{n} \times \mathrm{curl}\,\mathbf{v}. \tag{8.17}$$

Solution

Using (8.14), the i-th component of $\mathbf{n} \times \mathrm{curl}\,\mathbf{v}$ is

$$(\mathbf{n} \times \mathrm{curl}\,\mathbf{v})_i = \varepsilon_{ijk}\, n_j (\mathrm{curl}\,\mathbf{v})_k$$

$$= \varepsilon_{ijk}\, n_j \varepsilon_{krs}\, \partial v_s / \partial x_r,$$

by (8.15). But

$$\varepsilon_{krs} = -\varepsilon_{rks} = \varepsilon_{rsk},$$

because interchanging any pair of suffixes in the alternator changes its sign. Hence,

$$(\mathbf{n} \times \mathrm{curl}\,\mathbf{v})_i = \varepsilon_{ijk}\,\varepsilon_{rsk} n_j \partial v_s / \partial x_r$$

$$= (\delta_{ir}\delta_{js} - \delta_{is}\delta_{jr})\, n_j \partial v_s / \partial x_r$$

$$= n_j \frac{\partial v_j}{\partial x_i} - n_j \frac{\partial v_i}{\partial x_j}.$$

The i-th component of (8.17) is therefore

$$t_i(\mathbf{n}) = -pn_i + 2\mu n_j \frac{\partial v_i}{\partial x_j} + \mu(\mathbf{n} \times \mathrm{curl}\,\mathbf{v})_i$$

$$= -pn_i + \mu n_j \left(\frac{\partial v_i}{\partial x_j} + \frac{\partial v_j}{\partial x_i} \right). \tag{8.18}$$

But, we are given that

$$t_i(\mathbf{n}) = \sigma_{ij}n_j,$$

and substituting σ_{ij} from (8.16) an expression identical to (8.18) is obtained. This completes the verification.

Remark

As (8.17) is in vector form, it can be used to obtain the components of stress in a curvilinear coordinate system. For example, in cylindrical polar coordinates R, φ, z, the stress on an element whose normal is in the direction of the unit normal vector $\mathbf{e}_R$ is

$$\mathbf{t(e}_R) = -p\mathbf{e}_R + 2\mu\partial\mathbf{v}/\partial R + \mu\mathbf{e}_R \times \mathrm{curl}\,\mathbf{v}. \tag{8.19}$$

It is conventional to denote the cylindrical polar components of this stress vector by $\sigma_{RR}, \sigma_{R\varphi}, \sigma_{Rz}$ and hence

$$\mathbf{t(e}_R) = \sigma_{RR}\,\mathbf{e_R} + \sigma_{R\varphi}\,\mathbf{e}_\varphi + \sigma_{Rz}\,\mathbf{e}_z. \tag{8.20}$$

Using the methods of sections 4.13 and 4.14 [remember that $\mathbf{e}_R$ and $\mathbf{e}_\varphi$ are not *constant* vectors], it follows from (8.19) and (8.20) that

$$\sigma_{RR} = -p + 2\mu \partial v_R / \partial R$$

$$\sigma_{R\varphi} = \mu \left(\frac{\partial v_R}{R \partial \varphi} + \frac{\partial v_\varphi}{\partial R} - \frac{v_\varphi}{R} \right)$$

and

$$\sigma_{Rz} = \mu \left(\frac{\partial v_R}{\partial z} + \frac{\partial v_z}{\partial R} \right).$$

In a similar way, the components of stress on elements whose normals are in the directions of $\mathbf{e}_\varphi$ and $\mathbf{e}_z$ may be found.

Isotropic tensors of the fourth order

We shall now prove that any fourth order isotropic tensor U with components u_{ijkm} may be expressed as a sum of products of delta tensors in the form

$$u_{ijkm} = \lambda \delta_{ij} \delta_{km} + \mu \delta_{ik} \delta_{jm} + v \delta_{im} \delta_{jk}, \tag{8.21}$$

where λ, μ and v are arbitrary scalar invariants. The formula will be established by considering the effect of certain rotations of the coordinate axes. We observe first that, for U to be isotropic, under any rotation of the axes

$$u_{pqrs} = l_{pi} l_{qj} l_{rk} l_{sm} u_{ijkm}, \tag{8.22}$$

since the transformed components of U must be the same as those relative to the original axes.

Each of i, j, k and m can assume only the values 1, 2, or 3, and hence the components of U may be collected into the following four groups:

(i) the components u_{1111}, u_{2222}, u_{3333} in which all four suffixes are equal;
(ii) components which have three equal suffixes and the other one different (e.g. u_{1112});
(iii) components which have two distinct pairs of equal suffixes (typically, u_{1122}, u_{1212}, u_{1221});
(iv) components which have just one pair of equal suffixes and the other suffixes unequal (e.g. u_{1123}).

Consider a rotation of the axes $Ox_1 x_2 x_3$ through 180° about Ox_3. The direction cosines in this transformation are $l_{11} = l_{22} = -1$, $l_{33} = 1$ and $l_{ij} = 0$ when $i \neq j$. Hence, using (8.22),

$$u_{1113} = l_{1i} l_{1j} l_{1k} l_{3m} u_{ijkm} = -u_{1113}$$

and

$$u_{1123} = l_{1i} l_{1j} l_{2k} l_{3m} u_{ijkm} = -u_{1123}.$$

It follows that

$$u_{1113} = 0 \text{ and } u_{1123} = 0.$$

But these are typical components from groups (ii) and (iv), and hence if U is isotropic all components in those groups must vanish.

Consider next a rotation through $90°$ about the axis Ox_3 for which the transformation matrix is:

O	x_1	x_2	x_3
x_1'	0	1	0
x_2'	-1	0	0
x_3'	0	0	1

(8.23)

Using (8.22), under this rotation,

$$u_{1111} = l_{1i}\, l_{1j}\, l_{1k}\, l_{1m}\, u_{ijkm} = u_{2222}.$$

Similarly, by a cyclic interchange of the suffixes, $u_{2222} = u_{3333}$, and hence the components in group (i) are equal to each other. Also, under the rotation (8.23),

$$u_{1122} = l_{1i}\, l_{1j}\, l_{2k}\, l_{2m}\, u_{ijkm} = u_{2211},$$

(8.24)

and cyclic interchanges of the suffixes give

$$u_{2233} = u_{3322} \text{ and } u_{3311} = u_{1133}.$$

(8.25)

Now, applying the rotation represented by the transformation matrix

O	x_1	x_2	x_3
x_1'	0	1	0
x_2'	0	0	1
x_3'	1	0	0

,

we also have

$$u_{1122} = l_{1i}\, l_{1j}\, l_{2k}\, l_{2m}\, u_{ijkm} = u_{2233}.$$

(8.26)

Advancing the suffixes cyclically and combining the results so obtained with (8.24) and (8.25), it follows that

$$u_{1122} = u_{2211} = u_{2233} = u_{3322} = u_{1133} = u_{3311}.$$

This shows that all members of group (iii) with $i = j$ and $k = m$ are equal to each other. Similarly, it can be shown that all members with $i = k$ and $j = m$ are equal to each other, and that those with $i = m$ and $j = k$ are equal to each other. In summary, the stage now reached is that the tensor with components u_{ijkm} can only be isotropic if:

(a) $u_{1111} = u_{2222} = u_{3333}$;
(b) components such that $i = j \neq k = m$ are equal;
(c) components such that $i = k \neq j = m$ are equal;
(d) components such that $i = m \neq j = k$ are equal;
(e) all components not included under (a)—(d) vanish.

Consider the particular case in which all components in groups (c) and (d) are set equal to zero. A rotation of the axes through 45° about Ox_3 is represented by the transformation matrix:

	x_1	x_2	x_3
x_1'	$\frac{1}{2}\sqrt{2}$	$\frac{1}{2}\sqrt{2}$	0
x_2'	$-\frac{1}{2}\sqrt{2}$	$\frac{1}{2}\sqrt{2}$	0
x_3'	0	0	1

Under this rotation, (8.22) gives

$$u_{1111} = l_{1i} l_{1j} l_{1k} l_{1m} u_{ijkm} = \tfrac{1}{4}(u_{1111} + u_{2222} + u_{1122} + u_{2211}),$$

where the hypothesis that $u_{ijkm} = 0$ except when $i = j$ and $k = m$ has been used. But it has been shown that $u_{1111} = u_{2222}$ and $u_{1122} = u_{2211}$, and it follows therefore that $u_{1111} = u_{1122}$. Hence, all the non-vanishing components of this particular isotropic tensor have the same value, λ, say which may be expressed as

$$u_{ijkm} = \lambda \, \delta_{ij} \delta_{km} .$$

Similarly, by considering groups (c) and (d), in turn, two other fourth order isotropic tensors may be identified having components

$$u_{ijkm} = \mu \delta_{ik} \delta_{jm}, \quad u_{ijkm} = \nu \delta_{im} \delta_{jk},$$

respectively, where μ and ν are arbitrary scalar invariants. The most general isotropic tensor of order four is evidently a combination of all three of these and hence has components of the form

$$u_{ijkm} = \lambda \, \delta_{ij} \delta_{km} + \mu \delta_{ik} \delta_{jm} + \nu \delta_{im} \delta_{jk},$$

as stated earlier.

EXAMPLE 2

In linear elasticity theory, the material stress components σ_{ij} are assumed to be linear functions of the infinitesimal strain components E_{ij}. Given that σ_{ij} and E_{ij} are the components of second order symmetrical tensors and that the properties of the material are independent of direction, show that

$$\sigma_{ij} = \lambda E_{kk} \, \delta_{ij} + 2\mu E_{ij}, \tag{8.27}$$

where λ and μ are scalar invariants.

Solution

Denote the coefficient of E_{km} in the linear expression for σ_{ij} by u_{ijkm}, so that

$$\sigma_{ij} = u_{ijkm} E_{km}.$$

By the quotient rule (section 8.2, Exercise 8.4), the numbers u_{ijkm} are the components of a fourth order tensor and the assumption that the properties of the material are independent of direction indicates that this tensor is isotropic. Hence, by (8.21)

$$\sigma_{ij} = \lambda E_{kk} \delta_{ij} + \mu E_{ij} + \nu E_{ji}.$$

But $E_{ji} = E_{ij}$ and hence we may take $\nu = \mu$ giving (8.27).

EXERCISES

8.5 Using results from the text, prove that:

$$\delta_{ij} \varepsilon_{ijk} = 0,$$

$$\varepsilon_{ijk} \varepsilon_{rjk} = 2\delta_{ir},$$

$$\varepsilon_{ijk} \varepsilon_{ijk} = 6.$$

8.6 A second order symmetrical tensor has components s_{ij}. Show that $\varepsilon_{ijk} s_{ij} = 0$ for all values of k. Prove the converse, that if S is a second order tensor with components s_{ij} such that $\varepsilon_{ijk} s_{ij} = 0$, then S is symmetric.

8.7 Prove from first principles that $\delta_{ij} \delta_{km}$ are the components of a fourth order isotropic tensor.

8.8 Using theorems and results from the text, prove that $\delta_{ij} \varepsilon_{kmn}$ are the components of a fifth-order isotropic tensor.

8.9 Show that the only isotropic tensor of first order is the zero vector. [*Hint.* Consider a 180° rotation about one of the axes.]

8.10 Prove that the most general isotropic tensor of second order has components $\lambda \delta_{ij}$ where λ is an arbitrary scalar.

8.11 If A is a second order tensor with components a_{jk}, show from first principles that

$$b_i = \tfrac{1}{2} \varepsilon_{ijk} a_{jk} \qquad (i = 1, 2, 3)$$

are the components of a vector **b**. If the tensor A is anti-symmetrical verify that

$$\mathbf{b} = (a_{23}, a_{31}, a_{12}).$$

8.12 If A is a second order anti-symmetrical tensor with components a_{jk} and the vector $\mathbf{b}$ has components given by

$$b_i = \tfrac{1}{2} \varepsilon_{ijk} a_{jk} \qquad (i = 1, 2, 3),$$

show that $a_{rs} = \varepsilon_{irs} b_i$. Write down the matrix of components of A in terms of the components of $\mathbf{b}$.

8.13 Prove that

$$\varepsilon_{ijk} = \begin{vmatrix} \delta_{i1} & \delta_{i2} & \delta_{i3} \\ \delta_{j1} & \delta_{j2} & \delta_{j3} \\ \delta_{k1} & \delta_{k2} & \delta_{k3} \end{vmatrix}$$

and hence show that

$$\varepsilon_{ijk} \varepsilon_{rst} = \begin{vmatrix} \delta_{ir} & \delta_{is} & \delta_{it} \\ \delta_{jr} & \delta_{js} & \delta_{jt} \\ \delta_{kr} & \delta_{ks} & \delta_{kt} \end{vmatrix},$$

Deduce that

$$\varepsilon_{ijk} \varepsilon_{rsk} = \delta_{ir} \delta_{js} - \delta_{is} \delta_{jr}.$$

8.14 Using expression (8.14) for the vector product, prove the formula

$$\mathbf{a} \times (\mathbf{b} \times \mathbf{c}) = (\mathbf{a} \cdot \mathbf{c})\mathbf{b} - (\mathbf{a} \cdot \mathbf{b})\mathbf{c}.$$

8.4 TENSOR FIELDS

The definitions of scalar and vector fields given in section 4.3 extend readily to tensors of general order. If the components $a_{ij\ldots}$ of the tensor A are functions of the coordinates x_1, x_2, x_3 of the points in a given region (or, more generally, any point set), then A is called a **tensor function of position** or a **tensor field**. Tensors which occur in subjects such as fluid dynamics and elasticity theory usually constitute tensor fields and some knowledge of their properties is therefore essential.

The main new feature that we have to consider is the necessity of handling with ease multiple partial derivatives of the components $a_{ij\ldots}$. Such derivatives are often denoted thus:

$$\frac{\partial a_{ij\ldots}}{\partial x_p} = a_{ij\ldots,p}, \quad \frac{\partial^2 a_{ij\ldots}}{\partial x_p \partial x_q} = a_{ij\ldots,pq}, \quad \text{etc.}$$

Particular instances of frequent occurrence are:

$$\frac{\partial a}{\partial x_i} = a_{,i}, \quad \frac{\partial a_i}{\partial x_j} = a_{i,j}, \quad \frac{\partial a_{ij}}{\partial x_k} = a_{ij,k},$$

where i, j and k can usually range over the values 1, 2 or 3.

There is one important basic theorem, which we now state and prove.

Theorem

In an open region $\mathcal{R}$, let the components $a_{ij\ldots}$ (n suffixes) of a tensor A of order n be continuously differentiable functions of the coordinates x_1, x_2, x_3. Then

$$a_{ij\ldots,p} \qquad (i, j, \ldots, \quad p = 1, 2, 3)$$

are the components of a tensor field of order $n + 1$ called the **tensor gradient** of A.

Proof

Consider a rotation of the axes $Ox_1x_2x_3$ to new positions $Ox_1'x_2'x_3'$, as defined by (8.1). By the tensor law, the transformed components of A are

$$a_{rs\ldots}' = l_{ri}\, l_{sj} \ldots a_{ij\ldots}.$$

Hence

$$\frac{\partial a_{rs\ldots}'}{\partial x_q'} = l_{ri} l_{sj} \ldots \frac{\partial x_p}{\partial x_q'}\, \frac{\partial a_{ij\ldots}}{\partial x_p},$$

using the chain rule. But

$$x_p = l_{qp} x_q',$$

and it follows that

$$a_{rs\ldots,q}' = l_{ri}\, l_{sj} \ldots l_{qp} a_{ij\ldots,p},$$

which is the law of transformation of the components of a tensor of order $n + 1$. It is readily verified that the components $a_{ij\ldots,p}$ are unchanged by a translation of the axes (cf. part (a) of the proof that grad Ω is a vector field in section 4.4), and hence it follows that they form a tensor field of order $n + 1$ defined on the region $\mathcal{R}$.

The theorem extends to higher order derivatives. The reader should prove, for example, the corollary that if the components of A have second order derivatives which are continuous, then $a_{ij\ldots,pq}$ are the components of a tensor of order $n + 2$.

Tensor gradients of scalar fields

One example of a tensor gradient already familiar to us is the vector field grad Ω associated with the scalar Ω. In tensor terminology, Ω is a tensor of order zero and according to the theorem above the components

$$\partial\Omega/\partial x_i = \Omega_{,i} \quad (i = 1, 2, 3)$$

form a tensor field of order 1, i.e. a vector field; this is, of course,

$$\text{grad } \Omega = \left(\frac{\partial\Omega}{\partial x_1}, \frac{\partial\Omega}{\partial x_2}, \frac{\partial\Omega}{\partial x_3}\right).$$

By the corollary to the theorem, differentiating Ω twice with respect to the coordinates gives a second order tensor field with components $\Omega_{,ij}$. Contracting this over the two suffixes reduces the order by 2 (see section 8.2) and hence yields the scalar field

$$\Omega_{,ii} = \frac{\partial^2\Omega}{\partial x_i \partial x_i} = \frac{\partial^2\Omega}{\partial x_1^2} + \frac{\partial^2\Omega}{\partial x_2^2} + \frac{\partial^2\Omega}{\partial x_3^2},$$

which we recognise as the Laplacian $\nabla^2\Omega$.

The tensor gradient of a vector field

The gradient of the vector field with continuously differentiable components F_i is the second order tensor field with components $\partial F_i/\partial x_j$. This tensor may be expressed as a sum of a symmetrical and an anti-symmetrical part by setting

$$\partial F_i/\partial x_j = s_{ij} + a_{ij} \tag{8.28}$$

where

$$s_{ij} = \frac{1}{2}\left(\frac{\partial F_i}{\partial x_j} + \frac{\partial F_j}{\partial x_i}\right) \tag{8.29}$$

and

$$a_{ij} = \frac{1}{2}\left(\frac{\partial F_i}{\partial x_j} - \frac{\partial F_j}{\partial x_i}\right). \tag{8.30}$$

Two facts worthy of note are the following.

(i) Contracting the symmetrical part s_{ij} over the suffixes i and j gives

$$s_{ii} = \partial F_i/\partial x_i = \text{div } \mathbf{F}.$$

(ii) Using (8.15),

$$(\text{curl } \mathbf{F})_i = \varepsilon_{ijk}\, \partial F_k/\partial x_j$$

$$= \varepsilon_{ijk}\, s_{kj} + \varepsilon_{ijk}\, a_{kj}. \tag{8.31}$$

But $\varepsilon_{ijk}\, s_{kj} = 0$ (cf. Exercise 8.6, section 8.3), and hence taking $i = 1, 2$ and 3 in turn we find that

$$\text{curl } \mathbf{F} = 2(a_{32}, a_{13}, a_{21}) \tag{8.32}$$

$$= -2(a_{23}, a_{31}, a_{12}). \tag{8.33}$$

This shows that the curl of a vector field is expressible entirely in terms of the non-zero terms of the anti-symmetrical parts of its tensor gradient ($a_{ij} = 0$ when $i = j$, as is seen immediately from the definition (8.30)). Alternatively, we may say that the antisymmetrical part of the tensor gradient of a vector field $\mathbf{F}$ depends only on the components of curl $\mathbf{F}$. More explicitly, setting

$$\text{curl } \mathbf{F} = (c_1, c_2, c_3)$$

and comparing components with those in (8.32) and (8.33), the matrix of components of the anti-symmetrical part of the tensor gradient of $\mathbf{F}$ is

$$\begin{pmatrix} 0 & -\frac{1}{2}c_3 & \frac{1}{2}c_2 \\ \frac{1}{2}c_3 & 0 & -\frac{1}{2}c_1 \\ -\frac{1}{2}c_2 & \frac{1}{2}c_1 & 0 \end{pmatrix}.$$

EXAMPLE 3

In fluid dynamics, the vorticity is defined as $\omega = \text{curl } \mathbf{v}$, where $\mathbf{v}$ is the fluid velocity. If $\mathbf{r} = (x_1, x_2, x_3)$, and a_{ij} are the components of the anti-symmetrical part of the tensor gradient of $\mathbf{v}$, show that the i-th component of the vector field $\frac{1}{2}\omega \times \mathbf{r}$ is $a_{ij}x_j$.

Solution

Using (8.14) followed by (8.15), the i-th component of $\frac{1}{2}\omega \times \mathbf{r}$ is

$$\frac{1}{2}(\omega \times \mathbf{r})_i = \frac{1}{2}\varepsilon_{ikj}\omega_k x_j$$

$$= \frac{1}{2}\varepsilon_{ikj}\varepsilon_{krs}(\partial v_s/\partial x_r)x_j .$$

But an odd number of interchanges of the suffixes of the alternator changes its sign and an even number of interchanges leaves it unchanged. Using this fact followed by (8.13), it follows that

$$\frac{1}{2}(\omega \times \mathbf{r})_i = -\frac{1}{2}\varepsilon_{ijk}\varepsilon_{rsk}(\partial v_s/\partial x_r)x_j$$

$$= \frac{1}{2}(\delta_{is}\delta_{jr} - \delta_{ir}\delta_{js})(\partial v_s/\partial x_r)x_j$$

$$= \frac{1}{2}\left(\frac{\partial v_i}{\partial x_j} - \frac{\partial v_j}{\partial x_i}\right)x_j$$

$$= a_{ij}x_j.$$

This is the required result.

EXERCISES

8.15 Use tensor methods to establish the following identities:

(i) $\operatorname{div}(\mathbf{F} \times \mathbf{G}) \equiv \mathbf{G} . \operatorname{curl} \mathbf{F} - \mathbf{F} . \operatorname{curl} \mathbf{G};$

(ii) $\operatorname{curl}(\Omega \, \mathbf{F}) \equiv \Omega \, \operatorname{curl} \mathbf{F} - \mathbf{F} \times \operatorname{grad} \Omega ;$

(iii) $\operatorname{curl} \operatorname{curl} \mathbf{F} \equiv \operatorname{grad} \operatorname{div} \mathbf{F} - \nabla^2 \mathbf{F}.$

In each case, assume that all derivatives which are implied by these relations exist and are continuous.

8.5 THE DIVERGENCE THEOREM IN TENSOR FIELD THEORY

The gradient of a second order tensor field A with continuously differentiable components a_{ij} is a third order tensor with components $a_{ij,k}$. Contraction over the suffixes j and k gives a vector field whose i-th component is

$$a_{ij,j} = \frac{\partial a_{i1}}{\partial x_1} + \frac{\partial a_{i2}}{\partial x_2} + \frac{\partial a_{i3}}{\partial x_3}, \tag{8.34}$$

whilst interchanging i and j (for convenience) and then contracting over the suffixes j and k gives another vector field whose i-th component is

$$a_{ji,j} = \frac{\partial a_{1i}}{\partial x_1} + \frac{\partial a_{2i}}{\partial x_2} + \frac{\partial a_{3i}}{\partial x_3} \tag{8.35}$$

In the special case when A is symmetrical, the two vector fields will, of course, be identical since then $a_{ij} = a_{ji}$. If $\mathbf{F}$ is a vector field with continuous first derivatives, div $\mathbf{F}$ may be obtained by contracting the components $F_{i,j}$ of the tensor gradient of $\mathbf{F}$ with respect to the suffixes i, j, and hence (8.34) and (8.35) may be regarded as generalized divergences; they are, in fact, sometimes called the divergence of A with respect to the second suffix and the divergence of A with respect to the first suffix, respectively.

The divergence theorem proved in Chapter 6 may be extended to obtain corresponding results for the vector fields (8.34) and (8.35). Using our present notation, the divergence theorem (6.1) may be expressed as

$$\iint_S F_i n_i \, dS = \iiint_V F_{i,i} \, dV, \tag{8.36}$$

where $\mathbf{F} = (F_1, F_2, F_3)$ is a continuously differentiable vector field defined in a region V bounded by a simple closed surface S and $\mathbf{n} = (n_1, n_2, n_3)$ is the outward unit normal vector to S. To generalize the theorem, consider the three results (6.3), (6.4), and (6.5), from which (8.36) is derived. Replacing F_1, F_2, F_3, respectively by the continuously differentiable scalar functions a_{i1}, a_{i2}, a_{i3} and making an appropriate change of notation for the coordinates and unit vectors, we have

$$\iint_S a_{i1}\mathbf{e}_1 \cdot d\mathbf{S} = \iiint_V a_{i1,1} \, dV$$

$$\iint_S a_{i2}\mathbf{e}_2 \cdot d\mathbf{S} = \iiint_V a_{i2,2} \, dV$$

$$\iint_S a_{i3}\mathbf{e}_3 \cdot d\mathbf{S} = \iiint_V a_{i3,3} \, dV.$$

Addition of these results gives

$$\iint_S a_{ij}n_j \, dS = \iiint_V a_{ij,j} \, dV, \tag{8.37}$$

where $n_j = \mathbf{e}_j \cdot \mathbf{n}$. This is one of the generalized forms of the divergence theorem (8.36). Another form, corresponding to (8.35), is

$$\iint_S a_{ji} \, n_j \, dS = \iiint_V a_{ji,j} \, dV. \tag{8.38}$$

Further generalization is possible for tensors of higher order, but the cases we have mentioned are sufficient for many purposes.

EXERCISES

8.16 In the region V bounded by a simple closed surface S, the second order symmetrical tensor field with components σ_{ij} has continuous first order derivatives and is such that, throughout V, $\sigma_{ij,j} = 0$. On the bounding surface S, a vector field $\mathbf{F}$ is defined with i-th component $\sigma_{ij}n_j$, where $\mathbf{n} = (n_1, n_2, n_3)$ is the outward unit normal vector to S. Prove that:

$$\text{(i)} \iint_S \mathbf{F} \, dS = 0, \qquad \text{(ii)} \iint_S \mathbf{r} \times \mathbf{F} \, dS = 0,$$

where $\mathbf{r} = (x_1, x_2, x_3)$ is the position vector from the origin.

8.17 In the region V bounded by a simple closed surface S, the vector field $\mathbf{v} = (v_1, v_2, v_3)$ is continuously differentiable and the components of the symmetrical part of its tensor gradient are denoted by e_{ij}. If the tensor with components σ_{ij} is symmetrical and has continuous first order derivatives in V such that $\sigma_{ij,j} = 0$, show that

$$\iint_S v_i \sigma_{ij} n_j \, dS = \iiint_V \sigma_{ij} e_{ij} \, dV.$$

Representation theorems for isotropic tensor functions

9

9.1 INTRODUCTION

In theoretical studies of the behaviour of materials it is necessary to introduce postulates about the way in which materials respond to factors such as local deformations and local temperature gradients. These postulates naturally depend on the type of material under consideration and in continuum theories they are expressed mathematically in the form of certain relations between tensors called **constitutive equations**. A simple example of a constitutive equation is that relating the stress components σ_{ij} and the pressure p in an inviscid fluid, namely

$$\sigma_{ij} = -p\delta_{ij}.$$

Another example is the constitutive equation of linear elasticity theory, which is expressed in the form of the relation (8.27); we observe that it is a linear relation between the components σ_{ij} of the stress tensor and the infinitesimal strain components E_{ij}. Yet another example of a constitutive equation is Fourier's law of heat conduction, which may be expressed as

$$\mathbf{q} = -k \operatorname{grad} T,$$

where $\mathbf{q}$ is the heat conduction vector, T denotes temperature and k is the thermal conductivity.

Now a basic postulate which is applicable to a large class of materials is that the physical properties are independent of direction, or, in other words, that there are no preferred directions of response. Such materials are said to be **isotropic**. Mathematically, the requirement of isotropy implies that the relations defining the material properties should be invariant under a rotation of the coordinate axes, and this places certain restrictions on the form that the relations may take.

The main purpose of this chapter is to consider three types of relation which are of particular importance, and to derive theorems for their representation under the assumption of isotropy. Many generalizations of these theorems have been proved in recent years but their applications are largely at postgraduate level and lie outside the scope of the present book. Two of the cases considered involve second order symmetrical tensors and it is necessary to establish first some of the principal properties of such tensors.

9.2 DIAGONALIZATION OF SECOND ORDER SYMMETRICAL TENSORS

A classical theorem of great importance is that for any second order symmetrical cartesian tensor, it is always possible to choose axes, called **principal axes**, such that the matrix of components of the tensor is of the **diagonal form**

$$\begin{pmatrix} \lambda_1 & 0 & 0 \\ 0 & \lambda_2 & 0 \\ 0 & 0 & \lambda_3 \end{pmatrix}.$$

This section is devoted to proving this theorem.

Definition

A non-zero vector $\mathbf{v} = (v_1, v_2, v_3)$ is said to be an **eigenvector** of the second order tensor A, with components a_{ij}, if there exists a scalar λ such that

$$a_{ij}v_j = \lambda v_i \quad (i = 1, 2, 3); \tag{9.1}$$

λ is called the **eigenvalue** associated with the vector $\mathbf{v}$.

Written out in full, equations (9.1) are:

$$(a_{11} - \lambda)v_1 + a_{12} v_2 + a_{13} v_3 = 0$$

$$a_{21} v_1 + (a_{22} - \lambda)v_2 + a_{23}v_3 = 0$$

$$a_{31}v_1 + a_{32}v_2 + (a_{33} - \lambda)v_3 = 0.$$

A necessary and sufficient condition for these equations to have a solution in which at least one of v_1, v_2, v_3 is non-zero is that

$$\begin{vmatrix} a_{11} - \lambda & a_{12} & a_{13} \\ a_{21} & a_{22} - \lambda & a_{23} \\ a_{31} & a_{32} & a_{33} - \lambda \end{vmatrix} = 0. \tag{9.2}$$

This is a cubic equation for λ, called the **characteristic equation** of A. It follows that there are three eigenvalues $\lambda_1, \lambda_2, \lambda_3$ (though these may not be distinct), and corresponding to these eigenvalues there will be eigenvectors, which may be denoted by $\mathbf{v}^{(1)}, \mathbf{v}^{(2)}, \mathbf{v}^{(3)}$, respectively. It should be noted that the eigenvectors are arbitrary to the extent of a multiplicative scalar invariant factor. In other words, if $\mathbf{v}$ is an eigenvector and $\gamma \neq 0$ is a scalar invariant, then $\gamma\mathbf{v}$ is also an eigenvector.

Theorem 1

The eigenvalues of a second order tensor are independent of the coordinate system.

Proof

With respect to the axes $Ox_1'x_2'x_3'$, suppose that λ is an eigenvalue of the tensor A, and hence that

$$a_{rs}' v_s' = \lambda v_r'.$$

Expressing a_{rs}', v_s' and v_r' in terms of components relative to the axes $Ox_1 x_2 x_3$,

$$l_{ri}\, l_{sj}\, a_{ij}\, l_{sp}v_p = \lambda l_{rq}v_q .$$

Multiplying by l_{rt} and using the orthonormality conditions, we obtain

$$\delta_{it}\, a_{ij}\delta_{jp}\, v_p = \lambda\delta_{tq}v_q,$$

which reduces to

$$a_{tj}v_j = \lambda v_t.$$

Thus λ is also an eigenvalue in the coordinate system $Ox_1 x_2 x_3$, and the theorem follows.

Theorem 2

A real symmetric second order tensor has real eigenvalues.

Proof

Let λ be any eigenvalue of the tensor A. Multiplying (9.1) by the complex conjugate of v_i (which is denoted by $\bar{v}_i$),

$$\lambda v_i\bar{v}_i = a_{ij}v_j\bar{v}_i$$

$$= \tfrac{1}{2} a_{ij}v_j\bar{v}_i + \tfrac{1}{2}a_{ji}v_j\bar{v}_i \quad (\text{because } A \text{ is symmetrical})$$

$$= \tfrac{1}{2} a_{ij}(v_j \bar{v}_i + v_i \bar{v}_j).$$

The expression $v_j \bar{v}_i + \bar{v}_j v_i$ is real because it is the sum of conjugates, and since $v_i \bar{v}_i$ is real and non-zero, we conclude that λ is real.

Theorem 3

If A is a real symmetric second order tensor, it is possible to choose a set of principal axes, say $Ox_1'' x_2'' x_3''$, relative to which the components of A are

$$a_{11}'' = \lambda_1, \quad a_{22}'' = \lambda_2, \quad a_{33}'' = \lambda_3 \quad \text{and} \quad a_{ij}'' = 0 \text{ for } i \neq j,$$

where $\lambda_1, \lambda_2, \lambda_3$ are the eigenvalues of A.

Proof

Let λ_1 be an eigenvalue of A, and denote by $\mathbf{v}^{(1)}$ a corresponding eigenvector. Then

$$a_{ij} v_j^{(1)} = \lambda_1 v_i^{(1)}.$$

If $\mathbf{e}^{(1)}$ is the unit vector parallel to $\mathbf{v}^{(1)}$, it follows also that

$$a_{ij} e_j^{(1)} = \lambda_1 e_i^{(1)},$$

because $\mathbf{e}^{(1)} = \mathbf{v}^{(1)} / \left| \mathbf{v}^{(1)} \right|$.

Choose axes $Ox_1' x_2' x_3'$ such that Ox_1' is parallel to $\mathbf{e}^{(1)}$. Then, in the transformation matrix

	x_1	x_2	x_3
x_1'	l_{11}	l_{12}	l_{13}
x_2'	l_{21}	l_{22}	l_{23}
x_3'	l_{31}	l_{32}	l_{33}

$l_{11} = e_1^{(1)}, l_{12} = e_2^{(1)}$ and $l_{13} = e_3^{(1)}$.

Now

$$a_{rs}' = l_{ri} l_{sj} a_{ij},$$

and hence

$$a_{r1}' = l_{ri} l_{1j} a_{ij}$$
$$= l_{ri} e_j^{(1)} a_{ij}$$
$$= \lambda_1 l_{ri} e_i^{(1)}$$
$$= \lambda_1 l_{ri} l_{1i}$$

$$= \lambda_1 \delta_{r1}.$$

Since A is symmetrical in all coordinate systems (see section 8.2), it follows that, relative to the axes $Ox_1'x_2'x_3'$, the matrix of components of A is of the form

$$\begin{pmatrix} \lambda_1 & 0 & 0 \\ 0 & a_{22}' & a_{23}' \\ 0 & a_{32}' & a_{33}' \end{pmatrix}.$$

Suppose now that the axes $Ox_1'x_2'x_3'$ are rotated about Ox_1', and denote the axes in their new position by $Ox_1''\, x_2''\, x_3''$. Since Ox_1' and Ox_1'' coincide, the transformation matrix is of the form

	x_1'	x_2'	x_3'
x_1''	1	0	0
x_2''	0	l_{22}'	l_{23}'
x_3''	0	l_{32}'	l_{33}'

By considering the equations

$$a_{rs}'' = l_{ri}'\, l_{sj}'\, a_{ij}',$$

it is readily verified that this rotation of the axes changes only those components of A having suffixes 22, 23, 32, and 33, respectively. The completion of the proof therefore rests upon showing that it is possible to rotate the axes about Ox_1' in such a way that the set of scalars

$$\begin{pmatrix} a_{22}' & a_{23}' \\ a_{32}' & a_{33}' \end{pmatrix}$$

transforms to a set

$$\begin{pmatrix} \lambda_2 & 0 \\ 0 & \lambda_3 \end{pmatrix}.$$

Now this is the two-dimensional analogue of the theorem we require to prove, and by reasoning similar to that above (with all suffixes restricted to have two values instead of three), its proof depends in turn, upon proving the one-dimensional analogue. But the one-dimensional analogue is trivially true, because the matrix representation of the tensor is then a single scalar whose eigenvalue is itself. We conclude therefore that it is possible to choose axes $Ox_1''\, x_2''\, x_3''$ relative to which the tensor A has components given by the matrix

$$\begin{pmatrix} \lambda_1 & 0 & 0 \\ 0 & \lambda_2 & 0 \\ 0 & 0 & \lambda_3 \end{pmatrix}.$$

The diagonal elements are the eigenvalues of A, which, by virtue of Theorem 1, are independent of the choice of axes. This completes the proof.

Theorem 4

Let $Ox_1'x_2'x_3'$ be a set of rectangular cartesian coordinate axes such that Ox_1', Ox_2', Ox_3' are in the directions of the unit vectors $\mathbf{e}^{(1)}$, $\mathbf{e}^{(2)}$, $\mathbf{e}^{(3)}$, respectively. If $\mathbf{e}^{(1)}$, $\mathbf{e}^{(2)}$, $\mathbf{e}^{(3)}$ are eigenvectors of the second order tensor A, corresponding to the eigenvalues $\lambda_1, \lambda_2, \lambda_3$, respectively, then, relative to the axes $Ox_1'x_2'x_3'$, the components of A are $a_{11}' = \lambda_1$, $a_{22}' = \lambda_2$, $a_{33}' = \lambda_3$ and $a_{rs}' = 0$ if $r \neq s$.

Proof

Let A have components a_{ij} relative to the axes $Ox_1 x_2 x_3$. Then by hypothesis,

$$a_{ij}\, e_j^{(s)} = \lambda_s e_i^{(s)}, \quad (s = 1, 2, 3)$$

where the expression on the right-hand side is *not* summed over s.

The transformation matrix relating the coordinate systems $Ox_1 x_2 x_3$ and $Ox_1'x_2'x_3'$ has components $l_{ij} = e_j^{(i)}$. Hence, relative to the axes $Ox_1'x_2'x_3'$, A has components

$$a_{rs}' = l_{ri}\, l_{sj}\, a_{ij} = e_i^{(r)} e_j^{(s)} a_{ij}$$

$$= e_i^{(r)} \lambda_s e_i^{(s)}$$

$$= \delta_{rs} \lambda_s \quad \text{(not summed over } s),$$

which yields the required results.

EXAMPLE 1

Find three orthogonal eigenvectors of the tensor whose matrix of components is

$$\begin{pmatrix} 0 & 0 & 0 \\ 0 & 1 & 1 \\ 0 & 1 & 1 \end{pmatrix}.$$

Solution

The eigenvectors are obtained by finding the non-trivial solutions of the set of linear equations

$$-\lambda v_1 = 0$$

$$(1 - \lambda) v_2 + v_3 = 0 \tag{9.3}$$

$$v_2 + (1 - \lambda) v_3 = 0.$$

The characteristic equation is

$$\begin{vmatrix} -\lambda & 0 & 0 \\ 0 & 1-\lambda & 1 \\ 0 & 1 & 1-\lambda \end{vmatrix} = 0,$$

which reduces to

$$\lambda [(1 - \lambda)^2 - 1] = 0.$$

Solving, we obtain the three eigenvalues

$$\lambda = 2, \ \lambda = 0, \ \lambda = 0.$$

Taking $\lambda = 2$, equations (9.3) reduce to

$$v_1 = 0 \text{ and } v_2 = v_3.$$

Hence one eigenvector is

$$\mathbf{v}^{(1)} = (0, 1, 1).$$

Taking $\lambda = 0$, the first of equations (9.3) is satisfied for all values of v_1, and the other two equations reduce to $v_2 = -v_3$. Hence all vectors of the form $(v_1, v_2, -v_2)$ are eigenvectors orthogonal to $\mathbf{v}^{(1)}$. Thus we may choose, say,

$$\mathbf{v}^{(2)} = (0, -1, 1)$$

and

$$\mathbf{v}^{(3)} = (1, 0, 0).$$

These vectors are orthogonal to each other and orthogonal to $\mathbf{v}^{(1)}$, as required.

EXERCISE

9.1 Find the eigenvalues and eigenvectors of the tensor A whose components a_{ij} relative to rectangular cartesian axes $Ox_1 x_2 x_3$ are

$$\begin{pmatrix} 0 & 0 & 0 \\ 0 & 1 & 2 \\ 0 & 2 & 1 \end{pmatrix}.$$

Find an orthonormal set S of unit vectors such that, when referred to S, A becomes

$$\begin{pmatrix} 0 & 0 & 0 \\ 0 & -1 & 0 \\ 0 & 0 & 3 \end{pmatrix}.$$

9.3 INVARIANTS OF SECOND ORDER SYMMETRICAL TENSORS

By Theorem 1 of the previous section, the eigenvalues $\lambda_1, \lambda_2, \lambda_3$ of a second order symmetrical tensor A are independent of the coordinate system and they are therefore said to be **invariants** of A. However, it should be noted that the eigenvalues are not an ordered set, although they can be made so if they are distinct by requiring, for example, that the largest appear first and the smallest last.

Any function $f(\lambda_1, \lambda_2, \lambda_3)$ whose value is unaltered by interchanges of pairs of $\lambda_1, \lambda_2, \lambda_3$ is said to be **symmetric** in $\lambda_1, \lambda_2, \lambda_3$ and is an invariant of the tensor A. Such invariants play an important role in the development of theoretical continuum mechanics. Particular examples are the elementary symmetric functions

$$I_1 = \lambda_1 + \lambda_2 + \lambda_3 \,, \tag{9.4}$$

$$I_2 = \lambda_1 \lambda_2 + \lambda_2 \lambda_3 + \lambda_3 \lambda_1, \tag{9.5}$$

and

$$I_3 = \lambda_1 \lambda_2 \lambda_3; \tag{9.6}$$

they are called the **principal invariants** of A.

The characteristic equation (9.2) which determines the eigenvalues of A can be expressed as

$$(\lambda - \lambda_1)(\lambda - \lambda_2)(\lambda - \lambda_3) = 0$$

because its roots are $\lambda_1, \lambda_2, \lambda_3$. Expanding the left-hand side and using the definitions (9.4)–(9.6), we find that

$$\lambda^3 - I_1 \lambda^2 + I_2 \lambda - I_3 = 0, \tag{9.7}$$

which shows that the coefficients of the characteristic equation can be expressed in terms of the principal invariants of A.

Symmetric functions of the eigenvalues

Any symmetric function $f(\lambda_1, \lambda_2, \lambda_3)$ of the eigenvalues of A can be expressed as a function of the coefficients I_1, I_2, I_3 in the characteristic equation, for Cardan's solution of the cubic equation (9.7) identifies the roots $\lambda_1, \lambda_2, \lambda_3$ as functions of $I_1, I_2,$ and I_3. By substituting Cardan's expressions into $f(\lambda_1, \lambda_2, \lambda_3)$, and observing that the value obtained will be independent of the ordering of the roots because of the assumed symmetry, we obtain, as required

$$f(\lambda_1, \lambda_2, \lambda_3) = g(I_1, I_2, I_3).$$

EXERCISES

9.2 If λ_1 λ_2 and λ_3 are the roots of equation (9.7), express

$$\text{(i)} \; I = \lambda_1^2 \lambda_2 \lambda_3 + \lambda_1 \lambda_2^2 \lambda_3 + \lambda_1 \lambda_2 \lambda_3^2$$

and

$$\text{(ii)} \; F = \lambda_1^3 \lambda_2 \lambda_3 + \lambda_1 \lambda_2^3 \lambda_3 + \lambda_1 \lambda_2 \lambda_3^3$$

as functions of I_1, I_2, and I_3.

9.3 A second order symmetrical tensor A has components a_{ij}. Using only the tensors transformation law and the orthonormality conditions, show that the scalars

$$F_1 = a_{ii}, \quad F_2 = a_{ij} a_{ji}, \quad F_3 = a_{ij} a_{jk} a_{ki}$$

are invariants of A. By referring A to its principal axes express F_1, F_2 and F_3 in terms of the eigenvalues λ_1, λ_2 and λ_3 of A.
The principal invariants of A are I_1, I_2 and I_3. Verify that

$$I_1 = F_1, \quad I_2 = \tfrac{1}{2}(F_1^2 - F_2), \quad I_3 = \tfrac{1}{6}(2F_3 + F_1^3 - 3F_1 F_2).$$

9.4 REPRESENTATION OF ISOTROPIC VECTOR FUNCTIONS

Relative to the rectangular cartesian coordinate axes $Ox_1 x_2 x_3$, let the vectors **a** and **b** have components (a_1, a_2, a_3) and (b_1, b_2, b_3), respectively, and suppose that the components of **b** are functions of the components of **a**. Then

$$b_i = F_i(a_1, a_2, a_3) \quad (i = 1, 2, 3). \tag{9.8}$$

Suppose that, relative to the axes $Ox_1' x_2' x_3'$ whose orientations relative to the original axes $Ox_1 x_2 x_3$ are defined by the usual transformation matrix (8.1), the components of **a** and **b** are (a_1', a_2', a_3') and (b_1', b_2', b_3'), respectively, and that

$$b_i' = F_i(a_1', a_2', a_3'), \quad (i = 1, 2, 3), \tag{9.9}$$

where the functions F_i are the same as before. Then the dependence of the components of **b** on the components of **a** is unchanged by a rotation of the axes and **b** is said to be an **isotropic function** of **a**.
The condition of isotropy restricts severely the form that the functions F_i can take, as the following theorem shows.

Theorem

If the vector **b** is an isotropic function of the vector **a**, then it may be represented in the form

$$\mathbf{b} = \lambda(a)\mathbf{a} \tag{9.10}$$

where $\lambda(a)$ is a scalar invariant function of $a = |\mathbf{a}|$. Conversely, if (9.10) holds, then $\mathbf{b}$ is an isotropic function of $\mathbf{a}$.

Proof

Suppose that $\mathbf{b}$ is an isotropic function of $\mathbf{a}$. Choose axes $Ox_1 x_2 x_3$ such that $\mathbf{a}$ is parallel to Ox_1 and hence has components $(a_1, 0, 0)$, and let the corresponding components of $\mathbf{b}$ be (b_1, b_2, b_3). Then

$$b_1 = F_1(a_1, 0, 0), \quad b_2 = F_2(a_1, 0, 0), \quad b_3 = F_3(a_1, 0, 0). \tag{9.11}$$

Let the axes $Ox_1 x_2 x_3$ be rotated through 180° about Ox_1. The new axes $Ox_1' x_2' x_3'$ will then be such that Ox_1' is along Ox_1, and Ox_2' and Ox_3' are in the opposite directions to Ox_2 and Ox_3, respectively. Relative to the new axes, the components of $\mathbf{a}$ and $\mathbf{b}$ will therefore be $(a_1, 0, 0)$ and $(b_1, -b_2, -b_3)$, respectively, and since $\mathbf{b}$ is assumed to be an isotropic function of $\mathbf{a}$ we must have

$$b_1 = F_1(a_1, 0\ 0), \quad -b_2 = F_2(a_1, 0\ 0), \quad -b_3 = F_3(a_1, 0\ 0). \tag{9.12}$$

Comparing (9.11) and (9.12), it is seen that $b_2 = b_3 = 0$.

Consider next a rotation of the axes $Ox_1 x_2 x_3$ through 180° about Ox_2 to new positions $Ox_1' x_2' x_3'$. The axis Ox_1' will be opposite to Ox_1, and hence relative to the new axes, $\mathbf{a} = (-a_1, 0, 0)$ and $\mathbf{b} = (-b_1, 0, 0)$. Because of the assumption of isotropy,

$$-b_1 = F_1(-a_1, 0, 0)$$

and using (9.11) it follows that

$$F_1(-a_1, 0, 0) = -F_1(a_1, 0, 0).$$

Changing the sign of a_1 is therefore equivalent to changing the sign of F_1, and hence $F_1(a_1)$ must be expressible in the form $a_1 \lambda(|a_1|)$, where λ is a function of $|a_1|$ only. But as $a_2 = a_3 = 0$, $|a_1| = a$, and hence we conclude that

$$b_1 = \lambda(a)a_1, \quad b_2 = a_2 = 0, \quad b_3 = a_3 = 0.$$

Thus

$$\mathbf{b} = \lambda(a)\mathbf{a},$$

as required.

To prove that whenever the relation (9.10) holds $\mathbf{b}$ is an isotropic function of $\mathbf{a}$, consider the components

$$b_i = \lambda(a)a_i \quad (i = 1, 2, 3). \tag{9.13}$$

Multiplying each side of this equation by l_{ji} (defined by (8.1)) and summing over the three values of i gives

$$l_{ji} b_i = \lambda(a) l_{ji} a_i.$$

By the rule for transforming the components of a vector, this may be expressed as

$$b'_j = \lambda(a) a'_j \quad (j = 1, 2, 3) \tag{9.14}$$

which is the j-th component of (9.10) relative to the axes $Ox'_1 x'_2 x'_3$. The functional dependence of the components of **b** on the components of **a** is therefore unaffected by a rotation of the axes, and hence **b** is an isotropic function of **a**.

EXERCISE

9.4 If n is a constant, determine for which values of n the following vector function **c** of (a_1, a_2, a_3) is isotropic:

$$\mathbf{c} = (c_1, c_2, c_3)$$

where

$$c_1 = a_1^{n+1} + a_2^n a_1 + a_3^n a_1,$$
$$c_2 = a_1^n a_2 + a_2^{n+1} + a_2^n a_2$$

and

$$c_3 = a_1^n a_3 + a_2^n a_3 + a_3^{n+1}.$$

9.5 ISOTROPIC SCALAR FUNCTIONS OF SYMMETRICAL SECOND ORDER TENSORS

Relative to the rectangular cartesian coordinate axes $Ox_1 x_2 x_3$, let the second order symmetrical tensor A have components a_{ij} $(i, j = 1, 2, 3)$, and suppose that Ω is a scalar function of these components. For brevity, the relation may be expressed as

$$\Omega = \Omega(a_{ij}),$$

where it is to be understood that Ω is a function of all the components of A. Relative to new axes $Ox'_1 x'_2 x'_3$, suppose that the components of A are a'_{ij} and that

$$\Omega(a'_{ij}) = \Omega(a_{ij}). \tag{9.15}$$

This indicates that the dependence of Ω on the components of A is unchanged by a rotation of the axes, and Ω is therefore said to be an **isotropic function** of A. The theorem which follows shows that such isotropic scalar functions can be represented in a relatively simple form.

Theorem

If Ω is an isotropic scalar function of the second order symmetrical tensor A, then it may be represented in the form

$$\Omega = \Omega(I_1, I_2, I_3),\qquad (9.16)$$

where I_1, I_2, I_3 are the principal invariants of A (defined by relations (9.4)–(9.6)). Conversely, if (9.16) holds, then Ω is an isotropic function of A.

Proof

Suppose that Ω is an isotropic function of A. As A is a second order symmetrical tensor, Theorem 3 of section 9.2 shows that a principal set of axes $Ox_1 x_2 x_3$ may be chosen relative to which A has the diagonal form

$$\begin{pmatrix} \lambda_1 & 0 & 0 \\ 0 & \lambda_2 & 0 \\ 0 & 0 & \lambda_3 \end{pmatrix}. \qquad (9.17)$$

In general Ω depends on $a_{11}, a_{22}, a_{33}, a_{12}, a_{23}$ and a_{31} (because of the symmetry of A it is unnecessary to list a_{21}, a_{32} and a_{13}), and hence relative to the principal axes $Ox_1 x_2 x_3$,

$$\Omega = \Omega(\lambda_1, \lambda_2, \lambda_3, 0, 0, 0).$$

Consider a rotation to new axes $Ox_1' x_2' x_3'$, defined by the transformation matrix:

O	x_1	x_2	x_3
x_1'	0	1	0
x_2'	-1	0	0
x_3'	0	0	1

(9.18)

This rotation turns the axes through $90°$ about Ox_3. Using the tensor transformation rule, namely

$$a_{rs}' = l_{ri}\, l_{sj}\, a_{ij},$$

and noting that the components a_{ij} are as displayed in the matrix (9.17),

$$a_{11}' = l_{1i}\, l_{1j}\, a_{ij} = \lambda_2,$$

$$a'_{22} = l_{2i} \, l_{2j} \, a_{ij} = \lambda_1,$$

and

$$a'_{33} = l_{3i} \, l_{3j} \, a_{ij} = \lambda_3.$$

It is readily verified that $a'_{ij} = 0$ when $i \neq j$, and hence relative to the axes $Ox'_1 x'_2 x'_3$, A has the diagonal form

$$\begin{pmatrix} \lambda_2 & 0 & 0 \\ 0 & \lambda_1 & 0 \\ 0 & 0 & \lambda_3 \end{pmatrix}.$$

By the assumed condition of isotropy, it follows that

$$\Omega(\lambda_2, \lambda_1, \lambda_3, 0, 0, 0) = \Omega(\lambda_1, \lambda_2, \lambda_3, 0, 0, 0)$$

and hence interchanging λ_1 and λ_2 leaves the value of Ω unchanged. By a similar method it may be shown that interchanging any pair of the eigenvalues $\lambda_1, \lambda_2, \lambda_3$ leaves the value of Ω unchanged, and hence Ω is a symmetric function of λ_1, λ_2 and λ_3. From the remarks at the end of section 9.3, we conclude that Ω is a function of the principal invariants I_1, I_2, I_3 of A, as required.

To establish that when (9.16) holds Ω is an isotropic function of A is straightforward. For, I_1, I_2 and I_3 are invariant under a rotation of the axes and hence if (9.16) holds for one set of axes it holds for all sets obtained by a rotation. The dependence of Ω on the components of A is therefore invariant under a rotation of the axes and hence Ω is an isotropic function of A.

9.6 REPRESENTATION OF AN ISOTROPIC TENSOR FUNCTION

Let A and B be second order symmetrical tensors whose components relative to the axes $Ox_1 x_2 x_3$ are a_{km} and b_{ij}, respectively. Suppose that each of the components of B is a function of the components of A, and denote this by writing

$$b_{ij} = F_{ij}(a_{km}). \tag{9.19}$$

Relative to new axes $Ox'_1 x'_2 x'_3$, suppose that the components of A and B are a'_{km} and b'_{ij}, respectively, and that the relations

$$b'_{ij} = F_{ij}(a'_{km})$$

are satisfied, where the functions F_{ij} are as in (9.19). The dependence of the components of B on the components of A is then invariant under a rotation of the axes and B is said to be an **isotropic function** of A. As in the case of isotropic vector functions and isotropic scalar functions of a tensor, discussed in the previous two sections of this chapter, the condition of isotropy is naturally satisfied by only a limited class of functions. The final representation theorem that we prove shows the extent of the simplification that can be achieved.

Theorem

Let A be a second order symmetrical tensor whose principal invariants are I_1, I_2 and I_3. If the second order symmetrical tensor B is an isotropic function of A, then there exist scalar invariant functions α, β and γ of I_1, I_2 and I_3 such that the components a_{ij} of B can be represented in terms of the components a_{ij} of A in the form

$$b_{ij} = \alpha \delta_{ij} + \beta a_{ij} + \gamma a_{ik} a_{kj} \quad (i, j = 1, 2, 3). \tag{9.20}$$

Conversely, if (9.20) holds, then B is an isotropic function of A.

Proof

Suppose that B is an isotropic function of A and choose the axes $Ox_1 x_2 x_3$ such that A is of the diagonal form

$$\begin{pmatrix} \lambda_1 & 0 & 0 \\ 0 & \lambda_2 & 0 \\ 0 & 0 & \lambda_3 \end{pmatrix}. \tag{9.21}$$

Then

$$b_{ij} = F_{ij}(\lambda_1, \lambda_2, \lambda_3, 0, 0, 0). \tag{9.22}$$

Consider a rotation of the axes through $180°$ about Ox_1, defined by the transformation matrix

O	x_1	x_2	x_3
x_1'	1	0	0
x_2'	0	-1	0
x_3'	0	0	-1 .

Let the components of A and B relative to the new axes $Ox_1' x_2' x_3'$ be a_{ij}' and b_{ij}', respectively. Using the law for transforming the components of a second order tensor,

$$a_{11}' = l_{1i} l_{1j} a_{ij} = \lambda_1,$$

$$a_{22}' = l_{2i} l_{2j} a_{ij} = \lambda_2,$$

$$a_{33}' = l_{3i} l_{3j} a_{ij} = \lambda_3,$$

and

$$a'_{rs} = l_{ri}\, l_{sj}\, a_{ij} = 0 \quad \text{when } r \neq s.$$

Also,

$$b'_{12} = l_{1i}\, l_{2j}\, b_{ij} = -b_{12}$$

and

$$b'_{13} = l_{1i}\, l_{3j}\, b_{ij} = -b_{13}.$$

Now from (9.22),

$$b_{12} = F_{12}(\lambda_1, \lambda_2, \lambda_3, 0, 0, 0) \tag{9.23}$$

and

$$b_{13} = F_{13}(\lambda_1, \lambda_2, \lambda_3, 0, 0, 0). \tag{9.24}$$

Under the assumed condition of isotropy, the functions F_{12} and F_{13} are unchanged by a rotation of the axes, and hence relative to the axes $Ox'_1 x'_2 x'_3$,

$$-b_{12} = F_{12}(\lambda_1, \lambda_2, \lambda_3, 0, 0, 0) \tag{9.25}$$

$$-b_{13} = F_{13}(\lambda_1, \lambda_2, \lambda_3, 0, 0, 0). \tag{9.26}$$

Comparing these relations with (9.23) and (9.24), it is seen that $b_{12} = b_{13} = 0$. In a similar way, it may be shown that $b_{23} = 0$, and hence relative to the axes $Ox_1 x_2 x_3$, B is of the diagonal form

$$\begin{pmatrix} \mu_1 & 0 & 0 \\ 0 & \mu_2 & 0 \\ 0 & 0 & \mu_3 \end{pmatrix},$$

where μ_1, μ_2, μ_3 are the eigenvalues of B. One implication of the assumption of isotropy therefore is that the principal axes of A and B coincide.

Assume now that λ_1, λ_2 and λ_3 are *distinct*; the cases when two or more eigenvalues coincide are easier and are set later as exercises. Consider the following simultaneous equations in α, β and γ:

$$\mu_1 = \alpha + \beta\lambda_1 + \gamma\lambda_1^2$$

$$\mu_2 = \alpha + \beta\lambda_2 + \gamma\lambda_2^2 \tag{9.27}$$

$$\mu_3 = \alpha + \beta\lambda_3 + \gamma\lambda_3^2.$$

Since μ_1, μ_2 and μ_3 are components of the tensor B, they will be functions of λ_1, λ_2 and λ_3, and hence α, β and γ are functions of λ_1, λ_2 and λ_3. By Cramer's rule, the solution of equations (9.27) for α, β, γ may be expressed as

$$\alpha = \frac{\Delta_1}{\Delta}, \quad \beta = \frac{\Delta_2}{\Delta}, \quad \gamma = \frac{\Delta_3}{\Delta},$$

where

$$\Delta = \begin{vmatrix} 1 & \lambda_1 & \lambda_1^2 \\ 1 & \lambda_2 & \lambda_2^2 \\ 1 & \lambda_3 & \lambda_3^2 \end{vmatrix} \neq 0,$$

$$\Delta_1 = \begin{vmatrix} \mu_1 & \lambda_1 & \lambda_1^2 \\ \mu_2 & \lambda_2 & \lambda_2^2 \\ \mu_3 & \lambda_3 & \lambda_3^2 \end{vmatrix},$$

$$\Delta_2 = \begin{vmatrix} 1 & \mu_1 & \lambda_1^2 \\ 1 & \mu_2 & \lambda_2^2 \\ 1 & \mu_3 & \lambda_3^2 \end{vmatrix},$$

and

$$\Delta_3 = \begin{vmatrix} 1 & \lambda_1 & \mu_1 \\ 1 & \lambda_2 & \mu_2 \\ 1 & \lambda_3 & \mu_3 \end{vmatrix}.$$

Now in section 9.5, it was shown that the effect on the tensor A of the rotation of axes defined by the array (9.18) is to interchange λ_1 and λ_2; likewise the effect on B will be to interchange μ_1 and μ_2. The determinants Δ, Δ_1, Δ_2 and Δ_3 will therefore all change sign, but α, β and γ will be unchanged. By applying other suitable rotations of the axes, it follows similarly that interchanging any pair of λ_1, λ_2, λ_3 leaves α, β and γ unchanged, and hence α, β and γ are symmetric functions of λ_1, λ_2 and λ_3. From our earlier remarks in section 9.3 on symmetric functions of the eigenvalues, we conclude that α, β and γ can be expressed as functions of the principal invariants of A.

In the coordinate system $Ox_1 x_2 x_3$, the components a_{ij} and b_{ij} of A and B are zero when $i \neq j$ and the components $a_{11} = \lambda_1, a_{22} = \lambda_2, a_{33} = \lambda_3, b_{11} = \mu_1$, $b_{22} = \mu_2$ and $b_{33} = \mu_3$ are related by (9.27). Hence, relative to the axes $Ox_1 x_2 x_3$,

$$b_{ij} = \alpha \delta_{ij} + \beta a_{ij} + \gamma a_{ik} a_{kj} \qquad (i, j = 1, 2, 3). \tag{9.28}$$

But this is a relation between components of second order tensors with coefficients which are invariant under a rotation of the axes, and it follows that it will hold in all coordinate systems obtained by a rotation. Hence the first part of the theorem is proved.

The converse inference, that when (9.20) holds B is an isotropic function of A, is readily proved. For, if the relation (9.20) holds in a particular coordinate system $Ox_1 x_2 x_3$ then on rotating the axes to new positions $Ox_1' x_2' x_3'$ it becomes

$$b_{ij}' = \alpha \delta_{ij} + \delta a_{ij}' + \gamma a_{ik}' a_{kj}',$$

since α, β and γ are invariants. Hence the functional dependence of the components of B on those of A is invariant under a rotation of the axes and B is therefore an isotropic function of A.

EXERCISES

9.5 In the preceding theorem, suppose that two of the eigenvalues of A are equal and the third is different. Show that, under the assumption of isotropy, two of the eigenvalues of B must then coincide, and that the components of B can be represented in terms of the components of A by a relation of the form

$$b_{ij} = \alpha \delta_{ij} + \beta a_{ij}$$

where α and β are functions of the principal invariants of A.

State also the form of the representation theorem when all three eigenvalues of A are equal.

9.6 In classical fluid dynamics, the components σ_{ij} of the stress tensor are assumed to be *linear* isotropic functions of the rate of strain components e_{ij}. Given that $\sigma_{ij} = \sigma_{ji}$ and $e_{ij} = e_{ji}$ for all i and j, show that

$$\sigma_{ij} = (-p + \lambda e_{kk})\delta_{ij} + 2\mu e_{ij},$$

where p, λ and μ are scalar invariants.

Remark
The student of continuum mechanics should note that this relationship can be deduced from rather more basic assumptions. If the stress components are assumed to be functions of the tensor gradient of the velocity field, then it can be shown from certain fundamental invariance requirements that there can be dependence only on the symmetrical part of this tensor gradient (which is the rate of strain tensor) and that the relationship must be isotropic. For a full discussion of this and related matters, see, for example, Leigh, D. C. (1968) *Nonlinear Continuum Mechanics*, p. 145, McGraw-Hill, London.

Appendix A Determinants

Let $a_{11}, a_{12}, a_{21}, a_{22}$ be real or complex numbers or variables. We define

$$\begin{vmatrix} a_{11} & a_{12} \\ a_{21} & a_{22} \end{vmatrix} = a_{11}a_{22} - a_{12}a_{21} \tag{A1.1}$$

and call the expression on the left-hand side a **second order determinant**.
Similarly, a **third order determinant** is defined as

$$\begin{vmatrix} a_{11} & a_{12} & a_{13} \\ a_{21} & a_{22} & a_{23} \\ a_{31} & a_{32} & a_{33} \end{vmatrix} = a_{11}\Delta_{11} - a_{12}\Delta_{12} + a_{13}\Delta_{13} \tag{A1.2}$$

where

$$\Delta_{11} = \begin{vmatrix} a_{22} & a_{23} \\ a_{32} & a_{33} \end{vmatrix}, \tag{A1.3}$$

$$\Delta_{12} = \begin{vmatrix} a_{21} & a_{23} \\ a_{31} & a_{33} \end{vmatrix}, \tag{A1.4}$$

$$\Delta_{13} = \begin{vmatrix} a_{21} & a_{22} \\ a_{31} & a_{32} \end{vmatrix}. \tag{A1.5}$$

The determinants $\Delta_{11}, \Delta_{12}, \Delta_{13}$ are called the **minors** of the elements a_{11}, a_{12}, a_{13} respectively. The reader should observe that the minor of a_{11} is the second order determinant which remains when all the elements on the same row and on the same column as a_{11} are deleted; and similarly for the minors of a_{12} and a_{13}

The determinant, T say, on the left-hand side of (A1.2) can also be expanded, starting with the elements of the first column, giving

$$T = a_{11}\Delta_{11} - a_{21}\Delta_{21} + a_{31}\Delta_{31}, \tag{A1.6}$$

where $\Delta_{11}, \Delta_{21}, \Delta_{31}$ are the minors of a_{11}, a_{21}, a_{31} respectively. It is easily verified that the right-hand sides of (A1.2) and (A1.6) are identical. Expansions starting with the elements of the second or third rows or columns can also be performed, but we shall omit the details.

The reader should verify that if two rows or two columns of a determinant are equal, then the value of the determinant is zero.

Multiplication of determinants

The product of two second order determinants is given by the following rule:

$$\begin{vmatrix} a_{11} & a_{12} \\ a_{21} & a_{22} \end{vmatrix} \times \begin{vmatrix} b_{11} & b_{12} \\ b_{21} & b_{22} \end{vmatrix} = \begin{vmatrix} a_{11}b_{11} + a_{12}b_{21} & a_{11}b_{12} + a_{12}b_{22} \\ a_{21}b_{11} + a_{22}b_{21} & a_{21}b_{12} + a_{22}b_{22} \end{vmatrix}. \quad (A1.7)$$

The rule may be verified by expanding the expressions on the two sides.

The product of two third order determinants is given by the rule:

$$\begin{vmatrix} a_{11} & a_{12} & a_{13} \\ a_{21} & a_{22} & a_{23} \\ a_{31} & a_{32} & a_{33} \end{vmatrix} \times \begin{vmatrix} b_{11} & b_{12} & b_{13} \\ b_{21} & b_{22} & b_{23} \\ b_{31} & b_{32} & b_{33} \end{vmatrix}$$

$$= \begin{vmatrix} a_{11}b_{11} + a_{12}b_{21} + a_{13}b_{31} & a_{11}b_{12} + a_{12}b_{22} + a_{13}b_{32} & a_{11}b_{13} + a_{12}b_{23} + a_{13}b_{33} \\ a_{21}b_{11} + a_{22}b_{21} + a_{23}b_{31} & a_{21}b_{12} + a_{22}b_{22} + a_{23}b_{32} & a_{21}b_{13} + a_{22}b_{23} + a_{23}b_{33} \\ a_{31}b_{11} + a_{32}b_{21} + a_{33}b_{31} & a_{31}b_{12} + a_{32}b_{22} + a_{33}b_{32} & a_{31}b_{13} + a_{32}b_{23} + a_{33}b_{33} \end{vmatrix}.$$

$$(A1.8)$$

Again the rule can be verified by direct evaluation of the two sides, although the calculation is obviously tedious.

Transpose of a determinant

If the first, second, . . . rows of a determinant T are replaced by the corresponding columns, the **transpose** T' is obtained. Thus, if

$$T = \begin{vmatrix} a_{11} & a_{12} & a_{13} \\ a_{21} & a_{22} & a_{23} \\ a_{31} & a_{32} & a_{33} \end{vmatrix}, \quad (A1.9)$$

then

$$T' = \begin{vmatrix} a_{11} & a_{21} & a_{31} \\ a_{12} & a_{22} & a_{32} \\ a_{13} & a_{23} & a_{33} \end{vmatrix}. \quad (A1.10)$$

It may be verified by expansion that $T = T'$.

Interchange of two rows or columns

If two rows or two columns of a determinant are interchanged, its sign is changed. Thus, if the original value is a, the value after an interchange of rows or columns will be $-a$. The reader should prove this as an exercise.

Appendix B Expressions for grad, div, curl and ∇^2 in cylindrical and spherical polar coordinates

Cylindrical polar coordinates R, φ, z

$$\operatorname{grad} \Omega = \frac{\partial \Omega}{\partial R} \mathbf{e}_R + \frac{\partial \Omega}{R \partial \varphi} \mathbf{e}_\varphi + \frac{\partial \Omega}{\partial z} \mathbf{e}_z,$$

$$\operatorname{div} \mathbf{F} = \frac{1}{R} \frac{\partial}{\partial R} (R F_R) + \frac{\partial F_\varphi}{R \partial \varphi} + \frac{\partial F_z}{\partial z},$$

$$\operatorname{curl} \mathbf{F} = \frac{1}{R} \begin{vmatrix} \mathbf{e}_R & R\mathbf{e}_\varphi & \mathbf{e}_z \\ \dfrac{\partial}{\partial R} & \dfrac{\partial}{\partial \varphi} & \dfrac{\partial}{\partial z} \\ F_R & RF_\varphi & F_z \end{vmatrix},$$

$$\nabla^2 \equiv \frac{1}{R} \frac{\partial}{\partial R} \left(R \frac{\partial}{\partial R} \right) + \frac{1}{R^2} \frac{\partial^2}{\partial \varphi^2} + \frac{\partial^2}{\partial z^2}.$$

Spherical polar coordinates r, θ, φ

$$\operatorname{grad} \Omega = \frac{\partial \Omega}{\partial r} \mathbf{e}_r + \frac{\partial \Omega}{r \partial \theta} \mathbf{e}_\theta + \frac{\partial \Omega}{r \sin \theta \, \partial \varphi} \mathbf{e}_\varphi,$$

$$\operatorname{div} \mathbf{F} = \frac{1}{r^2} \frac{\partial}{\partial r} (r^2 F_r) + \frac{1}{r \sin \theta} \frac{\partial}{\partial \theta} (\sin \theta \, F_\theta) + \frac{1}{r \sin \theta} \frac{\partial F_\varphi}{\partial \varphi},$$

$$\text{curl }\mathbf{F} = \frac{1}{r^2\sin\theta}\begin{vmatrix} \mathbf{e}_r & r\mathbf{e}_\theta & r\sin\theta\,\mathbf{e}_\varphi \\ \dfrac{\partial}{\partial r} & \dfrac{\partial}{\partial\theta} & \dfrac{\partial}{\partial\varphi} \\ F_r & rF_\theta & r\sin\theta\,F_\varphi \end{vmatrix},$$

$$\nabla^2 \equiv \frac{1}{r^2}\frac{\partial}{\partial r}\left(r^2\frac{\partial}{\partial r}\right) + \frac{1}{r^2\sin\theta}\frac{\partial}{\partial\theta}\left(\sin\theta\frac{\partial}{\partial\theta}\right) + \frac{1}{r^2\sin^2\theta}\frac{\partial^2}{\partial\varphi^2}.$$

Appendix C The chain rule for Jacobians

If

$$\alpha = \alpha(u, v), \quad \beta = \beta(u, v)$$

and

$$u = u(u', v'), \quad v = v(u', v'),$$

then

$$\frac{\partial(\alpha, \beta)}{\partial(u', v')} = \frac{\partial(\alpha, \beta)}{\partial(u, v)} \times \frac{\partial(u, v)}{\partial(u', v')}.$$

Proof

Using suffixes to denote partial derivatives with respect to u, v, u', v'

$$\frac{\partial(\alpha, \beta)}{\partial(u, v)} \times \frac{\partial(u, v)}{\partial(u', v')} = \begin{vmatrix} \alpha_u & \alpha_v \\ \beta_u & \beta_v \end{vmatrix} \times \begin{vmatrix} u_{u'} & u_{v'} \\ v_{u'} & v_{v'} \end{vmatrix}$$

$$= \begin{vmatrix} \alpha_u u_{u'} + \alpha_v v_{u'} & \alpha_u u_{v'} + \alpha_v v_{v'} \\ \beta_u u_{u'} + \beta_v v_{u'} & \beta_u u_{v'} + \beta_v v_{v'} \end{vmatrix}$$

$$= \begin{vmatrix} \alpha_{u'} & \alpha_{v'} \\ \beta_{u'} & \beta_{v'} \end{vmatrix}, \quad \text{by the chain rule,}$$

$$= \frac{\partial(\alpha, \beta)}{\partial(u', v')}, \quad \text{as required.}$$

Answers to exercises

1 RECTANGULAR CARTESIAN COORDINATES AND ROTATION OF AXES

1.3 Four points, $(\pm \frac{1}{2} \sqrt{2}, \pm \frac{1}{2} \sqrt{2}, 0)$.

1.4 Eight points, $(\pm 3, \pm 2\sqrt{2}, \pm 2\sqrt{2})$.

1.5 Eight points, $(\pm \frac{1}{2}, \pm \frac{1}{2}, \pm \frac{1}{2})$.

1.6 (i) 5; (ii) $\sqrt{61}$.

1.7 $(-1, -1, 1)$; 1, 3, 0.

1.8 $3\sqrt{2}$.

1.10 $6/\sqrt{65}, 2/\sqrt{65}, 5/\sqrt{65}$.

1.11 $135°$.

1.12 $\frac{1}{3}\sqrt{3}, \frac{1}{3}\sqrt{3}, \frac{1}{3}\sqrt{3}$.

1.13 1, 1, 0.

1.16 $\cos^{-1} \frac{1}{3}$.

1.17 $(2, 8, -2)$.

1.19 $0, \frac{4}{3}$. The points O and N coincide.

1.20 $\begin{pmatrix} 1 & 0 & 0 \\ 0 & -1 & 0 \\ 0 & 0 & -1 \end{pmatrix}$; $(1, -1, -1)$.

1.23 $x = x' \sin \theta \cos \varphi + y' \cos \theta \cos \varphi - z' \sin \varphi$,
$y = x' \sin \theta \sin \varphi + y' \cos \theta \sin \varphi + z' \cos \varphi$,
$z = x' \cos \theta - y' \sin \theta$.

1.24 Let

$$l'_{11} = l_{22}l_{33} - l_{23}l_{32},$$

$$l'_{12} = l_{23}l_{31} - l_{21}l_{33},$$

$$l'_{13} = l_{21}l_{32} - l_{22}l_{31}.$$

Then

$$l_{11}l'_{11} + l_{12}l'_{12} + l_{13}l'_{13} = \begin{vmatrix} l_{11} & l_{12} & l_{13} \\ l_{21} & l_{22} & l_{23} \\ l_{31} & l_{32} & l_{33} \end{vmatrix}$$

$$= 1,$$

by (1.17) of the text. It follows from (1.9) that the angle between the lines with direction cosines l_{11}, l_{12}, l_{13} and $l'_{11}, l'_{12}, l'_{13}$ respectively is zero. Hence, since l_{11}, l_{12}, l_{13} are the direction cosines of Ox' relative to the axes $Oxyz$, so also are $l'_{11}, l'_{12}, l'_{13}$. Thus

$$l'_{11} = l_{11}, \quad l'_{12} = l_{12}, \quad l'_{13} = l_{13},$$

as required.

The two sets of three relations similar to those for l_{11}, l_{12}, l_{13} are obtained by cyclic permutation of the first suffix in each l_{ij}. Thus

$$l_{21} = l_{32}l_{13} - l_{33}l_{12},$$

$$l_{22} = l_{33}l_{11} - l_{31}l_{13},$$

$$l_{23} = l_{31}l_{12} - l_{32}l_{11},$$

and

$$l_{31} = l_{12}l_{23} - l_{13}l_{22},$$

$$l_{32} = l_{13}l_{21} - l_{11}l_{23},$$

$$l_{33} = l_{11}l_{22} - l_{12}l_{21}.$$

1.28 (a) 1, (b) -3, (c) -5.

2 SCALAR AND VECTOR ALGEBRA

2.2 (i) $(1, 2, 3)$; (ii) $(-1, -2, -3)$; (iii) $(1, 2, 2)$.

2.3 $\sqrt{26}$; $\sqrt{5}$.

2.4 $\frac{3}{5}$.

2.5 The axes Oz', Oz coincide; Ox', Oy coincide; and Oy' is oppositely directed to Ox. $(1, -2, 2)$.

2.7 $(1, 0, 0)$ and $(-1, 0, 0)$. No.

2.12 (i) $(1, 0, 4)$; (ii) $(-1, 0, -4)$.

2.13 The vectorial sum of the four consecutive displacements is zero.

2.15 When $\hat{\mathbf{a}} = \hat{\mathbf{b}}$, or $\mathbf{a} = \mathbf{0}$, or $\mathbf{b} = \mathbf{0}$.

2.17 No.

2.20 $a = -1, b = 3, c = 1$.

2.22 $\cos^{-1}(1/10)$.

2.23 (i) $3\sqrt{2}$; (ii) 0; (iii) $-\sqrt{2}$.

2.26 $\frac{1}{6}\sqrt{6}$.

2.27 The resolutes are 0, 24/5 and 0.

2.28 The diagonals of a rhombus intersect at right angles.

2.31 (i) A vertically upward unit vector; (ii) A vector of magnitude $\sqrt{2}$ pointing downwards; (iii) A unit vector pointing *NW*.

2.35 $\mathbf{u} = (2\lambda + 1, \lambda, -\lambda)$ where λ is an arbitrary real number.

2.36 $a = b = 1$.

2.37 Take $\mathbf{a} = \mathbf{b} = (1, 0, 0)$ and $\mathbf{c} = (0, 1, 0)$.

2.54 $|\mathbf{F} \times \mathbf{G}|/F^2$.

3 VECTOR FUNCTIONS OF A REAL VARIABLE. DIFFERENTIAL GEOMETRY OF CURVES

3.1 (i) Fig. (i); (ii) Fig. (ii); (iii) Fig. (iii); (iv) Fig. (iv); (v) Fig. (v).

3.2 $C = 1$; parabola; straight line.

3.3 (1, 0, 5).

3.7 (i) $\pi(-2\sin\pi t, \cos\pi t, 0), -\pi^2(2\cos\pi t, \sin\pi t, 0)$;

  (ii) $(1, 1, e^t), (0, 0, e^t)$;

  (iii) $t > 0$: (1, 1, 0) and (0, 0, 0);

     $t < 0$: (−1, 1, 0) and (0, 0, 0).

3.8 $(e^{-t}\cos t, e^{-t}\sin t, 0)$. See Fig. (vi).

3.13 **a** and **b** are such that $\mathbf{a} \times \mathbf{b}$ is a constant vector; and so **a** and **b** are parallel to a fixed plane and are such that the parallelogram with edges **a** and **b** has constant area.

3.14 (i) See Fig. (vii). This is a smooth, simple closed curve. (ii) See Fig. (viii). This is a smooth curve.

3.16 See Fig. (ix).

Fig. (i).

Fig. (ii).

Fig. (iii).

Fig. (iv).

Fig. (v).

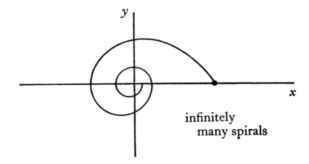

infinitely
many spirals

Fig. (vi).

Fig. (vii).

Fig. (viii).

Fig. (ix).

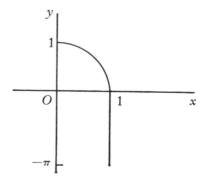

Fig. (x).

3.17 See Fig. (x).

3.18 $\mathbf{r} = (a\cos(s/c), a\sin(s/c), bs/c)$, where $c = (a^2 + b^2)^{\frac{1}{2}}$.

3.23 $\frac{1}{2}$.

3.26 $(-60, 10, -40)\ \text{ms}^{-1}$.

3.27 $(1, 2, 3)$ radians per second.

3.29 $x^2 + \left(y + \dfrac{u}{2\omega\cos\lambda}\right)^2 = \dfrac{u^2}{4\omega^2\cos^2\lambda}$.

3.31 $\mathbf{\Omega}$, $\mathbf{r}$ and $\mathbf{\rho}$ are coplanar.

4 SCALAR AND VECTOR FIELDS

4.1 Open.

4.3 $2(ax + hy)$; $2(hx + by)$; $2a$; $2b$; $2h$; $2h$.

4.5 $x/r, y/r, z/r$.

4.9 $(-2, 0, 2)$.

4.14 $27/\sqrt{6}$.

4.15 The direction of the vector (a, b, c).

4.16 $x/a^2, y/b^2, z/c^2$

4.18 2; $(-1, 0, -1)$; $(0, 1, 1)$.

4.19 $2 + 6z$; $\mathbf{0}$.

4.23 (i) $\nabla(\nabla \cdot \mathbf{F})$;   (ii) $\nabla \cdot (\nabla\Omega)$;   (iii) $\nabla \cdot (\nabla \times \mathbf{F})$;   (iv) $\nabla \times (\nabla\Omega)$;
     (v) $\nabla \times (\nabla \times \mathbf{F})$.

4.25 (i) $2(x^2, y^2, z^2)$; (ii) $(2x, 4y, 6z)$; (iii) $\mathbf{0}$; (iv) $3xyz$; (v) $(2, 2, 2)$; (vi) $\mathbf{0}$.

4.33 (i) $R = $ constant, a cylinder with axis Oz;
         $\varphi = $ constant, a half plane containing Oz;
         $z = $ constant, a plane parallel to the xy-plane.
    (ii) $r = $ constant, a sphere centre the origin;
         $\theta = $ constant, an infinitely long cone with vertex O, axis Oz;
         $\varphi = $ constant, a half-plane containing Oz.

4.36 $h_1 = h_2 = (\sinh^2\xi + \sin^2\eta)^{\frac{1}{2}}$, $h_3 = 1$; ellipses, hyperbolae.

4.39 $R\mathbf{e}_\varphi$; $r\sin\theta\,\mathbf{e}_\varphi$.

4.40 $R\mathbf{e}_R + z\,\mathbf{e}_z$.

4.41 $1, 1, 0$; $-r, 0, 0$; $-r\sin\theta, -r\cos\theta, r\cos\varphi$.

4.43 $\mathbf{e}_R + \frac{1}{2}\mathbf{e}_z$.

4.46 $-r^{-2}\cot\theta, -r^{-2}\text{cosec}^2\theta, 0$.

4.51 $\{R\cos^2\varphi - (1 + R^{-1})\sin^2\varphi\}\,\mathbf{e}_R + (1 + R^{-1})\sin\varphi\cos\varphi\,\mathbf{e}_\varphi$.

4.55 Substituting $d = ax_1^2 + by_1^2$ gives

$$e = ab(a - b)^2 x_1^2 y_1^2 / (a^3 x_1^2 + b^3 y_1^2).$$

Hence, when $a \neq b$, $0 < e < d$ and the sequence of iterations converges because it is monotone decreasing, and bounded below. When $a = b$ we have $e = 0$, giving convergence in one step. This is because the normal to a circle always passes through the centre.

4.56 Substituting $d = ax_1^2 + by_1^2 + cz_1^2$ gives

$$e = \frac{ab(a-b)^2 x_1^2 y_1^2 + bc(b-c)^2 y_1^2 z_1^2 + ca(c-a)^2 z_1^2 x_1^2}{a^3 x_1^2 + b^3 y_1^2 + c^3 z_1^2}$$

Hence, unless $a = b = c$, $0 < e < d$ and the sequence of iterations converges as in the previous exercise. When $a = b = c$ we have $e = 0$ because the normal to a sphere always passes through its centre.

4.57
$$A = \begin{pmatrix} 4 & 1 \\ 1 & 1 \end{pmatrix}, \mathbf{b} = \begin{pmatrix} 5 \\ 2 \end{pmatrix}$$

Fig. (xi). (a).

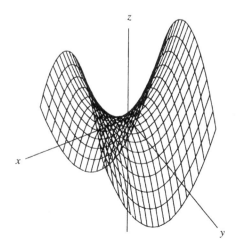

Fig. (xi). (b).

4.58 Near $(0, 0)$ the value of c can be either positive or negative, and we have a saddle point there. The lines of steepest descent are in fact the family of hyperbolae $k = xy$. This is because $\mathrm{grad}\, c = (x, -y)$ and $\mathrm{grad}\, k = (y, x)$ are orthogonal for all x and y. The two families of curves are shown in Fig. xi(a), and the arrows show the downhill direction along the curves of steepest descent. Starting from a point P on $y = -1$ takes you downhill along the side of the saddle, and eventually, unless $x_0 = 0$, you finish up travelling away from the origin. Fig. xi(b) shows a three-dimensional plot of the surface $z = x^2 - y^2$ which might help the reader to visualize the contours. Changing the minus sign so that $c = x^2 + y^2$ changes the contours to a family of concentric circles, when all lines of steepest descent pass through the origin.

5 LINE, SURFACE AND VOLUME INTEGRALS

5.1 $\frac{4}{5}\sqrt{5}(2 + \pi^2)a^2$.

5.3 $\frac{1}{2}63\sqrt{3}$.

5.4 πa^3.

5.5 $\frac{1}{12}(5\sqrt{5} - 1) + \frac{1}{3}\sqrt{3}$.

5.6 (i) $7/6$; (ii) $\mathbf{0}$; (iii) $\left(\dfrac{3}{4}, \dfrac{8\sqrt{2}}{15}, \dfrac{5}{12}\right)$.

5.7 $2\pi c(a, 0, \pi c)$.

5.8 $-2\pi a$.

5.9 6π.

5.10 $2\pi a(0, b, a)$; $2\pi^2 b(a^2 + b^2)^{\frac{1}{2}}\mathbf{k}$.

5.14 $\frac{1}{3}$.

5.15 $\frac{1}{2}(3e - 7)$.

5.17 $4(3 - \sqrt{2})/21$.

5.18 $\frac{1}{2}\pi(b^4 - a^4)$.

5.19 8π.

5.20 $-\frac{1}{2}\pi$.

5.21 $\frac{1}{4}\pi a^3 b$.

5.23 $\frac{1}{16}$.

5.24 $\pi^{\frac{3}{2}}$ (see Example 19, section 5.7).

5.25 $\frac{1}{6}$.

5.26 $\frac{4}{15}\pi abc(a^2 + b^2 + c^2)$.

5.28 $\frac{1}{2}(\sqrt{3}\cos\varphi, \sqrt{3}\sin\varphi, -1)$; $2r\, dr\, d\varphi$.

5.29 $2\displaystyle\int_0^{2\pi} (a^2\sin^2 u + b^2\cos^2 u)^{\frac{1}{2}}\, du$.

5.30 $\frac{1}{6}\pi(5\sqrt{5} - 1)$.

5.31 $z = \{3(x^2 + y^2)\}^{\frac{1}{2}}$.

5.33 $\frac{1}{8}$.

5.34 $6\pi a^3$.

5.35 (i) $\frac{1}{2}\pi$; (ii) $(0, 0, \frac{4}{3}\pi)$.

5.36 (i) $(\frac{1}{2}, \frac{1}{2}, 1)$; (ii) 1; (iii) $(\frac{1}{2}, -\frac{1}{2}, 0)$.

5.37 $(0, 0, -\frac{1}{9}\pi)$.

5.38 54.

5.39 $4\pi\{(a^2 + b^2)^{\frac{5}{2}} - a^5 - b^5\}/15b$.

5.41 $\pi a^5/15$.

5.42 $\frac{2}{3}$.

6 INTEGRAL THEOREMS

6.1 $12\pi a^5/5$.

6.3 **F** continuously differentiable in V; and V bounded by a simple closed surface S.

6.15 **F** continuously differentiable on S; and S is a simple open surface spanning the correspondingly oriented curve $\mathscr{C}$.

7 APPLICATIONS IN POTENTIAL THEORY

7.1 (i) 2; (ii) 1; (iii) 4.

7.4 $-\dfrac{1}{r}$; $\frac{1}{2}r^2$.

7.6 $\mathbf{A} = (0, \frac{1}{2}x^2 - yz, -xz)$.

7.7 $\psi = -\cos\theta$.

7.14 $\frac{2}{3}\pi$.

8 CARTESIAN TENSORS

8.1
$$\begin{pmatrix} a_{22} & -a_{21} & a_{23} \\ -a_{12} & a_{11} & -a_{13} \\ a_{32} & -a_{31} & a_{33} \end{pmatrix}.$$

8.12
$$\begin{pmatrix} 0 & b_3 & -b_2 \\ -b_3 & 0 & b_1 \\ b_2 & -b_1 & 0 \end{pmatrix}.$$

9 REPRESENTATION THEOREMS FOR ISOTROPIC TENSOR FUNCTIONS

9.1 $c_1(1, 0, 0), c_2(0, 1, -1), c_3(0, 1, 1)$, where either $c_1 = 1$ and $c_2 = c_3 = \pm\frac{1}{2}\sqrt{2}$, or $c_1 = -1$ and $c_2 = -c_3 = \pm\frac{1}{2}\sqrt{2}$.

9.2 (i) $I_1 I_3$; (ii) $I_3(I_1^2 - 2I_2)$.

9.3 $J_1 = \lambda_1 + \lambda_2 + \lambda_3, J_2 = \lambda_1^2 + \lambda_2^2 + \lambda_3^2, J_3 = \lambda_1^3 + \lambda_2^3 + \lambda_3^3$.

9.4 $n = 0$ and $n = 2$.

9.5 $b_{ij} = \alpha \delta_{ij}$.

Index